Physical Chemistry in Action

For further volumes:
http://www.springer.com/series/10915

Ian W.M. Smith • Charles S. Cockell •
Sydney Leach
Editors

Astrochemistry and Astrobiology

Springer

Editors
Ian W.M. Smith
The University Chemical Laboratory
University of Cambridge
Cambridge
United Kingdom

Charles S. Cockell
School of Physics and Astronomy
University of Edinburgh
Edinburgh
United Kingdom

Sydney Leach
Laboratoire d'Etude du Rayonnement
et de la Matière en Astrophysique (LERMA)
Observatoire de Paris-Meudon
Meudon
France

ISBN 978-3-642-31729-3 ISBN 978-3-642-31730-9 (eBook)
DOI 10.1007/978-3-642-31730-9
Springer Heidelberg New York Dordrecht London

Library of Congress Control Number: 2012949660

Printed on acid-free paper

Springer is part of Springer Science+Business Media (www.springer.com)

Preface

The related subjects of Astrochemistry and Astrobiology are in an era of explosive growth. Both are strongly multidisciplinary: thus, contributions to, and discussions of, these subjects involve scientists who might primarily describe themselves as astronomers, chemists, physicists, astrophysicists, molecular biologists, evolutionary biologists, etc. Proposed phenomena in both areas must be consistent with universal physico-chemical principles. A major purpose of this volume is to outline these physico-chemical principles and describe how they underpin our efforts to understand astrochemistry and to make predictions in astrobiology.

Our book comprises ten chapters, each written by an expert, or experts, in the subject matter of their chapter. Because the backgrounds of those entering the fields of astrochemistry and astrobiology can be very diverse, authors have all been asked to pitch their chapters at a level that should be understandable by this wide range of readership. Chapter 1 seeks to introduce those aspects of physical chemistry which are relevant to the discussions of various topics in astrochemistry and astrobiology to be found in the subsequent eight chapters. These later chapters can be roughly sub-divided into three (Chaps. 2, 3, and 4) that deal with topics within astrochemistry and five (Chaps. 5, 6, 7, 8, and 9) that examine aspects of astrobiology – interpreted broadly.

Major concerns in astrochemistry are (1) the identification of the molecules that exist in the cosmos and the characterisation of the physical conditions in those regions of the universe where the observed molecules are found; (2) laboratory measurements on the spectroscopy of potential molecules and on the rates and products of homogeneous and heterogeneous processes that may contribute to the formation and destruction of molecules, under the generally extreme interstellar conditions where molecules are found; (3) the creation of computer models that use laboratory data and seek both to reproduce what is found in the 'molecular universe', and to suggest what other molecules may be present and which processes appear to be especially important, so as to focus the efforts of laboratory scientists. These topics are dealt with, in turn in chapters by Maryvonne Gerin (in Chap. 2), Michael Pilling (in Chap. 3), and by Valentine Wakelam, Herma Cuppen and Eric Herbst (in Chap. 4).

In any search for life elsewhere in the universe, we only have one exemplar: life here on the Earth. Consequently, it makes sense to search for astronomical bodies which appear to offer living systems a similar environment. For planets outside the solar system, that is, exoplanets, this idea leads to the notion of a Habitable Zone – or more colloquially a 'Goldilocks Zone' – generally defined as the range of radii around a star where the temperature at a planet's surface will not be too high or too low for liquid water to exist. In addition, there may be other bodies, such as the Jovian moon, Europa, where liquid water might be stable. In Chap. 5, Lisa Kaltenegger describes the methods by which planets are detected and characterised, and how, in the future, it may be possible, using sensitive spectroscopic methods to search for biosignatures – signatures of life – in their atmospheres. The importance of water for life as we know it – and may know it – is the subject of Philip Ball's Chap. 6. He emphasises that water does not simply play a passive role in biochemistry, and considers the possibility that other solvents might support and encourage life. Of course, even in thinking about life here on Earth, we must be careful not to adopt too anthropogenic a view. Charles S. Cockell, in Chap. 7, considers the range of physical conditions – temperature, pressure, aridity, pH, etc. – which different life forms on our planet can tolerate.

In searching for – even contemplating – life on exoplanets, we may have the existence of life here on Earth to guide us, but we are severely handicapped by our ignorance about how earthly life came into being. In Chap. 8, Robert Pascal under the title 'Life, Metabolism and Energy', considers how any proposals for the emergence of life must be consistent with thermodynamic and kinetic constraints. Then, in Chap. 9, Irene Chen and her colleagues reflect on the possibility that our present biological world was preceded by one in which RNA, rather than DNA, played an important evolutionary role, and they also discuss the importance of the creation of cells and their linkage to lipids.

Sydney Leach's final chapter is different in kind from those that precede it. He writes a contemplative essay on each of the previous chapters in turn. In some places, he expands on the matters dealt with in earlier chapters, in others he inserts his own view of various topics, and in yet others he adds some extra material that might have been included earlier if space had allowed.

As the outline in the previous paragraphs indicates, the topics that are covered in the individual chapters proceed from those that might be classed as belonging to astrochemistry to those dealing with aspects of astrobiology. Nevertheless, every effort is made to bring together concepts in astrochemistry and astrobiology, which have traditionally been dealt with as separate areas of science.

Cambridge, UK
Edinburgh, UK
Meudon, France
September 2012

Ian W.M. Smith
Charles S. Cockell
Sydney Leach

Acknowledgement

We kindly attribute the name of this series to Professor Ian W.M. Smith.

Contents

Chapter 1
Aspects of Physical Chemistry

Ian W.M. Smith

Abstract Some of those topics in physical chemistry that are especially relevant to astrochemistry and astrobiology are introduced in this chapter. I start with some discussion of the chemical elements: their relative abundances, their electronic structure, and how chemical bonds are formed in simple molecules. This leads to a discussion of how changes between energy levels lead to molecular spectra that can be used to identify molecules at a distance – even the vast distances from Earth to astronomical objects. Having considered forces within molecules, I then discuss the weaker forces between molecules, including hydrogen bonding. The next section focuses on chemical reactions from both the standpoint of thermodynamics and that of chemical kinetics. Finally, some consideration is given to surface processes, which can occur on the dust particles found in the interstellar medium, and enzyme kinetics, which is of great importance in biology.

1.1 Introduction

It is frequently stated that Physics and Chemistry are 'universal', whereas biology, dependent as it is on the environment, may differ in different parts of the cosmos. The purpose of this volume is to provide an introduction to astrochemistry and astrobiology: especially the physico-chemical principles that underpin our efforts to understand astrochemistry and predict astrobiology. In this chapter, I shall briefly review several aspects of physical chemistry which play important roles in astrochemistry and astrobiology. The coverage of the topics that I have selected is necessarily very concise. For those wishing to probe the subjects more deeply, I have given page references to (a) the 9th edition of the book *Physical Chemistry* by Atkins and de Paula [1], and (b) the book *Physical Chemistry for the Biosciences* by Chang [2].

I.W.M. Smith (✉)
The University Chemical Laboratory, Lensfield Road, Cambridge CB2 1EW, UK
e-mail: iwms2@cam.ac.uk

I.W.M. Smith et al. (eds.), *Astrochemistry and Astrobiology*, Physical Chemistry in Action, DOI 10.1007/978-3-642-31730-9_1,

1.2 The Chemical Elements and Their Atomic Structure

Hydrogen, which has the simplest structure of all atoms and the chemical formula H, is the most abundant element in the Universe [3]. Along with smaller amounts of helium (He) and very small amounts of lithium (Li), hydrogen was generated in the aftermath of the big bang in which the Universe was created. The same processes created deuterium (D or ${}^{2}{}_{1}$H),[1] the isotope of hydrogen, which has an abundance of 1.56×10^{-4} relative to H in Earth's oceans. Whereas in a hydrogen atom, the nucleus consists of just a single proton, in deuterium the nucleus consists of a proton and a neutron held together by the residual strong force or nuclear force. All the other elements[2] were created later in the history of the Universe in the process referred to as stellar nucleosynthesis. Clearly, the early stars were created only from the elements created in 'big bang' nucleosynthesis; that is, H, D, He and Li.

Stellar nucleosynthesis occurs during the 'burning' of stellar fuel – that is, the fusion of nuclei in the centre of stars. The strong repulsion between nuclei that both carry a positive charge means that such nuclear reactions only occur at the extremely high temperatures and pressures found in the centre of active stars. As one important example, I cite the formation of carbon (C) nuclei which occurs in a two-step process. First, beryllium (Be) nuclei are created by the fusion of two He nuclei,

$$ {}^{4}_{2}\mathrm{He}^{2+} + {}^{4}_{2}\mathrm{He}^{2+} \rightarrow {}^{8}_{4}\mathrm{Be}^{4+}, \tag{1.1} $$

and then the reaction of the Be nucleus with a further He nucleus produces a carbon nucleus:

$$ {}^{8}_{4}\mathrm{Be}^{4+} + {}^{4}_{2}\mathrm{He}^{2+} \rightarrow {}^{12}_{6}\mathrm{C}^{6+}. \tag{1.2} $$

(To emphasise that these are reactions between atomic nuclei I include, as superscripts following the chemical symbol for the element, the total electric charge.)

Following their creation in the thermonuclear processes that might be termed 'stellar burning', the elements heavier than Li, principally carbon, nitrogen and oxygen, are dispersed into interstellar space by stellar winds or supernovae explosions that mark the death of certain stars. The abundances of the chemical elements have been estimated (with difficulty) by a number of authors. They do vary in different regions of the cosmos [3]. The abundances in the solar system are estimated from observations on the sun and on meteorites. Those given by Cameron [4] have been

---

[1] The notation ${}^{2}{}_{1}$H specifies the number of protons, 1, and the number of nucleons (protons + neutrons), 2. Any electrically neutral atom contains the same number of electrons and protons. To refer to an ion, a superscript after the chemical symbol is added. Thus a deuteron is signified by ${}^{2}{}_{1}\mathrm{H}^{1+}$. The number of protons in an elemental atom corresponds to its atomic number, Z.

[2] These elements, heavier than lithium, are all referred to by astronomers as 'metals' – somewhat to the amusement, or bemusement, of those trained as chemists.

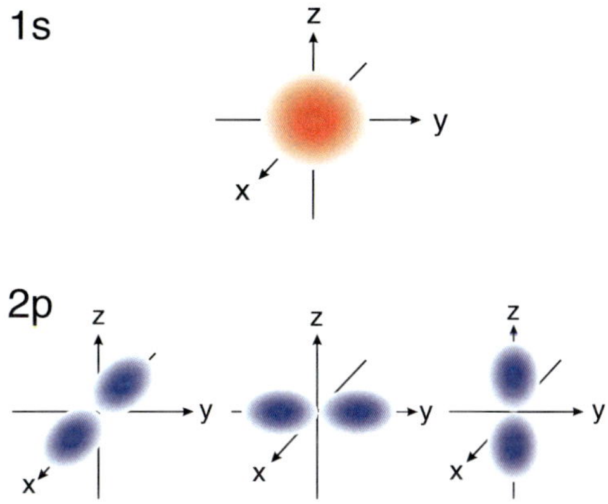

Fig. 1.1 Plots showing the angular distribution of the wavefunctions for the 1*s* and 2*p* orbitals of the hydrogen atom

Table 1.1 Partial listing of elemental abundances relative to that of (atomic) hydrogen [4]

H	**He**	**N**	**O**	**C**	**S**
1.00	9.8 (−2)	1.1 (−4)	8.5 (−4)	3.6 (−4)	1.6 (−5)
Si	**Fe**	**Na**	**Mg**	**P**	**Cl**
3.6 (−5)	4.7 (−5)	2.1 (−6)	3.8 (−5)	2.8 (−7)	3.0 (−7)

updated – and cited as 'cosmic abundances' by Greenwood and Earnshaw and are displayed in Fig. 1.1 of their book [3]. A list of the abundances of some of the most common elements (and those most involved in astrochemistry and astrobiology) are given in Table 1.1; these values are those estimated by Newson [5] and given on the website: www.astrophysicsspectator.com/tables/Abundances.html.

The full list of elemental abundances show that, very approximately and with some exceptions, the elemental abundances decay exponentially with atomic number, *Z*, at least up to about $Z = 40$. They also roughly reflect the nuclear binding energies. There are exceptions: for example, the abundances of Li, B and Be are much lower and that of Fe is higher than would be expected by assuming an exponential decay with *Z*. The anomalously high relative abundance of Fe is apparently associated with the unusually high stability of the nucleus of its commonest isotope ^{56}Fe.

The chemical nature of the elements depends on their 'atomic structure'; that is the distribution of the electrons around the nucleus in the 'atomic orbitals'. A detailed treatment of this subject is beyond the scope of this chapter, but it may be useful to give a simplistic discussion of this topic based on the Periodic Table of the elements, which is reproduced as Table 1.2. In the Periodic Table, the elements are arranged in increasing order of their atomic numbers as if they are on the lines of a book. Each line is termed a 'period'. However, there is not the same number of elements on each line; rather the arrangement is such that elements of similar chemical properties are positioned in the same vertical column or 'group'. For example, the alkali metals, Li, Na, K, Rb, Cs and Fr, fall in the first column (Group 1) in the table. They are characterised by low ionisation energies – that is, relatively small amounts of energy are required to expel an electron and form the singly charged ions: Li^+, Na^+, K^+, etc. These elements readily form salts, especially

Table 1.2 Periodic Table of the Elements, showing the electron configuration of the ground state of each atom

							1	2										
1[a]							**H**	**He**										
	1[b]	**2**					$1s^1$	$1s^2$					**13**	**14**	**15**	**16**	**17**	**18**
2	3	4											5	6	7	8	9	10
	Li	**Be**											**B**	**C**	**N**	**O**	**F**	**Ne**
	$[He]2s^1$	$[He]2s^2$											$2s^22p^1$	$2s^22p^2$	$2s^22p^3$	$2s^22p^4$	$2s^22p^5$	$2s^22p^6$
3	11	12	**3**	**4**	**5**	**6**	**7**	**8**	**9**	**10**	**11**	**12**	13	14	15	16	17	18
	Na	**Mg**											**Al**	**Si**	**P**	**S**	**Cl**	**Ar**
	$[Ne]3s^1$	$[Ne]3s^2$											$3s^23p^1$	$3s^23p^2$	$3s^23p^3$	$3s^23p^4$	$3s^23p^5$	$3s^23p^6$
4	19	20	21	22	23	24	25	26	27	28	29	30	31	32	33	34	35	36
	K	**Ca**	**Sc**	**Ti**	**V**	**Cr**	**Mn**	**Fe**	**Co**	**Ni**	**Cu**	**Zn**	**Ga**	**Ge**	**As**	**Se**	**Br**	**Kr**
	$[Ar]4s^1$	$[Ar]4s^2$	$3d^14s^2$	$3d^24s^2$	$3d^34s^2$	$3d^54s^1$	$3d^54s^2$	$3d^64s^2$	$3d^74s^2$	$3d^84s^2$	$3d^{10}4s^1$	$3d^{10}4s^2$	$4s^24p^1$	$4s^24p^2$	$4s^24p^3$	$4s^24p^4$	$4s^24p^5$	$4s^24p^6$
5	37	38	39	40	41	42	43	44	45	46	47	48	49	50	51	52	53	54
	Rb	**Sr**	**Y**	**Zr**	**Nb**	**Mo**	**Tc**	**Ru**	**Rh**	**Pd**	**Ag**	**Cd**	**In**	**Sn**	**Sb**	**Te**	**I**	**Xe**
	$[Kr]5s^1$	$[Kr]5s^2$	$4d^15s^2$	$4d^25s^2$	$4d^45s^1$	$4d^55s^1$	$4d^55s^2$	$4d^75s^1$	$4d^85s^1$	$4d^{10}$	$4d^{10}5s^1$	$4d^{10}5s^2$	$5s^25p^1$	$5s^25p^2$	$5s^25p^3$	$5s^25p^4$	$5s^25p^5$	$5s^25p^6$
6	55	56	57	72	73	74	75	76	77	78	79	80	81	82	83	84	85	86
	Cs	**Ba**	**La**	**Hf**	**Ta**	**W**	**Re**	**Os**	**Ir**	**Pt**	**Au**	**Hg**	**Tl**	**Pb**	**Bi**	**Po**	**At**	**Rn**
	$[Xe]6s^1$	$[Xe]6s^2$	$5d^16s^2$	$5d^26s^2$	$5d^36s^2$	$5d^46s^2$	$5d^56s^2$	$5d^66s^2$	$5d^76s^2$	$5d^96s^1$	$5d^{10}6s^1$	$5d^{10}6s^2$	$6s^26p^1$	$6s^26p^2$	$6s^26p^3$	$6s^26p^4$	$6s^26p^5$	$6s^26p^6$
7	87	88	89															
	Fr	**Ra**	**Ac**															
	$[Rn]7s^1$	$[Rn]7s^2$	$6d^17s^2$															

[a]These digits (increasing downwards) number the 'groups'.
[b]These digits (increasing across the table) number the 'periods'.
Elements with atomic numbers from 58 (Cerium) to 71 (Lutetium) are the lanthanides or 'rare earths'. The electronic configurations of Ce is $[Xe]4f^15d^16s^2$, that of element 59 Praseodymium (Pr) is $[Xe]4f^36s^2$. Thereafter the 4f orbitals fill to element 70 Ytterbium (Yb), which has the configuration $[Xe]4f^{14}6s^2$, and Lutetium, which has the configuration $[Xe]4f^{14}5d^16s^2$.
Elements with atomic numbers of 90 (Thorium) and greater are the actinides.

with the elements towards the right-hand side of the table such as the halogens in Group 17 – but not with the noble gases in group 18, He, Ne, Ar, etc.

The construction and the development of the Periodic Table was a work of genius, based on empirical knowledge of the chemical properties of different elements. However, a full *understanding* of the chemical behaviour of atoms – and therefore their place in the periodic table only came with the emergence of quantum mechanics (1, Chaps. 7 and 8; 2, Chap. 11). A detailed discussion of this topic is certainly beyond the scope of the present book, but it is relatively straightforward to summarise enough of the conclusions of quantum mechanics for current purposes.

A treatment of the hydrogen atom by quantum mechanics leads to the conclusion that its single electron can occupy any of a number of atomic orbitals. These orbitals are characterized by three quantum numbers given the symbols: n, l and m_l. Each orbital can accommodate two electrons which differ in their electron spin, m_s, which can only take the value + ½ or − ½. In the H atom, the energy associated with the electron depends only on the value of n, the principal quantum number, which can take integral values from 1 to infinity. This energy can be expressed by the simple formula: $E_n = -R_\infty/n^2$, relative to a zero corresponding to the removal of the electron from the influence of the positive proton (i.e., ionisation). The Rydberg constant, R_∞, corresponds to the ionisation energy of the H atom (13.60 eV). The value of the second quantum number l relates to the value of the angular momentum of an electron occupying a particular orbital: l can take the value $(n - 1)$, $(n - 2)$, down to zero. Thus if $n = 3$, l can have the values 2, 1 or 0. It is because the angular momentum is 'quantised'; that is, it can only take values corresponding to $\{l(l + 1)\}^{1/2}\,\hbar$ – where $\hbar$ is Planck's constant (h) divided by 2π - that only discrete values of the energy are allowed.

The orbitals with $l = 0$, 1, 2, 3, etc. are referred to as *s*-orbitals, *p*-orbitals, *d*-orbitals, *f*-orbitals, etc. Finally, m_l, which can take integral values running from $+ m_l$ to $- m_l$, refers to the spatial orientation of the orbital. An electron which occupies an orbital characterised by the same values of n and l, but different m_l, has the same energy – that is, these orbitals are degenerate – in the absence of an electric or magnetic field. There are $(2l + 1)$ orbitals for each value of l. This means that an *s*-orbital with a given principal quantum number can accommodate two electrons, whereas the *p*-orbitals with the same values of n can accommodate six electrons, *d*-orbitals with the same n ten electrons, and so on. As the value of the principal quantum number n increases the electron density is progressively displaced further from the nucleus. Orbitals with different values of l have different symmetries, and these can be represented by diagrams such as Fig. 1.1.

If the orbital energies given by the exact quantum mechanical solution for the H atom held for other atoms, we might expect to see the rows of the Periodic Table contain successively 2, 8, 18, 32 elements as orbitals with $n = 1, 2, 3, 4$ were filled with electrons. In reality, the relative energies of atomic orbitals (or the electrons occupying atomic orbitals) vary as the atomic number of the elements increases and, besides the attractive force between the nucleus and the electron in the H atom, there are also repulsive forces between the several electrons in the heavier atoms. Although the problem can no longer be exactly solved by the methods of quantum mechanics, it is apparent that the energies of orbitals with the same principal quantum number increase with increasing *l*. Consequently, it is helpful to think of the orbitals as occupying 'shells' of similar energy: thus, the *K* shell consists of just the 1*s* orbital, the *L* shell the 2*s* and 2*p* orbitals, the *M* shell the 3*s* and 3*p* orbitals and the *N* shell the 4*s*, 3*d* and 4*p* orbitals.

If we then take on board the facts that (a) each orbital can accommodate two electrons, which differ only in their spin quantum number m_s, and (b) the Aufbau principle, which states that to find the ground (or lowest energy) electron configuration the orbitals are filled in order of their ascending energy, we can arrive at the electron configurations given under the formulas of the elements given in the Periodic Table in Table 1.2. The outer – or valence – electrons largely determine the chemical behaviour of the elements and the nature of the bonding when molecules form.

Thus, if we start with a simplified consideration of diatomic molecules, we can identify two limiting kinds of behaviour. First, in the salt molecules formed from the alkali metal atoms of Group 1 and the halogen atoms of Group 17, for example NaCl, an electron is almost completely transferred from the halogen to the alkali metal and the attractive bonding forces are essentially ionic or electrostatic: the molecule is, to a good approximation, Na^+Cl^- and, and due to the separation of charge, this molecule will clearly possess an electric dipole moment. At the other extreme, two identical atoms will 'share' their valence electrons equally, hence forming a covalent bond. With two hydrogen atoms, two electrons are shared and a single covalent bond is formed; with two nitrogen atoms, six electrons, three from each atom are shared and a triple covalent bond results. The single bond in H_2 is the result of two electrons occupying a molecular orbital which can be thought of as arising from favourable overlap of the 1*s* atomic orbitals on the individual H atoms. This is classed as a σ orbital (or σ bond); the electrons occupying such orbitals do not give rise to electronic angular momentum around the interatomic axis. In N_2, there is again a σ bond; but, in addition there are two π bonds arising from the occupation of molecular orbitals arising from the overlap of *p* orbitals on the individual N atoms. The difference in strength of the single bond in H_2 and the triple bond in N_2 is demonstrated by the difference in their dissociation energies (D_0): 432.1 kJ mol^{-1} in the case of H_2, and 941.7 kJ mol^{-1} in N_2.

Of course, in homonuclear molecules like H_2 and N_2, the distribution of electrons must be symmetrical about a plane that bisects the internuclear axis and is perpendicular to it. Consequently, these molecules possess no electric dipole moment, which, as we shall see, has important consequences for their spectroscopy. On the other hand, heteronuclear molecules like CO, the second most abundant molecule in the universe, do possess an electric dipole moment.

The simple ideas that have been introduced in relation to diatomic molecules can be carried over to a brief discussion of bonding in polyatomic molecules: in particular, the distinction between σ bonds and π–bonds. Moreover, the structure of polyatomic molecules can be rationalised in terms of the atomic orbitals on neighbouring atoms that are used to create molecular orbitals: for example, if one *s* orbital and two *p* orbitals are used to create three bonding molecular σ-orbitals (as in the C atoms in C_2H_4), the three bonds are aligned 120° to one another.

The element carbon is exceptional and is, of course, essential to life as we know it here on Earth. It is also present in the majority of molecules that have been definitely identified in the interstellar medium. Its electronic configuration is $2s^2 2p^2$ and, by sharing its valence electrons, it is capable of forming four $\sigma-$bonds around each C atom, as in CH_4, C_2H_6, C_3H_8 and all the other alkanes. These alkanes can be referred to as saturated molecules, because they only involve $\sigma-$bonds, between neighbouring C atoms and between C and H atoms, and there is no possibility of further H atoms adding to the molecules. This is in contrast to unsaturated molecules, such as the alkenes, C_2H_4, C_3H_6, etc., and the alkynes like C_2H_2. These are frequently formulated as $H_2C{=}CH_2$, $H_3CC{=}CH$ and $HC{\equiv}CH$, emphasising the presence of double and triple bonds. In principle, these molecules can add hydrogen, thereby becoming saturated, but many of the organic molecules identified in the interstellar medium are unsaturated, despite H_2 being, by far, the most abundant molecule in most regions where interstellar molecules are found. It is also useful to draw a distinction between these molecules, in which all the electrons are 'paired', and those species, free radicals or simply radicals, which have unpaired electrons. Important examples from the viewpoint of astrochemistry include CN, C_2H and C_4H, as well as many atoms; for example, H, C and N. Generally, radicals exhibit high reactivity, especially with unsaturated molecules and with other radicals.

The ability of carbon to form four single bonds leads to its ability to create chiral molecules. These molecules, which are very important in biology, are said to be optically active, because they rotate the plane of polarised light. In general, a molecule is chiral if it does not have a centre of inversion or a mirror plane [1, p. 426]. These properties are achieved when a carbon atom in a molecule is bound to four different atoms or groups of atoms as in, for example, CHFClBr. A chiral molecule and its mirror image form an enantiomeric pair. Though they are mirror images of one another, such pairs of molecules cannot be superimposed on one another. The amino-acid alanine, $H_2NCH(CH_3)COOH$, is an example of a chiral molecule, whereas the closely related amino-acid glycine, H_2NCH_2COOH, is non-chiral.

I have pointed out the apparently unique role that carbon plays in the 'molecules of life' on Earth. A question that naturally arises is whether any other element might play the role that carbon plays under other conditions. The most likely candidate would seem to be silicon, since Si atoms have a similar electronic configuration, $[Ne]3s^2 3p^2$, to that of C atoms, $[He]2s^2 2p^2$. One empirical argument against this possibility – at least, on Earth-like planets – invokes the fact that, on Earth, silicon is approximately a thousand times more abundant than carbon but life here makes use of carbon, not silicon, in constructing pre-biotic molecules and beyond them, life forms.

A number of factors may contribute to the different chemical and biochemical behaviour of these two elements. One is the considerable difference in their oxides: CO_2, a rather unreactive gas, and SiO_2, most commonly found as a hard crystalline solid. CO_2 is, of course, inhaled or exhaled by terrestrial organisms, but it is hard to see how an organism could inhale or exhale SiO_2! Silane (SiH_4), the direct analogue of methane, does exist as a flammable gas. However, analogues of the larger alkanes and especially of the alkenes and alkynes are unstable; the smaller ones react vigorously with water and the large ones spontaneously decompose. In addition, silicon fails to bond to many of the other elements with which carbon forms bonds and which are necessary for metabolism. It appears that a basic reason for this difference in behaviour arises from the difference in atomic size between Si and C. Thus, the Si-Si bond length in disilane, Si_2H_6, is 0.2332 nm [6] compared with the C-C bond length of 0.1522 nm in C_2H_6 [7]. This increase in bond length means that any π bonds formed, for example, in Si_2H_4 or Si_2H_2 would be much weaker than the equivalent bonds in alkenes and alkynes. The difference between silicon and carbon, in particular the unique ability of carbon to form bonds with many other elements in molecular structures of enormous versatility, indicates that the formal regularities of the Periodic Table only tell part of the chemical story.

1.3 Energy Levels and Spectroscopy [1, Chaps. 12 and 13; 2, Chap. 14; 8]

The fact that the energy levels of atoms are 'discrete', as long as electrons are bound to the nucleus, has been introduced in Sect. 1.2. The situation is particularly simple in the case of the hydrogen atom since there is only one electron and the energy of each atomic state corresponds to that of the orbital which is occupied by the electron. Spectroscopy is the study of the interaction of electromagnetic radiation with atoms and molecules when changes or transitions occur between the quantised energy levels. Through Planck's famous relationship between the frequency (ν) of radiation and the energy of the corresponding photon (E_ν), $E_\nu = h\nu$, one can determine the spacing of the two energy levels between which the transition occurs. For example, the energies of the H-atom levels for which $n = 2$ and $n = 1$ are $E_n = -R_\infty/2^2$ and $E_n = -R_\infty/1^2$, respectively, so the difference in energy between them is, $(3/4)R_\infty$, and the frequency of the radiation associated with this change is $(3/4)R_\infty/h$. This transition can be observed in absorption, when external radiation of the correct frequency promotes the electron from the level with $n = 1$ to a level with $n = 2$, or emission, when the electron in the excited orbital $n = 2$ spontaneously falls into the $n = 1$ level and a photon is emitted. The photon which is absorbed or emitted corresponds to radiation at a wavelength of $\lambda = 121.6$ nm, which is often referred to as Lyman-α radiation. It is prevalent throughout the interstellar medium, but cannot be observed at the Earth's surface because it is absorbed by the terrestrial atmosphere. In certain regions of space, there is also emission from levels above $n = 2$ corresponding, for example, to the $n = 3 \rightarrow n = 2$ emission. This emission occurs at $\lambda = 656.27$ nm and is responsible for the pink colour of several regions of the cosmos.

The spectra of atoms are different for the different elements and their observation can therefore identify the presence of particular elements. An important example is the Fraunhöfer lines. These are observed as dark (that is, absorption) lines on the spectrum of the sun and the elements responsible for several of these lines were identified by noting their coincidence with the wavelengths emitted when the same elements were heated; for example, in a Bunsen flame. In the sun, these lines appear in absorption because the background source is hotter than the medium through which the radiation is passing, whereas they are observed in emission from a flame since the background is cooler than the sample.

Quantisation, or the existence of discrete energy levels, is not confined to the energies associated with the motion of electrons. Once we move to molecules, it is necessary additionally to consider the energy levels that are associated with other motions: in particular, with the overall rotation of the molecule and with the molecular vibrations in which the nuclei move their positions relative to one another. The interpretation of molecular spectra is facilitated by the fact that the gaps between energy levels associated with different kinds of motion are quite different, with the result that the spectra associated with them occur at different wavelengths or frequencies in the electromagnetic spectrum.

A further result is that it is generally possible to think of each *bound* electronic state of a molecule to possess its own manifold of vibrational energy levels and, for each of these vibrational states, there will be a manifold of more closely spaced rotational levels. The electronic potential energy (V) of each state of a molecule depends on the instantaneous nuclear geometry. For diatomic molecules (e.g., AB), V depends on the internuclear separation (r_{AB}), and this variation can be represented by a potential energy curve as shown in Fig. 1.2. This figure is also used to represent schematically some of the vibrational and rotational levels associated with this electronic state. For diatomic molecules, there is just one molecular vibration and two rotations about perpendicular axes through the centre-of-mass. With molecules comprised of more than two atoms ($\boldsymbol{N}$), it is impossible to represent how V depends on all the coordinates required to define the instantaneous geometry. If the molecule is non-linear, it possesses three moments of inertia and three overall molecular rotations, together with ($3\boldsymbol{N} - 6$) vibrations; linear molecules have two equivalent rotations (like a diatomic molecule) and ($3\boldsymbol{N} - 5$) vibrations.

Promoting *small* molecules from their lowest or ground electronic state to excited electronic states generally requires radiation in the ultraviolet region of the electromagnetic spectrum. However, observations of the spectra of astronomical sources reveal a large number of absorption features (see Fig. 1.3) referred to as the diffuse interstellar bands [9, 10]. Almost a 100 years after their discovery, the carriers of these bands remain uncertain, though they are generally attributed to neutral or ionised polyatomic molecules.

Because the Earth's atmosphere absorbs wavelengths below about 300 nm, [11] observations at lower wavelengths (for example, of H_2 and CO) generally require

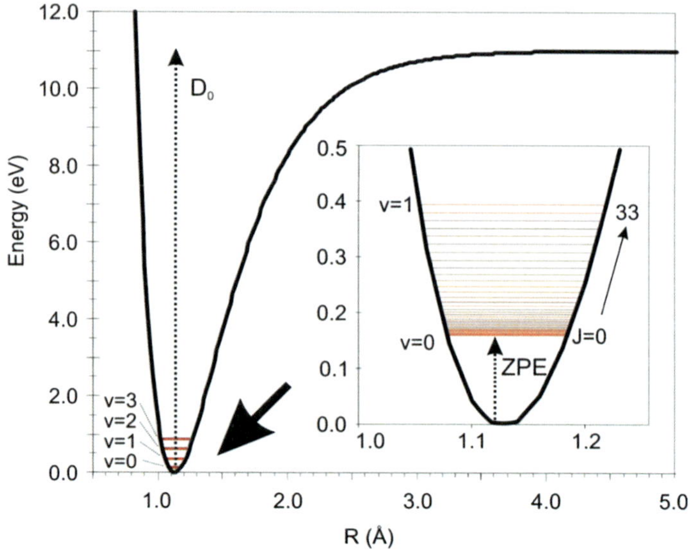

Fig. 1.2 The potential energy curve for the electronic ground state of CO. The *inset* shows (on a very different scale) the spacing of low-lying vibrational and rotational energy levels. v and J are the vibrational and rotational quantum numbers. ZPE shows the 'zero-point energy' between the minimum of the potential energy curve and the lowest quantum state and D_0 is the dissociation energy

Fig. 1.3 Picture showing the diffuse interstellar bands (Courtesy of NASA/JPL-Caltech)

that the spectrometers are deployed on space platforms. Exceptions to this general rule are the diatomic species CH, CH^+ and CN which absorb at wavelengths above 300 nm and were the first molecules to be identified in the interstellar medium [12] from observations of their absorption lines. Observations of the electronic absorption spectra of H_2 and CO between 91.5 and 111.2 nm are particularly important as they enable the relative abundances of these molecules, the commonest in the interstellar medium, to be established [13]. It is especially important in relation to H_2, since it has no rotational and only very weak vibrational spectra – for reasons that will soon become apparent. Consequently, CO, which is relatively easy to observe, is often employed as a marker for H_2 based on the relatively few but very important observations of the [CO]:[H_2] ratio.

The majority of the 160 or so gas-phase molecules[3] that have been positively identified in the interstellar medium, ranging from diatomics to molecules containing 13 atoms [14], have been observed via their rotational emission spectra; that is, by observation of the characteristic frequencies that are emitted as a molecule undergoes a transition from one rotational energy level to a second level lower in the manifold. The wavelengths associated with such transitions are sufficiently long that (a) they are not significantly scattered by the dust particles that are present in the dense interstellar clouds, where many of the interstellar molecules are found, and (b) the radiation is transmitted through the Earth's atmosphere, which is transparent at most of the wavelengths characteristic of these emissions.

For diatomic and other linear molecules, the energies of rotational levels are given (to a good degree of approximation) in terms of a rotational quantum number (N) and the moment of inertia (I_{AB}) by the expression:

$$E_N = N(N+1)\hbar^2/2I_{AB}. \tag{1.3}$$

The spacing, or energy difference, between the successive levels ($N + 1$) and N is therefore $2(N + 1)\,\hbar^2/2I_{AB}$. This can be converted to a frequency by dividing by h and to the reciprocal of a wavelength by further dividing by c, the speed of light. The wavelength of the $N = 1$ to $N = 0$ transition in CO is 2.6 mm.

Observation of rotational transitions not only identifies the molecules present in the interstellar medium but can also be used to infer both the abundance of that molecule and the physical conditions – the gas density and the temperature – in the regions from which the observed radiation is emitted. Assigning the observed frequencies to particular molecules involves a comparison with rotational spectra from laboratory experiments: mostly obtained using the technique of Fourier transform microwave spectroscopy [15]. Its application to potential interstellar molecules is far from straightforward, since the species of interest are frequently unstable and must be produced 'on the fly' [16]. Before definitely assigning the molecules responsible for frequencies observed from the interstellar medium, more than one coincidence with laboratory frequencies should be observed for each molecule, and allowance must be made for 'Doppler shifts' arising from motions of the source of radiation, which are particularly large for observations from galaxies other than our own.

Having identified a molecule, the next objective is to estimate its (relative) abundance – generally expressed relative to that of H_2. The first stage in this procedure is to measure the absolute strength of the transitions that are observed. Quantum mechanics demonstrates (a) that molecules without an electric dipole moment (like H_2) do not undergo *pure* rotational transitions, and (b) in molecules that have a dipole moment the transitions are limited by selection rules: for example, in linear molecules, N can only change by one; that is, $\Delta N = \pm 1$.

[3] This total does not include the isotopomers of several species that have also been observed.

Quantum mechanics also yields an expression for the Einstein spontaneous emission coefficient (the reciprocal of the mean radiative lifetime) [8]:

$$A_{(N+1)\to N} = \frac{64\pi^4\nu^3}{3(4\pi\varepsilon_0)hc^3}\,\mu^2_{\mathrm{D}(N+1),N} \tag{1.4}$$

where ν is the frequency of the transition and $\mu_{\mathrm{D}(N+1),N}$ is the transition dipole moment, which in the case of rotational transitions corresponds to the permanent electric dipole moment of the molecule. To take one example, the A coefficient for the $N = 1$ to $N = 0$ transition in CO is $2.16 \times 10^{-7}\ \mathrm{s}^{-1}$, corresponding to a mean radiative lifetime of *ca.* 54 days. However, it should be noted that, because of the ν^3 factor in (1.4), $A_{(N+1)\to N}$ increases rapidly with N.

To explain how the observations on different lines from the same molecule can be used to infer the gas density and temperature at the source, it is necessary to spend a moment explaining how molecules are distributed over their rotational levels in a situation where these populations are in equilibrium and, equivalently, a single temperature (T) describes the distribution. The distribution is then described by the Boltzmann distribution law [1, p. 567], which states that the population (N_i) in a particular level (i) depends on the energy of the level (E_i) and its degeneracy (g_i) – that is, the number of quantum states that possess the energy E_i - according to the expression:

$$N_i \propto g_i \exp(-E_i/k_{\mathrm{B}}T) \tag{1.5}$$

where $\propto$ represents 'is proportional to'. Based on this formula, the fraction (f_i) of an ensemble of molecules that occupy a particular level i is given by:

$$f_i = g_i\exp(-E_i/k_{\mathrm{B}}T)/\Sigma_i g_i\exp(-E_i/k_{\mathrm{B}}T) \tag{1.6}$$

where the denominator defines the partition function, q, associated with the manifold of levels that are being considered.

For the rotational levels of a linear molecule, $g_i = (2N + 1)$ and $E_i = N(N + 1)\hbar^2/2I_{\mathrm{AB}}$ so that:

$$f_N = (2N+1)\exp(-N(N+1)\hbar^2/2I_{\mathrm{AB}}k_{\mathrm{B}}T)/q_{\mathrm{rot}} \tag{1.7}$$

If the rotational levels are closely spaced compared with $k_{\mathrm{B}}T$, the summation sign in the denominator of the expression on the right-hand side of (1.6) can be replaced by an integration and q_{rot}, the rotational partition function, is given by

$$q_{\mathrm{rot}} = (2I_{\mathrm{AB}}k_{\mathrm{B}}T)/\hbar^2 \tag{1.8}$$

Equation (1.7) describes the distribution of an interstellar molecule over its rotational levels under two different equilibrium circumstances. First, at very low densities, as in the background interstellar medium, the distribution will be in equilibrium with the background radiation field, which corresponds to a temperature of 2.725 K. Second, at high densities, the distribution will be maintained by collisions and the temperature inferred from the rotational distribution corresponds to the translational temperature of the surrounding gas.

In practice, the distribution of a molecule over its rotational levels may not be described by a single temperature. This is because the A coefficient depends strongly on the rotational level, because of the ν^3 factor in (1.4), whereas the rate at which collisions re-distribute molecules amongst rotational levels is almost independent of N. As a result, by careful analysis of the rotational distribution inferred from the relative intensities of several lines, and possibly some modelling, both the gas density and the translational temperature can be estimated [17] (for further discussion, see Chap. 2).

As Fig. 1.2 shows, the spacing between successive vibrational levels in a molecule is generally greater than that between (low-lying) rotational levels but much less that that between neighbouring electronic states. The frequencies associated with transitions between neighbouring vibrational levels fall in the infrared part of the electromagnetic spectrum. The application of infrared spectroscopy to the study of interstellar molecules is hampered by absorption in the Earth's atmosphere. The value of infrared spectroscopy has been greatly enhanced in recent years by the deployment of spectrometers mounted on satellites, such as the *Infrared Space Observatory* (ISO), the *Spitzer Space Telescope* and the *Herschel Space Observatory*.

Both infrared absorption and infrared emission spectra have added to our knowledge of the molecules present in the cosmos. In the former case, a suitable star can be used as a background source. Such studies are useful in identifying small gas-phase molecules such as CO_2 and C_2H_2 which have no permanent electric dipole moment, and therefore no pure rotational transitions. The equivalent requirement in respect of vibrational transitions is that the vibrational motion must bring about a change in the electric dipole moment. This is true, for example, for both the asymmetric stretching mode of CO_2 and its bending vibration. However, homonuclear diatomic molecules, like H_2, N_2 and O_2, have no permanent electric dipole moment, nor does their vibrational motion generate one. Therefore they do not undergo rotational transitions, and their 'quadrupole-allowed' vibrational transitions are much weaker than those which cause changes in the electric dipole of a molecule.

So far I have only considered the spectroscopy of gas-phase species that are free to rotate. This is not true of molecules adsorbed onto dust grains. However, such species do exhibit infrared spectra resulting from vibrational transitions, although the infrared absorptions are broader and not always easy to assign [17]. Those that have been assigned with certainty include H_2O, CO and CO_2. The spectra of larger molecules become too weak and non-specific for proper identification.

Infrared emission from many regions of the interstellar medium is dominated by what are often called the unidentified infrared bands (UIBs).As shown in Fig. 1.3, these broad features are observed at several wavelengths between 3 and 15 μm [18]. They are attributed to vibrational transitions associated with the C–H and C–C stretching vibrations (at *ca.* 3.3 and 6–8 μm, respectively) and CH in-plane and CH out-of-plane bending modes (at *ca.* 8.5 and 10–15 μm, respectively) in large

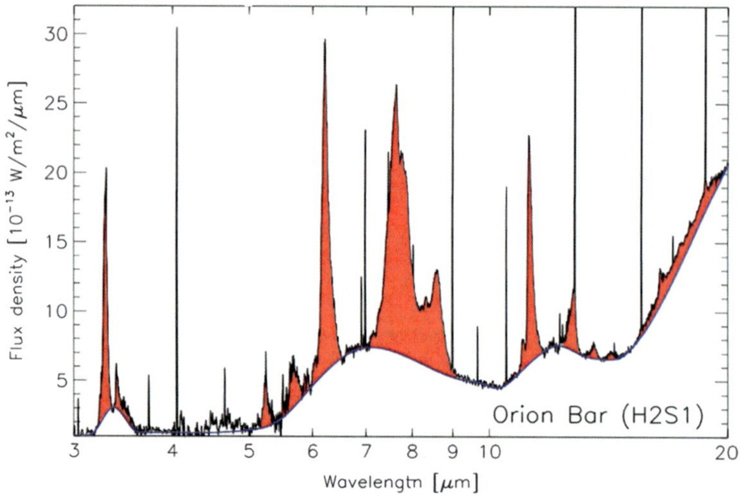

Fig. 1.4 The IR emission spectrum observed from the Orion bar (Adapted from Peeters [18])

aromatic molecules (i.e., polycyclic aromatic hydrocarbons: PAHs). These molecules are probably promoted to electronically excited states by the absorption of ultraviolet light from nearby stars and then undergo a process known as internal conversion by which the electronically excited molecules are converted by a radiationless transition to high vibrational levels of the electronic ground state. These molecules relax by undergoing successive vibrational transitions. The emission spectra are too broad and unstructured to identify which particular PAHs are responsible for the observed emission bands (Fig. 1.4).

1.4 Intermolecular Forces and Hydrogen Bonding [1, Chap. 17; 2, Chap. 13; 19, 20]

In Sect. 1.2, I made some brief remarks about chemical bonding – especially bonding in diatomic molecules. These comments were illustrated in Fig. 1.1 by a representative potential energy curve showing how the electronic energy varies with internuclear separation in the bound state of a diatomic molecule. In addition, I gave values of the dissociation energies for H_2 (432.0 kJ mol^{-1}) and for N_2 (941.7 kJ mol^{-1}), as examples of molecules held together by a single and triple bonds. In general, the strength of chemical bonds is reflected in their dissociation energies (D_0 – as shown in Fig. 1.2), which vary between *ca.* 100 and *ca.* 1,000 kJ mol^{-1}; for example, $D_0 = 148.9$ kJ mol^{-1} for I_2 and 1071.9 kJ mol^{-1} for CO.

As well as these strong forces *within* molecules, there are weaker attractive (and, at shorter internuclear separations, strong repulsive) forces *between* molecules; that is, intermolecular forces, frequently referred to as van der Waals forces. Again, the interaction between two non-bonding atoms can be represented by a potential energy curve but now the depth of the minimum (usually denoted by ε) is typically two orders-of-magnitude smaller than D_0 for a chemical bond; for example, in the interaction between two argon atoms, the minimum on the potential energy curve lies only 1.2 kJ mol^{-1} below the energy at large separation, and this minimum is found at an internuclear separation of 0.376 nm, much longer than a chemical bond. Despite their relative weakness, intermolecular forces are important in a number of ways. For example, they are the forces that cause gases to depart from ideal gas behaviour and they are responsible for condensation. The fact that the boiling points of the noble gases increase in the order helium to xenon is a reflection of the parallel increase in ε from 88 J mol^{-1} for He – He to 2.34 kJ mol^{-1} for Xe – Xe [21].

Gas phase reactions that are important at the very low temperatures (*ca.* 10 K) found in the cold cores of dense interstellar clouds must have no barrier along the path of minimum potential energy leading from reactants to products or, in other words, no activation energy. The rates of such reactions are controlled by the ability of the long-range intermolecular forces to bring the reactants into a close collision: a process generally referred to as 'capture'. For systems where the long-range potential only depends on the separation of the two reactants, it is relatively easy to estimate the rate coefficients for close collisions. One first needs to define an *effective* potential energy which takes into account that angular momentum must be preserved in the collision and the energy associated with orbital motion increases as the separation (R_{AB}) between the two colliding species (A and B) decreases:

$$V_{eff}(R_{\mathrm{AB}}) = \left[E_{trans}b^2/{R_{\mathrm{AB}}}^2\right] - C/{R_{\mathrm{AB}}}^{-n} \tag{1.9}$$

In this equation, the intermolecular attraction is assumed to be proportional to ${R_{\mathrm{AB}}}^{-n}$ and the first term on the right of the equation represents the energy associated with the orbital motion for collisions, where E_{trans} is the energy associated with relative motion of the colliding pair and b is the impact parameter, which is the closest distance the centres of the two particles would approach in the absence of intermolecular forces. One can obtain [21] an expression for the cross-section for close collisions by: (1) finding the value of R_{AB} ($R_{\mathrm{AB},max}$) at which $V_{eff}(R_{\mathrm{AB}})$ has its maximum value (by differentiating the right-hand-side of (1.9) and setting the result to zero); (2) finding the corresponding value of $V_{eff}(R_{\mathrm{AB}})$, that is, $V_{eff}(R_{\mathrm{AB},\,max})$, and (3) using these results to find the maximum value of b at which collisions with relative energy E_{trans} can 'surmount' the 'centrifugal barrier', $V_{eff}(R_{\mathrm{AB},\,max})$. Finally, one can obtain a rate coefficient for close collisions by multiplying the expression for the cross-section by the Maxwell-Boltzmann expression for the distribution of relative velocities and integrating the result.

We can distinguish two simple but important cases where the long-range attraction between two molecules (or atoms) varies as ${R_{\mathrm{AB}}}^{-n}$: (1) for an ion interacting with a polarisable molecule, $n = 4$, and (2) for two neutral molecules, dispersion forces give rise to an attraction with $n = 6$. These attractive forces increase the rate coefficient for close collisions above the value estimated for 'hard-spheres' by factors of *ca.* 5 for ion-molecule collisions (independent of temperature), and *ca.* 2 for neutral-neutral collisions at 298 K [21].

In the case where $n = 4$, the simple treatment outlined above, and generally referred to as the Langevin model, yields an expression for the rate coefficient that is independent of temperature and which can be calculated knowing the polarisability of the molecule (and the reduced mass associated with the collision). Because the charge-induced dipole attraction is strong, the model can be applied to many ion-molecule reactions with reasonable success. For the interaction between an ion and a polar molecule, the long-range energy depends also on a second term which depends on orientation: that is, the angle between the axis of the dipole and the line joining the centres of the two species. The effect of the charge-dipole interaction is to increase the rate of 'capture', especially at low temperatures [12]. The adaptation of these models to reactive collisions between ions and neutral molecules will be dealt with in more detail in Chap. 3.

When the two collision partners are electrically neutral, the long-range mutual attraction is much weaker. In addition to dispersion forces, responsible for the $-R_{AB}{}^{-6}$ term in the expression for the long-range potential, it is generally necessary to include a number of other contributions arising from the asymmetric distribution of electrical charge in the two species, to the long-range potential [19]. These include: dipole-dipole, dipole-quadrupole, dipole-induced dipole interactions. The strength of these attractions depends not only on the separation between the molecules but also on their orientation with respect to one another. A version of transition state theory [21, 22] suitable for estimating the rate coefficients for neutral-neutral 'barrierless' reactions has been described [23]. This method calculates contributions to the reactivity for collisions with specified collision energy and total angular momentum and averaging over the various possible orientations is included to allow for the angular dependence of the long-range intermolecular potential. It has been applied, with fair success, to reactions between pairs of free radicals, for which it is difficult to determine rate coefficients experimentally, especially at low temperatures.

Amongst the strongest intermolecular attractive forces are those between ions and polar molecules. They play an important role in the solution of salts in water [2, Sect. 5.7]; for example:

$$\mathrm{NaCl(s) + water \rightarrow Na^+(aq) + Cl^-(aq).} \tag{1.10}$$

The enthalpy change associated with this process can be estimated by constructing a thermochemical cycle and well-known values of the enthalpies associated with processes such as: (1) the formation of dissolved NaCl from Na(s) and ½Cl_2(g), (2) the sublimation of Na(s), (3) the ionisation of Na(g), (4) the dissociation of ½Cl_2(g), (5) the capture of an electron by Cl(g). The conclusion is that the enthalpy of hydration; that is, the enthalpy associated with

$$\mathrm{Na^+(g) + Cl^-(g) \rightarrow Na^+(aq) + Cl^-(aq)} \tag{1.11}$$

is $-783.4\ kJ\ mol^{-1}$. This is a measure of the strong ion-dipole interactions between one mole of the two ions with the water molecules that form hydration spheres around them. The large, negative, value of the enthalpy of hydration plays the predominant part in determining the high solubility of salts in water. Strictly, of course, it is also necessary to consider the entropy of hydration, since it is the Gibbs energy change that occurs on dissolving a salt which determines the equilibrium constant for (1.10) (see (1.20), below) and hence the solubility of the salt. In general, the entropy of hydration of gaseous ions is negative, reflecting the ordering of the water molecules around the dissolved ions when hydration occurs.

At the other extreme from the dissolution of salts in water, we can consider the creation of ideal solutions or mixtures. In these cases, the enthalpy of mixing is zero (or, in reality, close to zero) and it is the increase in entropy which thermodynamically drives the mixing process. The paradigm for ideal solutions is a mixture of two closely related compounds, say benzene and toluene. The intermolecular forces between benzene-benzene, toluene-toluene, and benzene-toluene pairs are very similar, so that there is no net change in the strength of the interactions as these two liquids are mixed and hence the enthalpy of mixing is very small. One result of this is that the partial vapour pressures of the two components are proportional to their mole fraction in the mixture (Raoult's Law) and it can be shown that the change in Gibbs energy results entirely from the positive change in entropy that occurs on mixing.

An important and interesting class of solutes are those classed as *amphiphilic*, meaning they contain both *hydrophobic* groups (often referred to as their tails) and *hydrophilic* groups (their heads). The hydrophobic tail is often a large alkyl group, whereas the hydrophilic head might be ionised or strongly polar. The balance between the water-hating and water-loving properties of parts of these molecules will depend, *inter alia*, on the length of the hydrocarbon tail, and the strength of the interaction between the head and the solvent water molecules. Amongst these molecules are soaps and surfactants which accumulate at an interface between, for example, oil and water phases. Where there is only the water solvent available, at a sufficient concentration (the critical micelle concentration) and temperature (the Krafft temperature), soap molecules can aggregate to form micelles, which are clusters of molecules containing *ca.* 100–1,000 molecules, in which the hydrophobic tails are in the centre of the micelles and the hydrophilic heads are on the periphery. With still longer hydrophilic tails, surfactant molecules, such as the carboxylic acid, $CH_3(CH_2)_{16}COOH$ (stearic acid), congregate at the surface of water with their hydrophilic heads within the water and the tails sticking out of the water. Such films lower the surface tension of the solute and demonstrate behaviour, in two dimensions, that is analogous to that of real gases in three dimensions.

In addition to forming micelles, amphiphiles can also form double layers or membranes, in which the hydrophobic tails in each component of the layer point towards one another and intermingle whilst the hydrophilic heads point outwards into the water or aqueous solution on either side of the membrane. In lipid bilayers, which are important in biology (see Chap. 9), the constituent molecule can be a phospholipid, such as phosphatidic acid, which has a negatively charged phosphate head and two long hydrocarbon tails.

Earlier in this section, I contrasted the interaction energies for a chemical bond with the much smaller energies associated with the intermolecular forces arising from dispersion forces and the non-symmetric charge distributions in a pair of interacting molecules. Of intermediate strength is the hydrogen bond. A hydrogen bond arises between a proton donor, say X–H, and a proton acceptor, Y, where X and Y are electronegative atoms such as F or O. Generally, as in the water dimer, there is a strong directional attraction of the X – H bond to an electron-rich region on Y. The strength of such bonds varies widely but 30 kJ mol^{-1} is a sensible upper limit to normal hydrogen bonds. Hydrogen bonds can form between molecules when X – H and Y are in different molecules and an intermolecular hydrogen bond forms, or within a single molecule; that is, intramolecular hydrogen bonds can form. Because the strengths of hydrogen bonds are comparable with thermal energies at room temperature (where $RT \approx 2.5$ kJ mol^{-1}), the bonds can make and break relatively easily under ambient conditions. As a result they are of prime importance in biology, not least in binding single strands of DNA into the double helix.

One simple piece of evidence for hydrogen bonding comes from a comparison of the boiling points of a range of hydride compounds. The boiling points of the hydrides of the group 14 (see Table 1.1) elements (i.e., CH_4, SiH_4, GeH_4 and SnH_4), in which hydrogen bonding is not possible, rise monotonically as one should expect – reflecting the increase in molar mass and polarisability. However, for the hydrides of groups 15, 16 and 17, the boiling points for the lightest compounds are anomalously high – a result that is attributed to the existence of intermolecular hydrogen bonding.

Much has been learnt about hydrogen bonding from the application of rotational and vibrational spectroscopy [24, 25] to small, hydrogen-bonded, clusters formed in the super-cold environment of molecular jets. In the case of water, accurate structural information has been obtained on the clusters $(H_2O)_n$ from the dimer ($n = 2$) to the hexamer ($n = 6$) and theoretical calculations have been performed [20] that agree well with the experimental observations [26]. However, it is not easy to transfer what has been learnt from these studies to the structure of liquid water where hydrogen bonding is doubtless responsible for its unusual, indeed unique, properties. Water is so important, as a solvent and as a participant in biology, that it is the subject of Chap. 6. Here, I note that the temperature range in which it is liquid, (273.15–373.15 K under a pressure of 1 bar), and therefore available as solvent to accommodate and promote chemical and biological processes, is rather narrow. This has led, in astrobiology, to the concept of a habitable zone or Goldilocks zone: that being the range of distances from a star where an Earth-like planet can maintain liquid water on its surface, that being a prerequisite for life – or, at least, life as we know it.

1.5 Chemical Reactions: Thermodynamics and Kinetics [1, Chaps. 2, 3, 21, and 22; 2, Chaps. 3, 4, 6, and 9; 22]

Chemical models of the interstellar medium (see Chap. 4) contain *ca.* 4,500 gas-phase reactions. The vast majority of these are bimolecular reactions; that is, they occur as the result of binary collisions between, for example, an ion and a neutral species, two neutral species, and ions or molecules with electrons. The rate of such elementary processes, expressed in terms of the change in concentration with time (t) of the reactant or product species, is proportional to the product of the concentrations of the two reactants. If the reactants are represented by **A** and **B** and the products by **C** and **D**, the reaction is:

$$\mathbf{A} + \mathbf{B} = \mathbf{C} + \mathbf{D} \tag{1.12}$$

and, for the rate of the reaction, we can write:

$$-d\,[\mathbf{A}]/dt = -d[\mathbf{B}]/dt = +d[\mathbf{C}]/dt = +d[\mathbf{D}]/dt = k_f(T)[\mathbf{A}][\mathbf{B}] \tag{1.13}$$

Here, the square brackets denote concentrations in units of molecule cm^{-3}, or simply cm^{-3}, so that $k_f\,(T)$, the rate constant or rate coefficient for the 'forward' reaction between **A** and **B**,[4] has units of cm^3 $\mathrm{molecule}^{-1}\ \mathrm{s}^{-1}$, or $\mathrm{cm}^3\ \mathrm{s}^{-1}$, and ($T$) is included to emphasise that rate coefficients generally depend on temperature.

Chemical reactions proceed towards *equilibrium*, when there is no further net chemical change. At this point, 'reactants' and 'products' will still be present and, at a microscopic level, reactants will continue to form products and products to form reactants, but the rates of these processes have become equal. For the reaction represented by (1.12), this equality can be represented by the equation:

$$k_f(T)[\mathbf{A}]_e[\mathbf{B}]_e = k_r(T)[\mathbf{C}]_e[\mathbf{D}]_e \tag{1.14}$$

Here, $k_f(T)[\mathbf{A}]_e[\mathbf{B}]_e$ and $k_r(T)[\mathbf{C}]_e[\mathbf{D}]_e$ are the equal rates of the forward and reverse reaction. The symbols $[\mathbf{A}]_e$, $[\mathbf{B}]_e$, $[\mathbf{C}]_e$ and $[\mathbf{D}]_e$ are the concentrations of the species at equilibrium. The equilibrium constant can be expressed, in terms of the concentrations at equilibrium, as:

$$K_c(T) = \frac{[\mathbf{C}]_e[\mathbf{D}]_\mathbf{e}}{[\mathbf{A}]_e[\mathbf{B}]_e} \tag{1.15}$$

Combining (1.14) and (1.15) shows that $K_c(T)$ is the ratio of the rate coefficients for the forward and reverse reactions:

---

[4] By convention, the 'forward' reaction is that proceeding from left to right in the chemical equation, and the 'reverse' reaction is that proceeding from right to left.

$$K_c(T) = k_f(T)/k_r(T) \tag{1.16}$$

The application of thermodynamics to chemical reactions enables equilibrium constants to be calculated from a knowledge of the macroscopic thermal, or microscopic molecular, properties of the reactants (**A** and **B**) and the products (**C** and **D**). Using statistical thermodynamics [20], the equilibrium constant for a reaction involving gas-phase species, can be expressed, in terms of the per unit volume partition functions, (q_i/V), for the reactants and products, by

$$K_c(T) = \frac{(q_C/V)(q_D/V)}{(q_A/V)(q_B/V)} \exp(-\Delta E_0/k_B T) \tag{1.17}$$

ΔE_0 is the difference between the energies of the zero-point energy levels[5] in the products and in the reactants. If the sum of the zero-point energies for **C** and **D** is lower than the sum of the zero-point energies for **A** and **B**, ΔE_0 is negative (the reaction is exoergic) and $K_c(T)$ is greater than one: frequently very much greater than one since $-\Delta E_0$ is often $\gg k_B T$. This is true for many reactions that occur in the very low temperature ($T \approx 10$ K) cold cores of dense interstellar clouds. As a result, in modelling the chemistry in these environments, it is generally only necessary to include reactions proceeding in an exoergic direction. Interesting exceptions to this rule are those reactions where isotope exchange is involved [21]; for example:

$$H_3^+ + HD = H_2D^+ + H_2 \tag{1.18}$$

In such cases, the values of $-\Delta E_0$ and K_c are determined by the difference between the sums of the zero-point energies for the products and reactants. For reaction (1.18), $(-\Delta E_0/k_B)$ is equal to *ca.* 187 K. Although this is small relative to the same quantity for most chemical reactions, it is nevertheless much larger than the temperature in the cold regions of dense interstellar clouds where many molecules are found. Consequently, the observed fractions of many deuterated molecules are far larger than would be expected simply on the basis of the cosmic abundance of deuterium relative to hydrogen.

In classical thermodynamics, as distinct from statistical thermodynamics, the equilibrium constant for a reaction involving only gas-phase species is expressed in terms of the 'reduced' partial pressures of the reactants and products at equilibrium; that is, the partial pressures divided by the standard pressure (p^o), which is 1 bar. In these terms, the equilibrium constant can be expressed as;

---

[5] Associated with each vibration in a molecule, there is 'zero-point energy' corresponding approximately to ½hv where v is the frequency of the vibration. Consequently, as shown in Fig. 1.2, the lowest vibrational energy level with the quantum number $v = 0$, has an energy ½hv above the minimum of the potential energy curve.

$$K(T) = \frac{(p_{e,\mathrm{C}}/p^{\mathrm{o}})(p_{e,\mathrm{D}}/p^{\mathrm{o}})}{(p_{e,\mathrm{A}}/p^{\mathrm{o}})(p_{e,\mathrm{B}}/p^{\mathrm{o}})} \tag{1.19}$$

It should be noted that $K(T)$ will always be dimensionless, even when there are unequal numbers of products and reactants in the balanced chemical equation. For the elementary reaction, of the type represented by (1.12), $K_c(T)$ will not only be dimensionless but will also have the same numerical value as $K(T)$.

Classical thermodynamics relates the equilibrium constant, $K(T)$, to the standard Gibbs energy for the reaction, $\Delta_{\mathrm{r}}G^{\mathrm{o}}$, by the equation:

$$RT \ln K(T) = -\Delta_{\mathrm{r}}G^{\mathrm{o}} \tag{1.20}$$

When $\Delta_{\mathrm{r}}G^{\mathrm{o}}$ is negative the reaction is said to be exoergonic and $K(T)$ is greater than 1, frequently much greater than 1, and when $\Delta_{\mathrm{r}}G^{\mathrm{o}}$ is positive the reaction is said to be endoergonic and $K(T)$ is less than 1. Using the Gibbs-Helmholtz equation that relates Gibbs energy to enthalpy and entropy, we can write:

$$ln\, K(T) = -\Delta_{\mathrm{r}}H^{\mathrm{o}}/RT + \Delta_{\mathrm{r}}S^{\mathrm{o}}/R \tag{1.21}$$

In these equations, the sign Δ_{r} denotes the difference between the sum of the specified thermodynamic quantity for the products and that for the reactants, the superscript ^{o} denoting that the species are in their thermodynamic standard state, for gases 1 bar. These quantities can be derived from 'thermal measurements': that is, measurements of heats evolved for particular reactions, molar specific heats and latent heats. Such quantities can be determined (and then tabulated) for relatively stable species. When $-\Delta_{\mathrm{r}}H^{\mathrm{o}} \gg RT$, the value of the first term on the right-hand-side of (1.21) is likely to dominate the second term, and $K(T)$ for such strongly exothermic reactions can be huge, meaning that $k_f(T) \gg k_r(T)$.

Although thermodynamics has its uses in respect of astrochemistry, it is clear that the abundances of interstellar molecules are not determined by thermodynamics but rather by kinetics [27]. The relatively large observed abundances of free radicals and unsaturated hydrocarbon molecules clearly support this statement. It is useful to start a discussion of kinetics by recalling (1.16), writing it in logarithmic form, and then differentiating that equation with respect to $(1/RT)$:

$$ln\, K_c(T) = ln\, k_f(T) - ln\, k_r(T) \tag{1.22}$$

$$d\, ln\, K_c(T)/d(1/RT) = d\, ln\, k_f(T)/d(1/RT) - dln\, k_r(T)/d(1/RT) \tag{1.23}$$

The Van't Hoff equation expresses how $K_c(T)$ depends on temperature; that is:

$$d \ln K_c(T)/d(1/RT) = \Delta_{\mathrm{r}}U^{\mathrm{o}}/RT^2 \tag{1.24}$$

Van't Hoff then argued that the individual rate coefficients, $k_f\,(T)$ and $k_r\,(T)$, are influenced by temperature by two different energies, E_f and E_r, whose difference is $\Delta_r U^o$, so that (1.20) becomes [28]:

$$\Delta_r U^o/RT^2 = E_f/RT^2 - E_f/RT^2 \tag{1.25}$$

Van't Hoff recognized that generally $\Delta_r U^o$ depends on temperature and therefore the energies E_f and E_r may also be temperature-independent: if they are not, then E_f and E_r correspond to the activation energies (E_{act}) for the forward and reverse reactions and the rate coefficients vary with temperature according to what is usually designated as the Arrhenius equation:

$$k(T) = A\exp(-E_{act}/RT) \tag{1.26}$$

Extending van't Hoff's argument, one can combine (1.20) and (1.22) to derive an equation which relates the rate coefficients to changes in *Gibbs energy*:

$$\ln k_f(T) - \ln k_r(T) = -\Delta_r G_c{}^o/RT = -\Delta_f G_{act,c}{}^o/RT + \Delta_r G_{act,c}{}^o/RT \tag{1.27}$$

indicating that the rate coefficients for the forward and reverse reactions depend on the Gibbs energies of activation – therefore involving entropic, as well as enthalpic, factors. This argument leads to the definition of the transition state for a reaction being the point along the pathway leading from reactants to products where the free energy has its maximum value. The relationship between $-\Delta_r G_c{}^o$, $\Delta_f G_{act,c}{}^o$, and $\Delta_r G_{act,c}{}^o$ is illustrated in Fig. 1.5.

It was apparently a student of Van't Hoff, D. M. Kooij, who first allowed for the temperature-dependence of $\Delta_r U^o$, E_f and E_r by proposing a modified form of (1.26):

$$k(T) = A'T^m\exp(-E'_{act}/RT) \tag{1.28}$$

This equation and slight variants of it are frequently referred to as the modified Arrhenius equation and sometimes as the Kooij equation, and are often used to express the temperature dependence of the rate coefficients in kinetic databases compiled for use in modelling atmospheric, combustion and astrochemical environments. For example, in KIDA (a Kinetic Database for Astrochemistry [29]), rate coefficients and their temperature dependences are expressed by the equation:

$$k(T) = \alpha(T/300)^\beta\exp(-\gamma/T) \tag{1.29}$$

For many reactions between free radicals and neutral *saturated* molecules, $k(T)$ increases with temperature and β and γ in (1.29) both have positive values. The reactions between CN and H_2 and C_2H and H_2, that is

$$CN + H_2 \rightarrow HCN + H \tag{1.30}$$

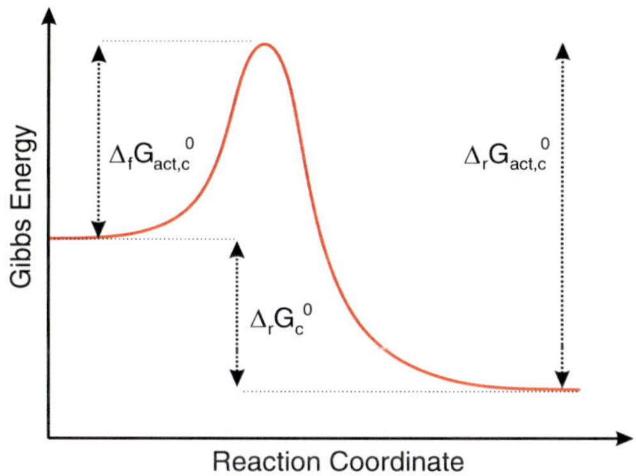

Fig. 1.5 Variation of Gibbs energy along the path for reaction, showing the relationship between the standard Gibbs energy for a reaction and the Gibbs energies of activation for the forward and reverse reactions

$$C_2H + H_2 \rightarrow HCCH + H, \tag{1.31}$$

provide examples [30]. For reaction (1.30), $\beta = 2.60$ and $\gamma = 960$ K, and for (1.31) $\beta = 2.32$ and $\gamma = 444$ K. The rate coefficients for these reactions are very small at low temperatures and they cannot play a role in the colder environments in the interstellar medium.

On the other hand, the rate coefficients for a number of reactions between strongly electronegative radicals (like CN) and unsaturated molecules increase as the temperature is lowered below 300 K [31]. For these reactions and those between two free radicals, it seems that there is no energy barrier preventing facile formation of a reactive intermediate which then decomposes to the final products. The values of the rate coefficients for such 'barrier-less' reactions are sometimes determined by 'capture'; that is, the bringing together of the reactants under the influence of long-range forces [see above, Sect. 1.4]. As discussed in Chap. 3, this is the mechanism for most reactions between ions and neutral molecules.

Up to this point, I have emphasised the application of thermodynamics to systems in the gas-phase. In solution, particularly in aqueous solutions where so much of biology occurs, the description of thermodynamic behaviour has to undergo some changes [1, Chap. 5; 2, Chaps. 5, 6 and 7]. In particular, it is impossible to apply statistical thermodynamics, an alternative definition of 'standard state' must be employed, and because the values of $\Delta_f H^o$ and S^o (and hence $\Delta_f G^o$) cannot be determined using the thermal properties of the species, they are *relative*, rather than *absolute*. In practice, the standard state is usually defined as that at unit activity, where activities can be thought of as concentrations corrected for non-ideal behaviour; that is departures from Henry's Law and Raoult's Law. The values of $\Delta_f H^o$, S^o and $\Delta_f G^o$ are evaluated relative to the same quantities for aqueous H^+ ions at unit activity; to put it another way, these three quantities are set equal to zero for aqueous H^+ ions, and the values for any other aqueous ion are found from measurements on reactions in which that ion and $H^+(aq)$ play a role.

These measurements are frequently made using electrochemical methods. The relationship between the Gibbs energy for the reaction ($\Delta_r G$) in an electrochemical cell and the potential (E) of the cell measured when no current is being withdrawn is

$$-\nu F E = \Delta_r G \tag{1.32}$$

The quantity F is the Faraday constant and has the value 96,485 C mol^{-1}. The cell reaction can be written as the difference between the reactions at the right-hand and left-hand electrodes. Thus for the well-known Daniell cell, Cu^{2+} ions are reduced at the right-hand electrode (the cathode):

$$\mathrm{Cu^{2+}(aq)} + 2e^- \rightarrow \mathrm{Cu(s)} \tag{1.33}$$

and, at the left-hand electrode (the anode), Zn^{2+} ions are oxidised:

$$\mathrm{Zn^{2+}(aq)} + 2e^- \leftarrow \mathrm{Zn(s)} \tag{1.34}$$

so that the overall cell reaction is:

$$\mathrm{Cu^{2+}(aq)} + \mathrm{Zn} \rightarrow \mathrm{Cu(s)} + \mathrm{Zn^{2+}(aq)}. \tag{1.35}$$

The integer ν in (1.29) refers to the number of electrons included in the equations representing the reduction and oxidation processes; that is, 2 in (1.33) and (1.34). E is, of course, independent of the choice of ν but $\Delta_r G$, an *extensive* quantity, does depend on it.

If E is evaluated with all the constituents in their standard states, then (1.32) yields $\Delta_r G^o$, the standard Gibbs energy for the cell reaction, so that combining (1.20) and (1.32):

$$RT \ln K = \nu F E^o \tag{1.36}$$

When the left-hand electrode in an electrochemical cell is a hydrogen electrode, where the reaction is

$$\mathrm{H^+(aq)} + e^- \rightarrow \tfrac{1}{2}\mathrm{H_2(g)} \tag{1.37}$$

The value of E measured under, or interpolated to, standard conditions, that is E^{o}, corresponds to the standard electrode potential for the right-hand electrode. (1.36) can be used to find the standard Gibbs energy change for the reaction occurring at this electrode, and measurements of how E varies with temperature yield values of $\Delta_r H^{o}$ and $\Delta_r S^{o}$. The value of E^{o} for any other cell can be calculated from the difference of the standard electrode potentials for the two electrodes.

The substances taking part in the 'half-reaction' at an electrode are frequently referred to as a redox couple and can generally be written as:

$$\mathrm{Ox} + \nu e^{-} \rightarrow \mathrm{Red} \tag{1.38}$$

In compilations of standard electrode potentials, it is usual to write the half-reactions in this way; that is with electrons on the left-hand side. For the processes represented by (1.33) and (1.34), the standard electrode potentials at 298 K are 0.34 V and -0.78 V, so that E^{o} for the cell is 1.10 V and $\Delta_r G^{o}$ for the overall reaction written in (1.35) is -101.1 kJ mol^{-1}.

The conventional standard state for aqueous hydrogen ion (unit activity, pH $= 0$) corresponds to very strongly acidic conditions and is inappropriate to normal biological conditions. Therefore, in biochemistry it is common to adopt pH $= 7$, that is a hydrogen ion activity of 10^{-7}, as the standard state. This difference between the definitions of the standard state is only important when a reaction involves H^{+}(aq) ions, as in:

$$\mathrm{A} + \nu\mathrm{H}^{+}(\mathrm{aq}) \rightarrow \mathrm{P} \tag{1.39}$$

Of course, such reactions are important in biochemical systems. The difference in the standard free energy of reaction using the biological standard state of $a = 10^{-7}$ and the usual chemical one of $a = 1$ is $\nu\, RT \ln 10^{-7}$, which amounts to -39.93 kJ mol^{-1} for $\nu = 1$ and $T = 298$ K. This type of reaction is more spontaneous (has a higher negative value of the standard Gibbs energy) at pH 0 than at pH 7. An example of such a reaction is the interconversion of the reduced and oxidised forms of nicotinamide adenine dinucleotide (NADH and NAD^{+}), which is a co-enzyme (see Sect. 1.6) found in all living cells [2, Chap. 6],

$$\mathrm{NADH} + \mathrm{H}^{+} \rightarrow \mathrm{NAD}^{+} + \mathrm{H}_2 \tag{1.40}$$

For the oxidation of NADH to NAD^{+}, the change in standard Gibbs energy using pH $= 0$ as the standard state is -21.8 kJ mol^{-1} but is + 18.1 kJ mol^{-1} for pH 7.

In biochemistry, a main function of NAD^{+} is its involvement in electron transfer reactions. Thus, the conversion of NADH to NAD^{+} releases two electrons that can then be used to reduce O_2 to water:

$$\mathrm{NAD}^{+} + \mathrm{H}^{+} + 2e^{-} \rightarrow \mathrm{NADH} \tag{1.41}$$

$$\frac{1}{2}O_2 + 2H^+ + 2e^- \rightarrow H_2O \tag{1.42}$$

The redox potentials for these two steps (referred to the biological standard state) are -0.32 V and 0.816 V, respectively; so that overall for

$$NADH + H^+ + \frac{1}{2}O_2 \rightarrow NAD^+ + H_2O \tag{1.43}$$

the change in standard Gibbs energy is -219 kJ mol^{-1}. In other words, reaction (1.43) is strongly exoergonic.

It is important to appreciate that endoergonic reactions can play a role in biochemical systems when they are coupled to exoergonic reactions so that the change in Gibbs energy for the overall process is negative. Many important biochemical processes involve a complex sequence of such reactions. The exoergonic reaction is frequently the hydrolysis of ATP (adenosine 5'-triphosphate) to ADP^+ (adenosine 5'-diphosphate) + HPO_4^{2-} and the coupled reactions are catalysed by enzymes (see Sect. 1.6).

Just as in the gas-phase, thermodynamics tells only part of the story in respect of reactions in solution: kinetics also plays its part. An important additional consideration is that, in solution, if a bimolecular reaction is intrinsically fast as, for example, in acid–base neutralisation, the rate-determining process can be the diffusion of the reactants through the solvent before they encounter one another. If the reaction occurs every time the reactants (say, **A** and **B**) meet and they are assumed to be spheres with radii r_A and r_B, it can be shown that the rate coefficient (k_D) for the diffusion-controlled reaction is given by:

$$k_D = 4\pi N_A (r_A + r_B)(D_A + D_B) \tag{1.44}$$

where N_A is the Avogadro constant and D_A and D_B are the diffusion coefficients of **A** and **B**. If Stoke's law is assumed to govern the diffusion of the reactants and they are also assumed to be the same size, the expression for the rate coefficient reduces to

$$k_D = 8RT/3\eta \tag{1.45}$$

where η is the coefficient of viscosity of the solvent.

1.6 Surfaces, Interfaces and Catalysis

Much interesting chemistry, and biology, occurs not in a single phase (gas, liquid or solid) but at a surface or an interface between phases [1, Chap. 23; 2, Chap. 10]. The term 'surface' is generally used when one of the phases is gaseous, the other liquid or solid; whereas 'interface' is generally used for the boundary between two condensed phases; for example, two solutions. I begin by considering some aspects of surface chemistry, which are relevant, *inter alia*, to the chemistry that might occur on the dust grains that are found in those cold dark regions of the interstellar medium that are richest in molecules.

When molecules cling to the surface of a solid, we speak of adsorption. It is useful to begin by distinguishing two kinds of adsorption according to the strength of the attraction between the adsorbed molecules and the solid surface. If the interaction is weak, arising from van der Waals forces (see Sect. 1.4), it is customary to speak of physical adsorption or physisorption. The enthalpy change when a molecule is physisorbed is comparable to the enthalpy of condensation of the molecule and is typically *ca.* 20 kJ mol^{-1}. Physisorbed molecules retain their integrity – that is, no bonds are broken – and they are likely to be quite mobile on the surface.

By contrast, in chemical adsorption or chemisorption, the molecules are bound to the surface by forming a chemical bond and the magnitude of the enthalpy of adsorption will be much larger than in physisorption, of the order of the strength of a chemical bond; *ca.* 200 kJ mol^{-1}. In general, and again in contrast to the situation for physisorption, chemisorption is 'activated': that is, the rate at which molecules are chemisorbed will increase with increasing temperature. Because new bonds are created in chemisorption, the bonds within the adsorbed molecules are generally weakened, with the result that chemisorption can lower the activation energy for certain chemical reactions. Hence chemisorption can create the conditions necessary for the heterogeneous catalysis of chemical reactions and this is important in a number of industrial processes. The classic example of heterogeneous catalysis is the Haber-Bosch process for the production of ammonia from nitrogen and hydrogen:

$$N_2(g) + 3H_2(g) + \text{catalyst} \rightarrow 2\ NH_3(g) + \text{catalyst} \tag{1.46}$$

The catalyst[6] is usually of iron promoted with K_2O, CaO and Al_2O_3. The physical conditions (150–250 bar pressure and a temperature between 300 °C and 550 °C) are chosen to strike a balance between maximising both the yield of NH_3 (thermodynamics) and the rate of production of NH_3 (kinetics).

The operating conditions for the Haber-Bosch process are far from the physical conditions in those regions in the interstellar medium where the majority of molecules are found. On the other hand, the low temperatures in cold, dark molecular clouds do 'encourage' the occurrence of physisorption, and it is also necessary to recognise that many of the species that may be physisorbed are unsaturated molecules or radicals for which low energy reaction pathways may exist on dust grains.

---

[6] A catalyst accelerates the rate of reaction without changing the position of equilibrium and without itself being changed during the course of reaction. As the equilibrium constant is not changed, it means that any catalyst must accelerate the rates of forward and reverse reactions in equal proportions. The changes in rate occur because the catalyst lowers the Gibbs energies of activation for the forward and reverse reactions.

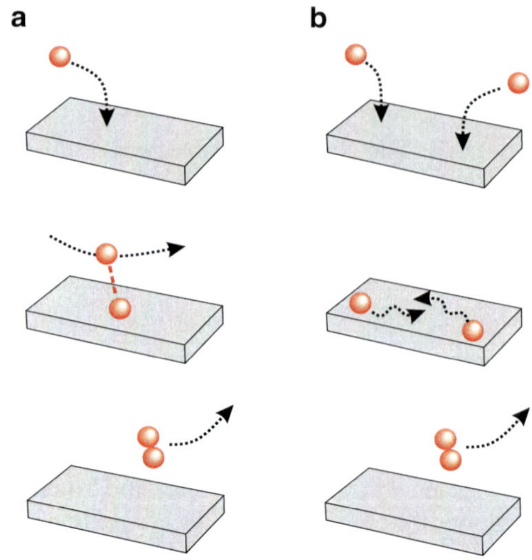

Fig. 1.6 Cartoons illustrating the (**a**) Eley-Rideal and (**b**) Langmuir-Hinshelwood mechanisms for reactions at surfaces

In surface chemistry, adsorption isotherms describe the equilibrium situation. However, just as in the consideration of the gas-phase chemistry in the interstellar medium, it is the *kinetics* of surface processes which are more relevant. Two mechanisms for surface-catalysed reactions can be distinguished and are illustrated by the cartoons in Fig. 1.6. In the Eley-Rideal mechanism, it is assumed that reaction occurs when a species (say, ***A***) from the gas-phase impacts on a species (say, ***B***) that is adsorbed on the surface. At significant surface coverage, the rate of reaction will be proportional to the product of the fraction of the surface covered in ***B*** ($\theta_{\boldsymbol{B}}$) and the pressure ($p_{\boldsymbol{A}}$) of the species ***A***, which will be proportional to the rate of collisions of ***A*** with unit area of the solid surface. An alternative picture is encapsulated in the Langmuir-Hinshelwood mechanism. Here it is assumed that reaction occurs in encounters between species both of which are adsorbed on the surface. Then the rate of reaction will be proportional to the product of the fractions of the surface covered by ***A*** and by ***B***; that is proportional to $\theta_{\boldsymbol{A}}\theta_{\boldsymbol{B}}$.

The role of surface-catalysed reactions in determining the abundances of molecules observed in the interstellar gas remains a topic of intense research (see Chap. 4). It is accepted that in cold, dark interstellar clouds the recombination of hydrogen atoms to form H_2 occurs on the surface of dust grains, causing H_2 to be the dominant form of hydrogen in these regions of the interstellar medium. In this process, the surface serves to stabilise the freshly made molecules by removing at least some of the energy that is released when the chemical bond forms. In the interstellar medium, even in dense clouds, the gas density is far too low for this mediating influence to be provided in ‘three-body collisions’ in the gas-phase. Because of its high vapour pressure, even at the very low temperatures in these environments, H_2 formed on surfaces can escape into the gas-phase. It is less clear if other molecules that are formed on dust grains can similarly escape from grains in the cold cores of interstellar clouds, or if they only escape either as the clouds warm up under the influence of gravitational collapse or as a result of heating by shocks. It is possible that newly formed molecules can utilise the energy released as they form to overcome the energy binding them to the surface and hence escape into the gas-phase. Chemical models of interstellar clouds, which incorporate surface chemistry as well as the freezing-out of molecules when they strike the cold surfaces of dust grains, have now been developed and they are considered in Chap. 4.

Heterogeneous catalysis is one of the three main categories of catalysis, the others being homogeneous catalysis and enzymatic catalysis. The study of homogeneous catalysis, particularly acid–base catalysis, has a venerable history starting with Kirchhoff's studies of the conversion of starch to glucose two centuries ago and it is important in organic chemistry and biochemistry. The word 'homogeneous' implies that the reactants and catalyst are all in the same phase, most generally in aqueous solution. An example is the bromination of acetone which is subject to catalysis by both acids and bases. In acid solutions, the rate-determining step is the formation of the enol form:

$$\mathrm{CH_3C(O)CH_3 + HA \rightarrow [\mathit{CH_3}C^+(OH)CH_3) + A^-] \rightarrow H_2C = C(OH)CH_3 + HA} \tag{1.47}$$

followed by the fast bromination of the enol:

$$\begin{aligned}\mathrm{H_2C} &= \mathrm{C(OH)CH_3 + Br_2 \rightarrow [CBrH_2C^+(OH)CH_3) + Br^-]} \\ &\rightarrow \mathrm{CBrH_2C(O)CH_3 + H^+ + Br^-}\end{aligned} \tag{1.48}$$

It is interesting to note that the overall reaction generates hydrogen ions so that it is said to be autocatalytic: that is, its rate increases as the reaction proceeds due to the increase in the hydrogen ion concentration (a decrease in the pH of the solution) arising from the reaction.

Clever as mankind has been in devising catalysts for boosting the industrial production of important chemicals, enzymatic catalysis, which has evolved in nature, is truly remarkable. Not only do enzymes hugely increase the rate of natural processes (by factors of the order of 10^6–10^{18} [2, p. 363]) but they achieve this effect with great selectivity; that is, the desired reaction proceeds much faster than in the absence of the enzyme, but other closely related reactions are not accelerated by the enzyme.

An enzyme (denoted here by ***E***) is generally a protein which contains one or more 'active sites' to which a reactant molecule can bind. In a sense, enzymatic catalysis is intermediate between homogeneous and heterogeneous catalysis in that the active site or sites are on the surface of the enzyme but the enzyme and reactant molecules are in the same solution phase. In the first step of the catalysed reaction, the reactant molecule, usually referred to as the substrate (***S***), binds to an active site on the enzyme, in a process which is reversible and which generally utilises intermolecular forces, of the kind considered in Sect. 1.4, to form an enzyme-substrate complex (***ES***). As in other kinds of catalysis, the role of the enzyme is to

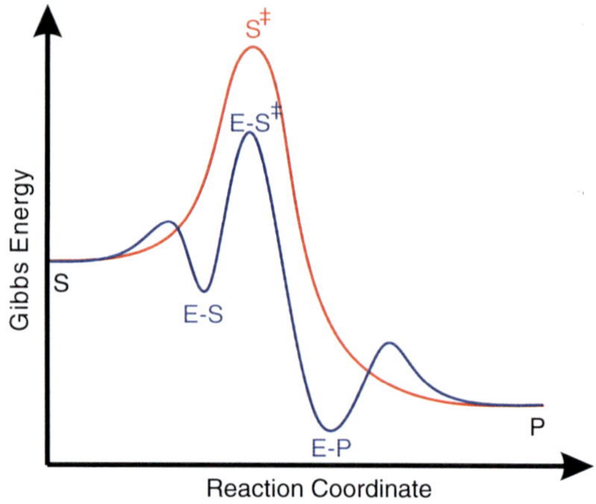

Fig. 1.7 The variation of Gibbs energy for an enzyme-catalysed reaction (in *blue*) showing the decrease in the Gibbs energy of activation compared with the uncatalysed reaction (in *red*)

reduce the free energy of activation. Consequently, the binding between the enzyme and substrate in the transition state ($\boldsymbol{ES}^{\ddagger}$) must be stronger than in the initially formed enzyme-substrate complex, and covalent bonding may be utilised to achieve this binding. The consequent lowering in Gibbs energy of activation compared with the uncatalysed reaction is illustrated in Fig. 1.7.

The product that is formed may itself bind to the enzyme, but not strongly, so that reaction proceeds to the product (***P***) of the reaction. Symbolically, these steps may be written:

$$\boldsymbol{E} + \boldsymbol{S} = \boldsymbol{ES} \rightarrow \boldsymbol{ES}^{\ddagger} \rightarrow \boldsymbol{EP} \rightarrow \boldsymbol{E} + \boldsymbol{P} \tag{1.49}$$

Here, the $=$ sign signifies that the first step of the reaction is reversible and the $\rightarrow$ s that reaction proceeds predominantly, in the forward direction, through the transition state, $\boldsymbol{ES}^{\ddagger}$, and the product-enzyme complex, ***EP***, so that the kinetics can be treated via the simplified scheme:

$$\boldsymbol{E} + \boldsymbol{S} \underset{k_{-1}}{\overset{k_1}{=}} \boldsymbol{ES} \overset{\mathrm{k}_2}{\rightarrow} \boldsymbol{E} + \boldsymbol{P} \tag{1.50}$$

An analysis of this kinetic scheme was proposed by Michaelis and Menten and later modified by Briggs and Haldane [2, p. 367–370]. The latter full treatment makes use of the steady-state approximation, which states that very soon after the start of the reaction the rate of formation of ***ES*** will be equal to its rate of loss, so that:

$$k_1[\boldsymbol{E}][\boldsymbol{S}] = k_{-1}[\boldsymbol{ES}] + k_2[\boldsymbol{ES}] \tag{1.51}$$

Recognising that the enzyme is present both as the enzyme–substrate complex, ***ES***, and as uncomplexed enzyme, ***E***, the concentration of ***E*** can be written in terms of the total enzyme concentration, $[\boldsymbol{E}_0] = [\boldsymbol{E}] + [\boldsymbol{ES}]$, so that:

$$k_1([\boldsymbol{E_0}] - [\boldsymbol{ES}])[\boldsymbol{S}] = (k_{-1} + k_2)[\boldsymbol{ES}] \tag{1.52}$$

and

$$[\boldsymbol{ES}] = \frac{k_1[\boldsymbol{E}_0][\boldsymbol{S}]}{k_1[\boldsymbol{S}] + k_{-1} + k_2} \tag{1.53}$$

The rate of reaction (v_0) is clearly the rate of conversion of ***ES*** to ***E*** + ***P***, so that it can be written as:

$$v_0 = k_2\,[\boldsymbol{ES}] = \frac{k_2 k_1[\boldsymbol{E}_0][\boldsymbol{S}]}{k_1[\boldsymbol{S}] + k_{-1} + k_2} = \frac{k_2[\boldsymbol{E}_0][\boldsymbol{S}]}{[\boldsymbol{S}] + K_{\mathrm{M}}} \tag{1.54}$$

where $K_{\mathrm{M}} = (k_{-1} + k_2)/k_1$ and is generally known as the Michaelis constant. The maximum rate of reaction ($V_{\max}$), that is achieved at high [S], is equal to $k_2[\boldsymbol{E}_0]$ and K_{M} corresponds to the concentration of substrate at which the rate of reaction half its maximum value. In order to find the values of $V_{\max}$ and K_{M} for a particular reaction, (1.54) is usually re-written in a form that creates a linear plot. A number of procedures have been adopted, most commonly:

$$\frac{1}{\mathrm{v}_0} = \frac{K_M}{V_{\max}[\boldsymbol{S}]} + \frac{1}{V_{\max}} \tag{1.55}$$

Finally, it is useful to note that the catalytic efficiency reaches a maximum when the formation of the enzyme-substrate complex ***ES*** is rate-determining. This corresponds to the situation where $k_2 >> k_{-1}$, so that every ***ES*** which is formed is converted to product, none re-dissociates to ***E*** + ***S***. However, the rate cannot exceed the rate of the collisions of ***E*** and ***S***, and this upper limit is determined by the rate of diffusion in the solution. Experiments on the kinetics of enzyme-catalysed reactions demonstrate that a number of them achieve this state of 'catalytic perfection' [2, p. 372].

1.7 Some Concluding Remarks

In this chapter, I have sought to introduce some of those topics in physical chemistry which are especially relevant when considering areas of astrobiology and astrochemistry. In the next eight chapters of this book, some of these 'astro-topics' are discussed in some detail, with emphasis being placed on the physicochemical principles that guide our understanding in these fields. In all of these chapters, use will be made of some of the topics in physical chemistry that have been introduced in the present chapter.

As stated earlier in the preface, the next three chapters deal with subjects that are traditionally assigned to astrochemistry; that is, with the identification of molecules, mainly small molecules, in the interstellar medium and the physical conditions in those regions of space where they are found, and with the chemical processes which lead to the formation and destruction of these molecules. The application of spectroscopy, at several wavelengths, is central to the former and an understanding of chemical kinetics – both homogeneous and heterogeneous – underpins the latter. These subjects bring together laboratory scientists, both experimentalists and theorists, astronomical observers, and modellers.

Spectroscopy will also be vital if and when we can search the atmospheres of potentially habitable planets for the presence of molecules that indicate the existence of life, a topic discussed in Chap. 5. The discussion of intermolecular forces, especially hydrogen bonding, in Sect. 1.4 serves as an introduction to Chap. 6, which is devoted to the role in biochemical systems of a molecule, water, whose universality on Earth might blind us to its remarkable properties. Quite a lot of this introductory chapter has been devoted to thermodynamics. The role and importance of thermodynamics when we consider what conditions might lead to and sustain life are particularly brought out in some of the later chapters of this book. The forces between dissolved species and their solvent and between molecules that are at the boundary of solubility and therefore can form micelles and lipid bilayers were introduced in Sect. 1.4. These species and their properties re-emerge in Chap. 9 as does the topic of enzyme catalysis introduced in Sect. 1.6.

Acknowledgements I am very grateful to Dr. Els Peeters for providing Fig. 1.4 and to Dr. Peter Barnes for his assistance in preparing the other diagrams for this article.

References

1. Atkins PW, de Paula J (2010) Physical chemistry, 9th edn. Oxford University Press, Oxford
2. Chang R (2005) Physical chemistry for the biosciences. University Science, Sausalito
3. Greenwood NN, Earnshaw A (1997) Chemistry of the elements, 2nd edn. Butteworth-Heinemann, Oxford, chaps. 1 and 2
4. Cameron AGW (1973) Abundances of elements in solar system. Space Sci Rev 15:121–146
5. Newson HE (1995) Composition of the solar system, planets, meteorites, and major terrestrial reservoirs. In: Ahrens TJ (ed) Global earth physics: a handbook of physical constants. American Geophysical Union, Washington, DC
6. Duncan DL, Harvie JL, McKean DC, Cradock S (1986) The ground state structures of disilane, methyl silane and the silyl halides, and an SiH bond length correlation with stretching frequency. J Mol Struct 145:225–242
7. Harmony MD (1990) The equilibrium carbon-carbon single bond length in ethane. J Chem Phys 93:7522–7523
8. Bernath PF (1995) Spectra of atoms and molecules. Oxford University Press, Oxford

9. Herbig GH (1995) The diffuse interstellar bands. Annu Rev Astron Astrophys 33:19–73
10. Jenniskens P, Désert F-X (1994) A survey of the diffuse interstellar bands (3800–8680 Å). Astron Astrophys Suppl Ser 106:39–78
11. Wayne RP (2000) Chemistry of atmospheres. Oxford University Press, Oxford
12. McKellar A (1940) Evidence for the molecular origin of some hitherto unidentified interstellar lines. Publ Astronom Soc Pac 52:187; (b) Adams WS (1941) Some results with the Coudé spectrograph of the Mount Wilson observatory. Astrophys J 93:11; (c) Douglas AE, Herzberg G (1941) CH^+ in interstellar space and in the laboratory. Astrophys J 94:381
13. Burgh EB, France K, McCandliss SR (2007) Direct measurement of the ratio of carbon monoxide to molecular hydrogen in the diffuse interstellar medium. Astrophys J 658:446
14. Wakelam V, Smith IWM, Herbst E, Troe J, Geppert W, Linnartz H et al (2010) Reaction networks for interstellar chemical modelling: improvements and challenges. Space Sci Rev 156:13–72
15. Balle TJ, Flygare WH (1981) Fabry-Perot cavity pulsed Fourier-transform microwave spectrometer with a pulsed nozzle particle source. Rev Sci Instrum 52:33–45
16. Thaddeus P, McCarthy MC, Travers MJ, Gottlieb CA, Chen W (1998) New carbon chains in the laboratory and in interstellar space. Faraday Discuss 109:121–135
17. Herbst E, van Dishoeck EF (2009) Complex organic interstellar molecules. Ann Rev Astron Astrophys 47:427–480
18. Peeters E (2011) Astronomical observations of the PAH emission bands. In: Joblin C, Tielens AGGM (eds) PAHs in the universe. EAS Publication Series 46:13–27
19. Rigby M, Smith EB, Wakeham WA, Maitland GC (1986) The forces between molecules. Clarendon, Oxford
20. Buckingham AD, Del Bene JE, McDowell SAC (2008) The hydrogen bond. Chem Phys Lett 463:1–10
21. Smith IWM (1980) Kinetics and dynamics of elementary gas reactions. Butterworths, London
22. Pilling MJ, Seakins PW (1995) Reaction kinetics. Oxford University Press, Oxford
23. Georgievskii Y, Klippenstein SJ (2005) Long-range transition state theory. J Chem Phys 122:194103-1–194103-17
24. Legon AC, Millen DJ (1987) Directional character, strength, and nature of the hydrogen bond in gas-phase dimers. Acc Chem Res 20:39–46
25. (a) Miller RE (1990) Vibrationally induced dynamics in hydrogen-bonded complexes. Acc Chem Res 23:10–16. (b) Nesbitt DJ (1988) High-resolution infrared-laser spectroscopy of weakly bound molecular complexes. Chem Rev 45:843–870
26. Liu K, Cruzan JD, Saykally RJ (1996) Water clusters. Science 271:929–933
27. Klemperer W (2011) Astronomical chemistry. Ann Rev Phys Chem 62:173–184
28. Smith IWM (2007) The temperature-dependence of elementary reaction rates: beyond Arrhenius. Chem Soc Rev 36:1–15
29. Wakelam V, Herbst E, Loison JC, Smith IWM, Chandrasekaran V, Pavone B et al (2012) A kinetic database for astrochemistry (KIDA). Astrophys J Supplt 199:21 and www.kida.obs.u-bordeaux1.fr
30. Baulch DL, Bowman CT, Cobos CJ, Cox RA, Just Th, Kerr JA et al (2005) Evaluated kinetic data for combustion modelling: supplement II. J Phys Chem Ref Data 34:757–1397
31. Smith IWM, Sage AM, Donahue NM, Herbst E, Quan D (2006) The temperature-dependence of rapid low temperature reactions: experiment and prediction. Faraday Discuss 133:137–156
32. McClelland BJ (1973) Statistical thermodynamics. Chapman and Hall, London
33. Herbst E, Millar TJ (2008) The chemistry of cold interstellar cloud cores. In: Smith IWM (ed) Low temperatures and cold molecules. Imperial College Press, London, pp 1–54

Chapter 2
The Molecular Universe

Maryvonne Gerin

Abstract This chapter presents a description of the interstellar medium. It starts with a summary of the interstellar medium structure and how the various phases are related to each other. The emphasis is put on molecular clouds, and on their densest regions, the dense cores, which are the birth place of stars. The evolution of matter during the star formation process and its observable consequences, especially in term of chemical composition is presented. The next section is dedicated to the constituents of the interstellar medium, with separate presentations of the gas species and the dust grains. Methods used by astronomers to derive useful information on the structure, temperature, ionization rate of interstellar environments as well as magnetic fields are briefly described. The last part of the chapter presents the telescopes and their instruments used for studying the interstellar medium across the electromagnetic spectrum.

2.1 Introduction

The formation of the first galaxies is now understood in the large scale context of the evolution of the Universe. Starting from the first seeds evidenced as tiny fluctuations in the Cosmic Microwave Background (CMB), the combined actions of expansion and gravity led to the growth of large scale structures, in the form of sheets and filaments of denser material, surrounded by large voids. Baryons condensed in the filaments to form the first stars and galaxy embryos. The first stages of this evolution were dominated by dark matter since the dark matter haloes were far larger in size and mass than individual galaxies, and therefore dominated the gravitational potential. Conversely, the last steps in the formation of galaxies and stars within these dark matter haloes were governed by the non-linear physics

M. Gerin (✉)
LERMA (Observatoire de Paris, CNRS and ENS), Paris Cedex 05 FR 75321, France
e-mail: maryvonne.gerin@lra.ens.fr

I.W.M. Smith et al. (eds.), *Astrochemistry and Astrobiology*, Physical Chemistry in Action,
DOI 10.1007/978-3-642-31730-9_2,

of the visible matter. This includes the complex cooling and heating effects of neutral and molecular gas, the action of magnetic fields on the large scale flows of matter, as well as the important dynamical and radiative feedback effects created by newly formed stars. Indeed as soon as a first generation of stars had been formed, the so-called Population III stars, the interstellar medium became enriched in heavy elements (compared to the initial composition resulting from the primordial nucleosynthesis shortly after the Big Bang), leading to more possibilities for cooling and heating the gas, and an easier synthesis of molecules.

The presence of molecules in high redshift massive galaxies [1] as well as in some absorption systems [2] is remarkable given the young age of the Universe at such redshifts: for instance $z = 6$ corresponds to less than 10^9 years after the Big Bang while the age of the Sun and the Solar System is 4.6×10^9 years. It demonstrates the very strong similarities of processes occurring along the whole evolutionary path, and the need for a rather fast formation of dust grains and molecules, at least in the most massive and densest systems. Stars contribute to the enrichment of the interstellar medium by ejecting matter outside of their envelope. This stellar matter, composed both of molecular gas and dust grains, contains the elements synthesized in the stellar core that are transported to the stellar surface and envelope by strong convective motions. In astrophysics, the term 'heavy elements' refers to those elements heavier than boron, which are only synthesized in stars. Therefore they include carbon, oxygen, and nitrogen for instance.

2.2 The Life Cycle Interstellar Medium

2.2.1 *Phases of the Interstellar Medium*

In the Milky Way, the interstellar medium has distinct phases that result from the interplay of radiation, large scale flows, magnetic fields and turbulence in the gas. Those phases are close to pressure equilibrium, with a median pressure[1] of $p \sim 3{,}500$ K cm^{-3}, which implies that the hottest phases have the lowest average particle densities. Table 2.1, adapted from Draine [3] presents the properties of the main interstellar phases.

The structure of the interstellar medium phases, and especially how they are spatially organized, is not fully elucidated. Large scale surveys of tracers of the neutral, ionized and coronal gas have led to determination of the global filling factors, and of the average physical conditions as listed in Table 2.1. The coronal gas, with temperatures in excess of 10^5 K is mostly formed by the bubbles created in

[1] It is customary to quote interstellar pressures as the product of the temperature and particle density, since the equation of state of ideal gas applies to these very dilute media (PV = nRT).

Table 2.1 Summary of interstellar phases (Adapted from Draine [3])

Phase	T/K	n/cm^{-3}	Volume (%)	Mass (%)
Coronal gas	$10^{5.5}$–10^7	0.004	50	<1
Warm ionized medium	$\sim 10^4$	0.3	10	23
Warm neutral medium	~5,000	0.6	40	36
Cold neutral medium	40–100	~30	1	24
Diffuse molecular gas	30–100	30–300	0.1	0.17
Dense molecular cores	6–50	10^3–10^6	0.01	<0.05
HII regions	10^4	10^2–10^4	0.01	<0.01

supernovae explosions and other energetic events. Because of its extreme temperature, this component represents a small fraction of the mass, but fills a significant fraction of the Galaxy volume. Most of the mass of ionized gas resides in the Warm Ionized Medium (WIM), a widely distributed component of the interstellar gas. This medium is kept ionized by the combined ionizing radiations of the massive stars of stellar types O and B, and to a lesser degree from other stellar populations such as white dwarfs.

Dynamical models (e.g. [4, 5]) have been constructed to study the formation of these phases and their interplay. The whole pattern is dynamical, with supernovae explosions and stellar winds from massive stars creating bubbles of hot gas expanding through the diffuse medium, accompanied by large scale motions and shock waves.

2.3 The Structure of the Neutral Interstellar Gas

While the warm ionized phases occupy most of the volume of the galaxy, they correspond to a relatively small fraction of the total mass. For star formation as well as for understanding molecular complexity, it is natural to focus on the phases where the hydrogen is in neutral form, either atomic or molecular.

Large scale surveys have established the overall spatial distribution and structure of these phases. The neutral atomic gas can exist in two stable phases, called the Warm Neutral Medium (WNM) and the Cold Neutral Medium (CNM) following the pioneering work of Field et al. [6]. These phases are roughly in pressure equilibrium with a median pressure of ~3,500 K cm^{-3} [7]. Observations of the HI hyperfine transition at 21 cm have confirmed the presence of these stable phases [8], but have also revealed that a significant fraction of the gas resides in the thermally unstable region of the pressure/density diagram. As for the ionized phases, it appears that the neutral gas is not in a static equilibrium, and continuously evolves from one phase to the other. Audit and Hennebelle [9] among others have performed extensive simulations of the dynamics of such a bistable medium.

In the cold neutral medium, hydrogen is not uniquely present in atomic form. Indeed, molecular hydrogen, together with other molecules with a small fractional abundance (of the order of 10^{-8}) are known to be present in the diffuse gas. The fraction of hydrogen in molecular form, $f(H_2) = 2N(H_2)/(N(HI) + 2\,N(H_2))$ varies from less than 10% in low extinction regions ($A_V \leq 0.1$ mag), up to nearly 100% in translucent gas with extinctions of a few magnitudes [10].

The overall structure of the diffuse neutral gas therefore appears to be governed by the interplay of radiation, turbulence and magnetic fields, with its specific equation of state. The structure of this diffuse gas is best studied by the HI 21 cm emission, and by the far infrared and sub-millimetre dust emission between 100 and 900 μm. The typical temperature of dust grains in these regions is 18 K, hence they radiate mostly at far infrared wavelengths. The IRAS satellite, followed by more powerful space missions (ISO, Spitzer, Herschel), have produced sensitive and large scale images of the dust thermal radiation at far infrared wavelengths, leading to a good statistical characterization of the structure. It is remarkable that the structure follows the same statistical properties from spatial scales of 8° down to 30″ (~15 to ~0.015 pc, where one parsec (pc) corresponds to 3.08×10^{16} m or about 3.26 light years) in the Polaris diffuse cloud [11]. Large scale flows induced by dynamical events, combined with turbulence, stir the gas and induce local compressions where cold neutral medium cloudlets can form. Numerical simulations have been performed to study the formation and evolution of the atomic gas (e.g. [9]). The simulated structures share many statistical properties with the interstellar gas, demonstrating that the main physical ingredients are well captured by such simulations.

Gravity becomes important in understanding the formation of more massive structures, the molecular clouds. Indeed, giant molecular clouds (GMCs) are the most massive structures in the Galaxy, with masses up to 10^7 $M_\odot$. The giant molecular clouds are located at the upper end of the hierarchical organization of molecular clouds, and follow the same scaling laws, with their mass (M) varying approximately as the square of their radius (R), and their internal velocity dispersion (σ) as the square root of their radius [12]. These scaling laws can be understood as revealing the interplay of the main acting forces, gravity, turbulence and magnetic fields. Indeed, for a system in virial equilibrium it is expected that the gravitational energy (scaling as $G\,M^2/R$) is balanced by the kinetic energy (scaling as $M\sigma^2$), which leads to the near constancy of the factor $M\sigma^2/R$. These scaling laws are reminiscent of the scaling laws for incompressible turbulence, that relate the scale size l and the local velocity dispersion at this scale δv as $l \sim \delta v^{\alpha}$. The exponent α that depends on the properties of turbulence is expected to be 1/3 for incompressible turbulence as first predicted by Kolmogorov [13]. In the interstellar medium, the scaling deviates from pure incompressible turbulence behaviour, as expected from the more complex nature of the interstellar medium (see [14] for a recent introduction).

2.3.1 *Dense Interstellar Cores*

2.3.1.1 Low Mass Cores

While the mean density remains moderate in the diffuse interstellar medium and in molecular clouds, with typical figures ranging between 100 and 1,000 cm^{-3}, high density regions, i.e. regions with densities larger than 10^4 cm^{-3} exist, but they occupy a small fraction of the total volume (Table 2.1). Such high density regions are called "dense cores", since they have specific properties compared to the diffuse and medium density medium: in addition to their higher density, they are associated with larger total gas column densities, smaller velocity dispersions, and, before stars form, cooler temperatures than their surrounding medium.

The excellent imaging capabilities of the Herschel space telescope have revealed, with unprecedented detail, the structure of molecular clouds and the regions where dense cores are formed. It is now clear that dense cores form within the ubiquitous filamentary structure of interstellar clouds, as localized density and column density maxima. There is a threshold in gas column density for dense core formation, estimated to be at extinctions of seven magnitudes from the analysis of either deep extinction images [15] or dust sub-millimetre emission maps [16]. With such a threshold, the dense cores will be commonly well shielded from the far ultra violet (FUV) radiation, with some exceptions in massive star forming regions or in photodissociation regions (PDRs) where dense molecular gas becomes directly exposed to FUV radiation.

It is now well established that dense cores are the birthplaces of stars. Most cores have relatively low masses, comparable or slightly larger than the mass of the sun, and will be able to form low mass stars like the sun only. The dense cores localized in the nearby molecular clouds such as the Taurus and Perseus complexes have been identified and catalogued through extensive surveys in the last decades. Dense cores are cold objects, with inner temperatures of about 10 K. Their typical size is 0.1 pc (3.08×10^{15} m, or 20,000 AU), and the typical line widths of molecular lines range between 0.2 and 0.6 km s^{-1}, a figure comparable with the thermal velocity dispersion of molecular hydrogen at a kinetic temperature of 10 K (0.2 km s^{-1}). The dark cloud Barnard 68 illustrated in Fig. 2.1 is considered as a model object of this class. Located at a distance of 125 pc, it appears as a dark patch hiding the stellar background. Although not completely axisymmetric, it can be well described as a spherical object, with a smoothly varying density profile that can be well fitted with a so-called Bonnor-Ebert profile, of the class of self-similar solutions of self gravitating isothermal gas spheres. This class of profiles is characterized by an inner plateau of nearly constant density, and a steep density decrease at the edge, scaling as r^{-2}, r being the local radius. In the case of Barnard 68, the maximum density is $n \sim 2.5 \times 10^5$ cm^{-3}. Barnard 68 does not host any young star or proto-stellar object, and seems to be in quasi-static equilibrium. Other dense cores show more concentrated profiles (e.g. LDN 1544 [18]) indicating that they are on the verge of collapse. Dense cores without sign of star formation are

Fig. 2.1 Image of the Barnard 68 dark cloud at visible wavelengths (*left*) and combining with near infrared (J, H and K coded with *red colour*) (*right*). This dense interstellar core blocks the light of the background stars, creating a dark patch. The dust grains responsible for this visible extinction are less effective in the infrared, enabling the detection of the brightest stars (Courtesy ESO, J. Alves [17])

called pre-stellar cores, and represent the initial conditions for the birth of stars, while those already hosting newly formed young stellar objects are called proto-stellar cores. Pre-stellar cores usually are colder and less turbulent than proto-stellar cores.

2.3.1.2 High Mass Cores

Finding the counterparts of the low mass dense cores for high mass stars has proven to be more difficult for several reasons

- The massive star forming regions are statistically more distant than the low mass star forming regions, leading to limited spatial resolution and a more difficult recognition of small compact objects.
- Massive stars are less numerous than low mass stars implying a smaller number of progenitors and the need to survey large areas in the sky to build statistically significant samples.
- Massive stars evolve significantly faster than low mass stars, and have a strong effect on their surrounding environment due to the combined effect of strong winds and radiation. Their birthplace must therefore be identified at a very early stage, before these negative feedback effects have had time to develop.

The best method for finding birth places of massive stars has been to use extensive surveys of large areas of the Galactic plane, either at sub-millimetre wavelengths targeting the dust thermal emission, or at mid infrared wavelengths targeting the PAH emission and the dust extinction. Indeed, the most massive cores are recognized as compact sub-millimetre sources and/or by maxima of the infrared extinction, two diagnostics of large column densities approaching 100 magnitudes of visual extinction, or H_2 column densities larger than 10^{23} cm^{-2}. The objects detected in the infrared are called "Infrared dark clouds" to mimic the name of the classical "dark clouds" such as Barnard 68. First recognized in the ISO imaging data [19], their study has rapidly developed using the subsequent infrared satellites WISE and Spitzer. As of today, the most recent catalogue contains over 10,000 objects [20]. A careful and lengthy work of cross identification and validation must be performed, after a particular object has been detected, to determine its nature. Although the search for massive pre-stellar cores is a rapidly evolving field, it is already well accepted that a fraction of the IR dark clouds belongs to this category, while other IR dark clouds already harbour massive proto-stars and are therefore at a later stage of evolution. The massive proto-stellar cores are typically located at larger distances than the local pre-stellar cores, because they are associated with larger molecular clouds that are relatively rare in the solar neighbourhood. They have larger masses too, typically of a few tens solar masses, sufficient to enable the formation of at least one massive star. Otherwise, their physical conditions and chemical composition are fairly similar to those of pre-stellar cores, with a tendency towards slightly warmer temperatures.

2.3.2 *Young Stellar Objects and Their Environment*

After the onset of gravitational collapse of a dense core, and when a first stellar embryo is formed, the physical conditions and chemical composition of the surrounding envelope are heavily modified. The increase of temperature due to the infrared radiation produced by the accreting object, combined with shocks and outflows and eventually with energetic radiation (FUV, X-rays) produced by the young proto-stars lead to a strong increase of the temperature in a small zone around the proto-star. This leads to the destruction of ice mantles, either through thermal evaporation or through non-thermal processes like sputtering by shocks or photo-desorption. The presence of relatively complex organic species such as methyl formate ($HCOOCH_3$) or dimethyl ether ($(CH_3)_2O$) as well as a tenfold increase of the abundance of formaldehyde (H_2CO) or methanol (CH_3OH) are clear signs of this phenomenon as explained in the review by Herbst and van Dishoeck [21].

The large number of complex molecules leads to a rich spectrum in the millimetre and sub-millimetre wavelength range, with numerous spectral lines. Two object classes are defined, that depend on the mass of the associated proto-star: high mass sources are called "hot cores", while the term "hot corinos" refers to solar type proto-stars. Indeed, because the time scales for low mass and high mass star formation are different, with high mass stars evolving significantly more rapidly than low mass stars, it is expected that clues on the chemical mechanisms will be found by studying both categories. Among the important differences, one must take into account the longer evolutionary time scale for low mass star

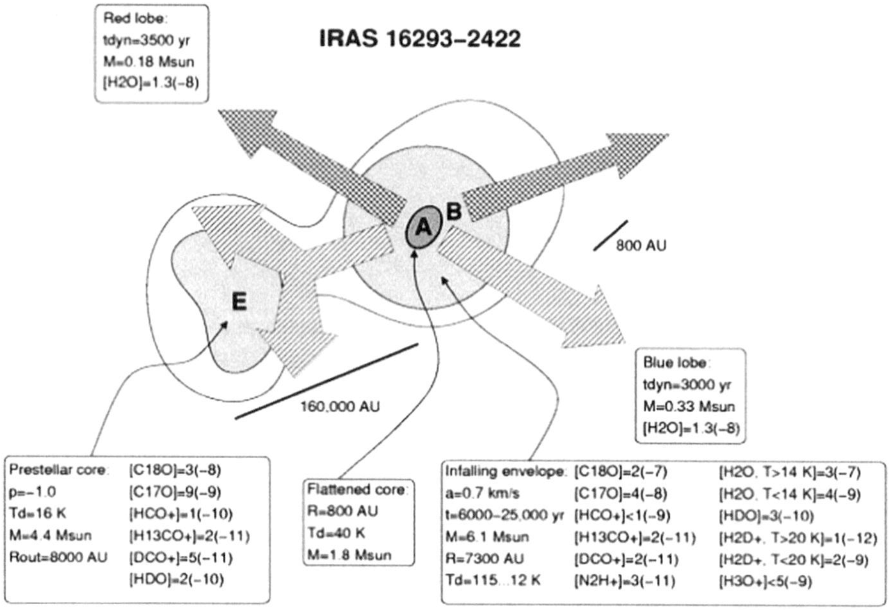

Fig. 2.2 Sketch of the environment of the low mass proto-star IRAS16293-2422 [23]. A binary system with sources A and B is located at the centre of a cold dense core. Sources A and B have distinct chemistries with a richer composition of organic species in source A. This system is associated with two molecular outflows, in the East–west and North-East-South-West directions represented by *arrows*. The former outflow is compressing a second dense core to the East (core E), triggering a new generation of star formation

formation, and the smaller envelope mass that leads to a lesser degree of processing of the species produced in the ice mantles in low mass proto-stars, compared to the hot cores in high mass star forming regions. The degree of deuterium fractionation is significantly larger in hot corinos compared to hot cores with relatively easy detections of doubly deuterated formaldehyde and even triply deuterated methanol in the prototypical source IRAS16293-2422 [22].

A key characteristic of young proto-stars is the presence of bipolar molecular outflows and jets as shown in Figs. 2.2 and 2.6. These collimated structures have a strong impact on the dense core where the proto-star is born, by stirring and disrupting the quiet cocoon. As explained below, the jets and outflows are connected with the accretion of matter on the central object, and contribute to the outward transport of energy and angular momentum that is necessary for energy and angular momentum conservation. These jets and outflows also have interesting chemical properties: because sputtering is efficient in shocks where dust grains are bombarded with atoms and molecules from the gas phase, grain mantles, and even a fraction of the grain cores, are destroyed in the molecular outflows. This leads to a specific chemistry, with significant enrichment of the shocked gas in some molecular species such as silicon monoxide (SiO) and water vapour (H_2O). Molecular outflows show up as broad "wings" in the molecular line profiles, tracing the presence of material at different velocities (either more positive or red-shifted, more negative or blue-shifted, or both) than the dense core, with a roughly symmetrical pattern of blue-shifted emission on one side of the proto-star and red-shifted emission on the other side (Fig. 2.2). Such a pattern can be analysed to deduce the expansion velocity, the opening angle as well as the orientation of the bipolar outflow on the plane of the sky. While molecular outflows are easily detectable in rotational lines of carbon monoxide (CO), care must be taken to separate the outflow signal from the emission associated with the dense molecular core. Other species are used that provide a better contrast, silicon monoxide or water vapour. The presence of strong emission from silicon monoxide, a species directly linked to the destruction of the silicates cores, as well as from water vapour, the main constituent of grain mantles, are two strong arguments supporting the destruction of mantles in such shocks. Molecular outflows are therefore good places to study the influence of shocks on the structure of the interstellar medium, as well as physical processes, especially the sputtering of ice mantles and grain cores.

The presence of magnetic fields has a strong influence on the process of star formation, because it introduces an asymmetry in the collapse and permits an efficient angular momentum transfer in the first stages of the gravitational collapse. Recent theoretical studies (e.g. [24] and references therein) based on extensive MHD simulations, have shown that the gravitational collapse proceeds with the formation of a flattened structure (a "pseudo-disk") and the launching of outflows. The fragmentation into several components, as well as the formation of a Keplerian disk depends on one key parameter, the mass to flux ratio $\mu = (M_0/\Phi)/(M/\Phi)_c$, relating the core mass M_0, the magnetic flux $\Phi = \pi B R_0^2 = \pi B (3\ M_0/4\pi\rho)^{2/3}$ and the critical value of the mass to magnetic flux ratio $(M/\Phi)_c = (c_1/3\pi)\ \surd(5/G)$ where c_1 is a numerical constant of about 0.53 [25]. Large values of μ (>20) correspond to a behaviour approaching the pure hydrodynamical case, while small values of μ (<0.1) correspond to a behaviour dominated by the magnetic field. Typical values deduced from the observations correspond to $\mu \sim 2$ to 5, a case where the magnetic field is dynamically important but not dominant. These simulations also show that it is important to include a realistic equation of state, including localized cooling or heating of the matter. Therefore it is expected that knowledge of the chemical composition, especially the species having a largest contribution to the thermal balance, will be included in future development of these investigations.

2.3.3 *Circumstellar Disks*

In the classical view of star formation [21, 26], young proto-stars are surrounded by an accretion disk, also called a circumstellar disk, which contributes to the feeding of the proto-star before it reaches its final mass. The so called class 0 proto-stars are deeply embedded and radiate mostly in the far infrared and sub-millimetre spectral range. At this stage, proto-stars are actively accreting, as testified by the presence of jets and molecular outflows. These outflows contribute to the release of energy and

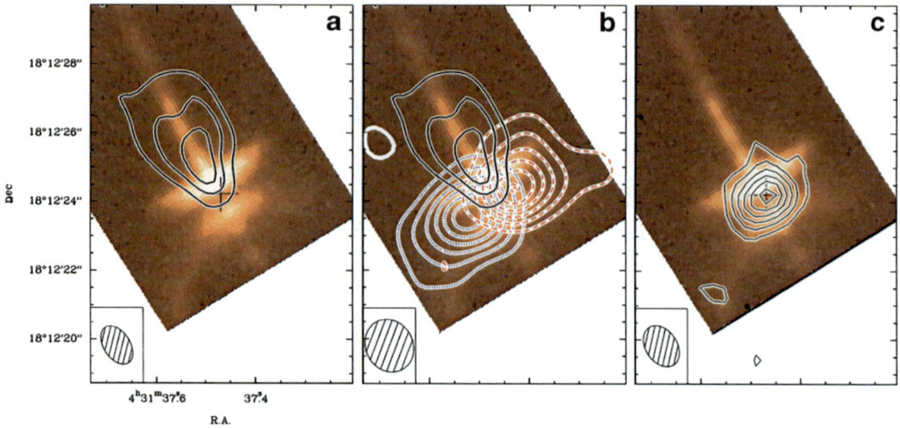

Fig. 2.3 The HH 30 system ([27]). The background image, taken with the Hubble Space Telescope (HST [28]), shows an edge-on disk traced by the *dark bar*, a jet perpendicular to the disk and scattered light from the embedded proto-star. The *left panel* presents the ^{12}CO($J = 2 \rightarrow 1$) emission at large positive and negative velocities relative to the dense core narrow emission. This high velocity ^{12}CO emission follows the narrow jet. The *middle panel* presents the ^{13}CO ($J = 2 \rightarrow 1$) emission in two velocity intervals indicated with *blue* and *red contours* for the emission approaching us/receding from us. This velocity pattern is consistent with Keplerian rotation around a solar mass star. The *right panel* presents the continuum emission due to dust grains in the circumstellar disk. The spatial resolution of the millimetre observations is $\sim 1''$ [27]

angular momentum outward that balances the losses of angular momentum and gravitational energy in the accretion process. However at this stage the presence of a Keplerian accretion disk is difficult to establish because most of the continuum and line radiation is associated with the dense core and the proto-stellar envelope. The presence of disks in the youngest proto-stars is therefore still debated.

Later stages in the proto-stellar evolution are called class I and class II, according to the accretion/ejection activity and the ratio of the total luminosity to the stellar luminosity, class II objects being more evolved, having a smaller accretion rate and a spectral energy distribution dominated by the proto-star radiation. Keplerian accretion disks are clearly present in class I and class II objects. In these objects, the spectral energy distribution becomes increasingly dominated by radiation from the (proto) stellar object, at visible and infrared wavelengths.

The presence of a disk can be detected in high angular resolution images, using adaptive optics systems or interferometers to reach sub-arcsec angular resolution (Fig. 2.3). These methods give access to the disk size, orientation and, coupled with spectroscopy, to its rotation pattern through the analysis of the variations of the line profiles with the position. An indirect detection of the presence of a disk can be established through analysis of the spectral energy distribution as the dust grains in the disk are significantly colder than the proto-star atmosphere and therefore produce an excess emission at longer wavelengths with respect to the close to black body radiation of the central object.

Circumstellar disks evolve together with their central object, becoming thinner and less massive through the combined actions of accretion onto the central object, and of dispersion of the disk triggered by the stellar radiation and winds/outflows. Dust grains evolve simultaneously with the gas: they settle to the disk mid-plane and start to grow in size through coagulation, first reaching centimetre sizes, and possibly up to boulder size or even larger objects. It is thought that planet embryos form at this period, which could start as early as 1 million years after the formation of the stellar embryo, and could last a couple of 100 million years. Planets gravitationally interact with their parental disk, creating density waves and gaps in the disk in which they evolve. This is especially true for the most massive ones that can radially migrate inward. The migration phenomenon is thought to be very common and also applies to the early phase of our solar system, as proposed in the so-called "Nice-Model" (e.g. [29] and references therein), in which the radial migration of proto-Jupiter and proto-Saturn led to a pronounced redistribution of the orbits of the small bodies, and could have caused an intense bombardment of the Earth and Moon surfaces.

While gas disks get rapidly dispersed, young stars and their planet embryos are still embedded in a tenuous disk of small dust particles continuously created during the frequent collisions of the large bodies (asteroids, comets, planet embryos, etc.). These dust disks are called "debris disks" to emphasize the fact that the dust does not originate from the ISM but is continuously replenished. The prototypical system surrounding the nearby star β Pictoris was discovered by the IRAS satellite (Fig. 2.4). It harbours an edge-on disk detected through the light scattered by the dust particles and extending up to hundreds of astronomical units. The small change in disk inclination between the inner and the outer regions has been interpreted as resulting from the gravitational interaction with a massive planet. This planet has been later discovered by high resolution imaging [30]. It may also cause the fall of comets/asteroids on to the star that were discovered earlier as variable absorption features in the visible stellar spectrum.

While β Pictoris is a young system (12 million years), with a relatively strong IR excess and bright scattered light emission from the disc, older systems like our Sun are still surrounded by a very faint dust disc. The study of proto-stars and their circumstellar discs, from the earliest phases of class 0 systems, to the debris discs, provides the necessary keys for understanding the formation and first evolutionary phases of our solar system. This is true for the dynamical evolution of the disc structure, as well as for the composition of the solid and gaseous matter. Conversely, information on the early evolution on the solar system, deduced from the analysis of the meteorites or from the study of the less evolved objects in the solar system, provides important constraints for understanding planet formation. It must be noted that the dynamical evolution of discs is closely related to the evolution of the solid and gaseous matter inside them. Dust grains provide shielding from the energetic radiation from the proto-star, and cold surfaces for freezing the volatiles like water, carbon monoxide, carbon dioxide, methanol, etc., onto grain mantles in the disk mid-plane. The presence of solid water is particularly important in the context of planet formation theory. Volatiles are more protected from the FUV

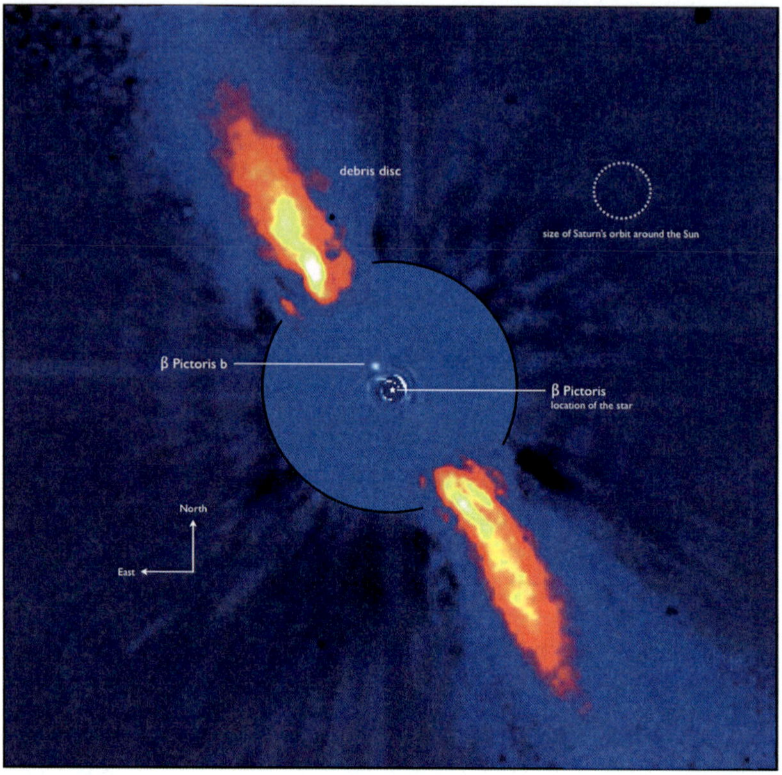

Fig. 2.4 The β Pictoris system in near infrared light. The faint reflected light radiation from the debris disc is revealed after a very careful subtraction of the much brighter stellar halo. The outer part of the image shows the dust disc, as observed in 1996 with the ADONIS instrument on the ESO 3.6 m telescope; the inner part of the image has been obtained with the NACO instrument on the ESO Very Large Telescope. The newly detected source is more than 1,000 times fainter than β Pictoris, aligned with the disc, at a projected distance of eight times the Earth-Sun distance. Both parts of the image were obtained on ESO telescopes equipped with adaptive optics [30]

radiation in grain mantles than in the gas phase, hence an active organic chemistry can proceed in the ice mantles, possibly forming the first building blocks of the organic species detected in comets and meteorites. The study of circumstellar discs in the context of planet formation is a very active field of research. Because circumstellar discs are relatively small objects a few arcsec across (Fig. 2.3), especially their inner region where planets form, the knowledge of the disk chemical composition is currently limited to the strongest spectral features available. With the upcoming of more powerful instruments, in terms of sensitivity and angular resolution like ALMA, VLT and soon E-ELT, tremendous progress is expected in the near future.

2.3.4 Photodissociation Regions and Shocks

Energy is fed in the interstellar medium either through impinging radiation from the surrounding stars or through stirring the gas by dynamical processes. Therefore, it is very important to understand (1) the coupling of matter with radiation on the one hand, and (2) the specific chemistry induced by shocks. In both cases, the modelling must involve the description of the heating and cooling processes of matter, which requires a detailed understanding of the relevant physical and chemical processes controlling the composition, electron fraction and thermal balance.

In addition, the modelling must include a macroscopic description of the system in consideration, its geometry, and the associated physical conditions.

The name "Photodissociation regions" or "PDRs" refers to the illuminated edges of dense molecular gas as well as the lower density diffuse and translucent interstellar medium, as both environments are bathed by far ultraviolet photons. One of the most famous examples, the edge of the horsehead nebula is illustrated in Fig. 2.5. In such regions the heating is dominated by the interaction of these far ultraviolet photons with matter, especially with the dust grains through the photo-electric effect. Indeed, FUV photons can eject an electron from an inner shell of an atom bound in a dust grain. This energetic electron with typically 1 eV of kinetic energy, subsequently heats the gas by colliding with other particles. The process is most effective for small dust grains, like the PAHs and very small grains. The cooling is dominated by radiation of abundant species in the gas phase such as atomic oxygen **OI**, ionized carbon **CII**, $\mathbf{H_2}$ and CO (Table 2.2). FUV photons also have a profound influence on the gas composition and on the dust structure. As the ionization energy of carbon is lower than that of hydrogen, ionized carbon is present in the neutral gas and can participate in the chemistry, contributing to the formation of many carbon bearing species. FUV photons can also induce the desorption of species frozen onto grain mantles, contributing to the destruction of these mantles and the release of frozen molecules into the gas phase. Finally molecules are destroyed by photons. During the destruction process, molecules can access excited states that produce specific emission lines. This is especially true for molecular hydrogen, H_2, whose photodissociation involves absorption involving its electronic transitions. The excited H_2 molecule can either dissociate or be de-excited through line emission and collisional de-excitation. The former process destroying H_2 is relatively inefficient, with radiation being preferred to dissociation in about 90% of the cases. The line emission produced during the de-excitation process can be detected in the near infrared (e.g. Fig. 2.5) and used as a clean probe of the interaction of FUV radiation with molecular gas.

State of the art models of photodissociation regions today involve hundreds of chemical species, thousands of chemical reactions, and tens of physical and chemical processes in order to accurately describe the physical and chemical structure of a piece of gas illuminated by FUV photons. This degree of sophistication cannot yet be implemented in 3D dynamical models. Most PDR models are therefore steady-state and one dimensional, with ongoing developments towards simplified three

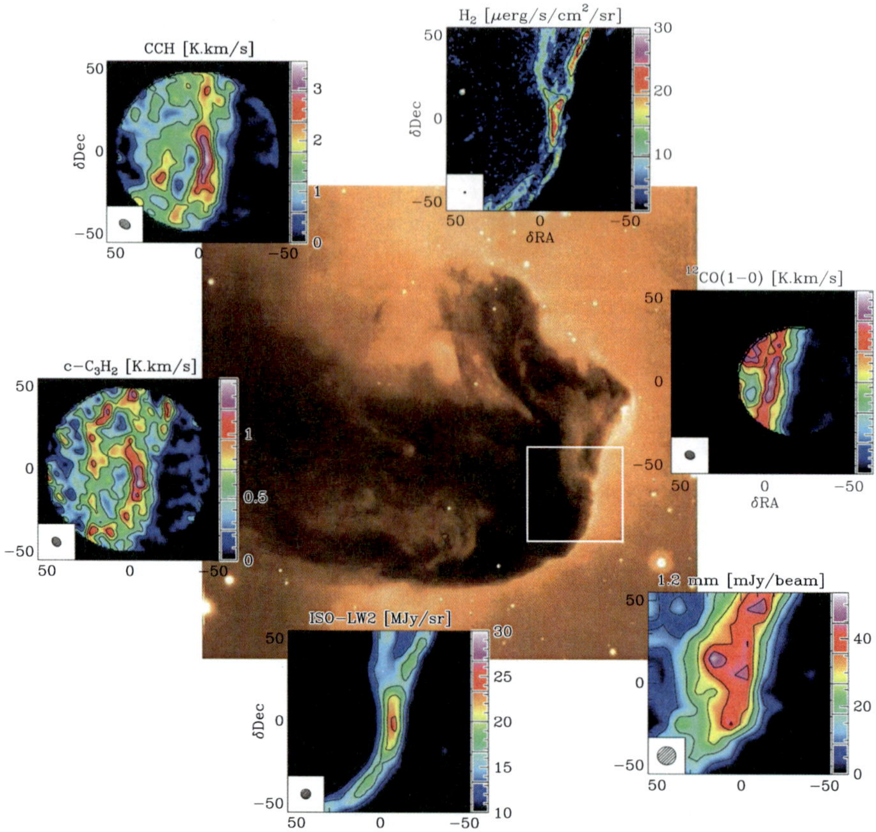

Fig. 2.5 The horsehead nebula, a dark cloud illuminated on its right edge by FUV radiation from the nearby O9 star σOri. This edge is a template source for studying the physical and chemical processes in UV irradiated interstellar matter. The background image is taken in the visible at ESO. The *small colour* pictures show the distribution in various spectroscopic tracers [31]

Table 2.2 Main gaseous cooling lines

Species	Transition	Wavelength/μm	T/K
H_2	Rotational lines	28, 17, 12, 9.3	200–1,000
Oxygen [OI]	Fine structure	63, 145	100–400
Ionized carbon [CII]	Fine structure	158	50–300
Neutral carbon [CI][a]	Fine structure	610, 370	20–100
CO[b]	Rotational lines	124–2,600	10–200
H_2O[c]	Rotational lines	538– ~60	50–500

[a]Weaker contribution compared to [CII].
[b]Stronger contribution than [CI], saturated lines.
[c]More important in shocks than in PDRs.

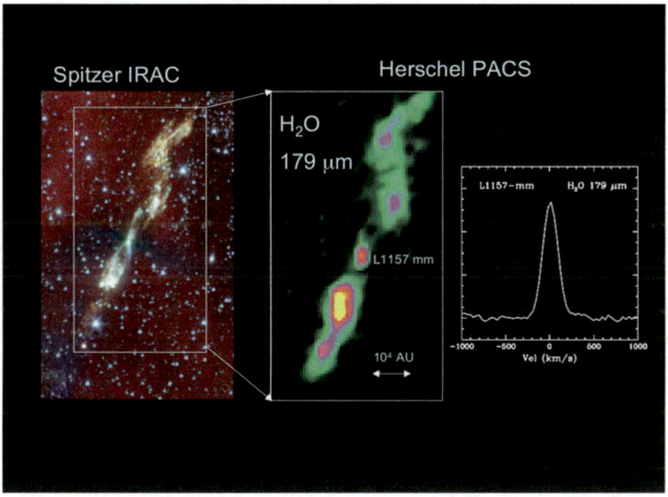

Fig. 2.6 The molecular outflow created by the class 0 proto-star L1157 viewed in the infrared by the Spitzer (*left*) and in the far infrared with Herschel (*right*) satellites. The Spitzer image taken with the IRAC camera is dominated by emission from the rotational lines of H_2. The Herschel image taken with the PACS instrument at 179 μm, the wavelength of one of the strongest lines of water vapour ($2_{1,2} - 1_{0,1}$) shows the excellent agreement between the H_2 and water vapour morphology [35]

dimensional geometries. The PDR code developed in the Meudon Observatory [32] is accessible on line (http://pdr.obspm.fr). It is however possible to implement a very coarse treatment of the chemistry and the interaction of matter with radiation in magnetohydrodynamic (MHD) simulations. The Heidelberg team has published pioneering results, showing the interplay of chemistry and dynamics clearly [33]. While both approaches have their pro and cons, it is foreseeable that they will be more often combined in the near future, thanks to a rapid increase in computer capabilities and the development of more advanced numerical codes.

Shocked regions are relatively frequent in the interstellar medium. Among the main shock sources are the jets from young stars creating molecular outflows (Fig. 2.6), and the shocks induced by supernovae explosions the most studied. The latter shocks can reach very large velocities, up to 100 km s^{-1} or higher, and have typical speeds of a few tens of km s^{-1}. Slower shocks are also present, as consequences of the large scale streaming motions along the spiral arms, stirring of the interstellar medium by stellar winds, distant supernovae shocks, and resulting turbulent motions. Understanding the shock physics is therefore as important as understanding the PDR physics for a global description of the interstellar medium. Because the interstellar medium is penetrated by magnetic fields, the shock structure can be very different from pure hydrodynamical shocks. As explained in the review by Draine and Mc Kee [34], in relatively slow shocks, a magnetic precursor is formed ahead of the shock, that informs the gas upstream as to the arrival of the shock. As the ionization fraction is very low, less than 10^{-7} in shielded molecular gas, the ionized and neutral fluids are decoupled by this magnetic precursor, with the neutral fluid being essentially unaffected while the ions are accelerated and heated. This decoupling induces a velocity drift between the ions and the neutrals, which leads to a delayed acceleration and heating of the neutrals through the ion-neutral friction. Since the gas can cool at the same time, the overall velocity and density pattern remains continuous in the shock reference frame. These shocks are therefore labelled "C-shocks" for "Continuous shocks". When the shock velocity becomes too large, the magnetic precursor cannot play its role and the shock becomes discontinuous as a pure hydrodynamical shock. These shocks are labelled "J-shocks" with "J" referring to the presence of "Jumps". This description is very schematic. In realistic cases, shocks may not be stationary, and have a more complex morphology than the ideal case of a one-dimensional shock front with the magnetic field oriented perpendicular to the shock front. For a more complete introduction to PDR and shock physics, the reader is invited to consult specialized references such as [34].

2.4 Constituents of the Interstellar Medium

The analysis of radiation across the electromagnetic spectrum has allowed astronomers to uncover the composition of the neutral interstellar medium, its spatial variation and how it relates to the local environment (physical conditions) as well as the overall evolutionary stage of the object.

2.4.1 Neutral Gas

Table 2.3 presents a summary of the molecules detected in the interstellar medium or in the envelope of evolved stars, most notably the bright carbon star **IRC +10216**. Species in which a hydrogen atom has been substituted by a deuterium atom are listed as well, as they are extremely valuable tools to understand the chemical evolution of the star-forming interstellar matter. The list has been established using on line data bases such as the Cologne Data base for Molecular Spectroscopy (CDMS http://www.astro.uni-koeln.de/cdms/molecules), the splatalogue (http://www.splatalogue.net) and The Astrochymist (http://astrochymist.org/).

This list has been obtained through dedicated observations across the electromagnetic spectrum, from the far ultra violet for H_2 or N_2, down to centimetre wavelengths for NH_3 or HI. The most effective detection method is the analysis of the rotational spectrum of molecules, in the centimetre to sub-millimetre wavelength spectral range because extremely sensitive detectors are available with high spectral resolution capabilities. However, this method is biased towards species with a permanent electric dipole moment that allow the efficient emission (or absorption) of rotational lines (see Chap. 1). Symmetrical species like N_2, C_2 and alike cannot be

Table 2.3 List of detected interstellar and circumstellar molecules, radicals and ions, grouped by the number of atoms (*N*) they contain. Species detected with UV, visible or infrared spectroscopy are indicated in italics

N	Observed species
2	*H_2*, *HD*, OH, O_2, OH^+, CH, CH^+, NH, HF, HCl, SH, SH^+, CO, CO^+ CS, SO, SO^+, SiO, *C_2*, CF^+, NO, O_2, PN, SiS, *N_2*, HCl^+, SiC, AlF, $AlCl$, $NaCl$, KCl, SiN, CP, PO, AlO, CN^-
3	*$H_3{}^+$*, *C_3*, C_2H, C_2O, C_2S, HCO, HCO^+, HOC^+, CH_2, O_2H_2O, H_2O^+, NH_2, HCN, HNC, HCS^+, H_2S, N_2H^+, OCS, SO_2, CO_2, HNO, N_2O, H_2Cl^+, $MgCN$, $MgNC$, $NaCN$, c-SiC_2, $SiCN$, $AlNC$, $SiNC$, HCP, CCP, $AlOH$, KCN, $FeCN$, O_2H
4	NH_3, H_3O^+, *C_2H_2*, H_2CO, *CH_3*, $HCNH^+$, H_2CN, H_2CS, HC_2N, c-C_3H, l-C_3H, C_3O, C_3S, l-C_3H^+, C_3N, $HNCO$, $HOCN$, $HCNO$, $HNCS$, $HSCN$, $HOCO^+$, $HOOH$, c-SiC_3, C_3N^-, PH_3
5	H_2C_2O, *C_5*, C_4H, C_4H^-, c-C_3H_2, l-C_3H_2, H_2CCN, *CH_4*, HC_3N, HC_2NC, HNC_3, HC_2NH, $HCOOH$, NH_2CN, $H2COH^+$, $HCOCN$, C_4Si, SiH_4
6	CH_3OH, C_5H, l-H_2C_4, *C_2H_4*, CH_3CN, CH_3NC, CH_3SH, HC_3NH^+, HC_2CHO, NH_2CHO, c-H_2C_3O, C_5N, C_5N^-, *l-HC_4H*, l-HC_4N, H_2CCNH
7	C_6H, C_6H^-, CH_2CHCN, CH_3CCH, HC_5N, CH_3CHO, CH_3NH_2, c-C_2H_4O, H_2CCHOH
8	CH_3C_3N, $HC(O)OCH_3$, CH_3COOH, CH_2OHCHO, C_7H, H_2C_6, *l-HC_6H* CH_2CHCHO, CH_2CCHCN, NH_2CH_2CN
9	CH_3C_4H, $(CH_3)_2O$, C_8H, C_8H^-, CH_3CH_2CN, CH_3CH_2OH, HC_7N, CH_3CONH_2, C_3H_6
10	CH_3C_5N, $(CH_3)_2CO$, $(CH_2OH)_2$, CH_3CH_2CHO
11	HC_9N, CH_3C_6H, C_2H_5OCHO
12	*C_6H_6*, $CH_3OC_2H_5$(?), C_3H_7CN
≥ 13	$HC_{11}N$, *C_{60}*, *C_{70}*
	Deuterated species
2	HD, ND
3	H_2D^+, D_2H^+ HDO, D_2O, HDS, D_2S, DCN, DNC, DCO^+, N_2D^+
4	NH_2D, ND_2H, ND_3, $HDCO$, D_2CO, $HDCS$, D_2CS, c-C_3D
5	C_4D, c-C_3HD, DC_3N
6	CH_2DOH, CD_2HOH, CD_3OH, CH_3OD, CH_2DCN

detected through their pure rotational pattern although they can be very abundant. When a suitable bright background source is present, absorption spectroscopy either in the UV-visible (searching for the electronic transitions of the molecules) or in the infrared (aiming at the vibrational transitions) can be performed. Because of the limited number of suitable sources, the weakness of the vibrational transitions, and lower spectral resolution, the number of species accessible through these techniques is restricted to relatively abundant species. Furthermore, it is important to note that the chemical composition is closely related to properties of the environment. All molecules listed in Table 2.3 will not be present in a single source. Some species like CO are widely present, while others like SiO are formed in specific events and can be used to locate these events. The analysis of the chemical composition of a given source therefore provides most of the clues on its structure, gas column density, gas density, electron fraction, magnetic field as well as its dynamics, through the analysis of the detected spectral lines.

Among the molecules detected so far in the interstellar medium, some species have been more widely used than others as physical and chemical probes. Some of the more commonly used species are described below. This choice is somewhat subjective and care must be taken when analysing molecular spectral lines, to compare with detailed physical and chemical models before drawing definitive conclusions.

2.4.1.1 Density and Temperature Determination

The local gas density cannot be directly measured as detected signals always are integrated along the line of sight. However it can be deduced from analysis of the degree of excitation of molecules, i.e. from the comparison of populations of different energy levels, which results from competition of radiative and collisional excitation and de-excitation (see Chap. 1). The most abundant collision partner is molecular hydrogen H_2. Collisions with hydrogen and helium atoms, and with electrons, can also be important especially in diffuse regions. The role of collisional excitation is best illustrated with the notion of critical density n_{cr}, which corresponds for a given molecular transition to the minimum gas density for exciting the molecule and producing an emission line. For a two level system at temperature T, and assuming H_2 as the sole collision partner, the balance between collisional excitation and emission of radiation can be written:

$$n_u A_{ul} + n_u n(H_2) C_{ul} + n_u I_\nu B_{ul} = n_l n(H_2) C_{lu} + n_l I_\nu B_{lu} \tag{2.1}$$

where n_u and n_l refer to the populations in the upper and lower energy levels of the system, A_{ul} and B_{ul} are the Einstein coefficients for spontaneous and induced emission, C_{lu} and C_{lu} the collisional rates for excitation/de-excitation, $n(H_2)$ the gas density and I_ν the local radiation intensity. The transition frequency ν is related to the energy level as

$$\nu = (E_u - E_l)/h \tag{2.2}$$

and the total number of molecules n_{mol} is $n_l + n_u$. The detailed balance implies,

$$C_{ul}/C_{lu} = B_{ul}/B_{lu} = g_l/g_u e^{(E_u - E_l)/k_B T} \tag{2.3}$$

Therefore equation (2.1) can be rewritten, neglecting the radiative coupling terms:

$$n_u/n_l = 1/((A_{ul}/n(H_2)C_{ul}) + (g_l/g_u)e^{h\nu/k_B T}) \tag{2.4}$$

This formula shows that when the collisions are frequent, the first term in the denominator becomes small compared to the radiative de-excitation, and the level populations approach thermodynamic equilibrium at the kinetic temperature T, namely $n_u/n_l = g_u/g_l\ e^{-h\nu/k_B T}$. On the contrary, when the de-excitations are dominated by radiative processes, the population in the upper level is very small and $n_u/n_l \sim n(H_2)C_{ul}/A_{ul}$. The change of behaviour occurs close to the critical density defined as $n_{cr} = A_{ul}/C_{ul}$. This concept can be used for multilevel systems, using the balance between the various excitation and de-excitation routes. The concept of critical density is useful for determining the order of magnitude of the gas density probed by a given molecular transition.

Ammonia (NH_3) and formaldehyde (H_2CO) are among the best species for probing the physical conditions as they have many accessible lines coupling levels of different excitation energies, and their collisional cross sections have been accurately computed [36, 37].

The kinetic temperature cannot be directly measured from the observations but can be deduced from the analysis of the line profiles and molecular excitation as the level populations are also sensitive to the temperature. For an accurate determination of the physical conditions, including the kinetic temperature, the full set of the statistical equilibrium calculations must be solved. There are some freely available numerical codes, such as RADEX ([38], http://www.strw.leidenuniv.nl/moldata/radex.html). They make use of extensive data bases of collisional cross sections, LAMDA (Leiden Atomic and Molecular DAtabase http://www.strw.leidenuniv.nl/moldata/), and BASECOL (http://basecol.obspm.fr/), that are now part of the European initiative VAMDC (Virtual Atomic and Molecular Data Center http://www.vamdc.eu/).

2.4.1.2 Ionization

The ionization fraction is an important parameter of the physics of the interstellar medium, as it controls the gas coupling with the magnetic fields. Because the main charge carriers are electrons, they are not directly accessible through spectroscopy and other methods must be used to probe the total ion content. The most common methods take advantage of the variations of the chemical composition of the gas as a function of the ionization fraction, and involve measurement of the abundances of different molecular ions (e.g. HCO^+, HOC^+, N_2H^+) or radicals (e.g. atomic carbon) that are particularly sensitive to the presence of electrons. The key point in these methods is the necessity to understand thoroughly the processes regulating the abundances of these molecular ions to be able to accurately extract the dependency on the ionization. For instance [39] discuss the variation of the ionization fraction across the edge of the horsehead nebula.

A related issue is the determination of the ionizing rate of the neutral gas produced by cosmic rays ζ. These energetic particles have a very long mean free path and can travel far from their birth place. They can ionize atomic and molecular hydrogen forming H^+ and $H_2{}^+$ respectively. $H_2{}^+$ rapidly reacts with molecular hydrogen to form the more stable ion $H_3{}^+$. Since H and O have almost identical ionization energies of 13.598 eV (H) and 13.618 eV (O) H^+ can transfer its charge to atomic oxygen, forming O^+ that reacts with molecular hydrogen to form OH^+. Then other hydrogen abstraction reactions lead to H_2O^+ and $\mathbf{H_3O^+}$. $H_3{}^+$, and possibly the oxygen ions, can be therefore be used as probes of the cosmic ray ionization rate ζ [40, 41].

2.4.1.3 Magnetic Field Measurement

As explained below, magnetic fields induce a small modification of the energy levels of molecules, through the Zeeman effect, which is particularly important for paramagnetic species like O_2 or SO, or radicals with an unpaired electron like CN, CCH or OH. With the typical strength of interstellar magnetic fields, from a few tens to a few hundreds μGauss, the signature of interstellar magnetic fields in the line profile is very weak and difficult to separate from other effects leading to deformations of the line profiles, especially systematic instrumental effects. In order to calibrate the instrumental effects, molecular species exhibiting hyperfine transitions with different sensitivities to the magnetic fields such as CN are usually preferred for probing the magnetic field intensity (see Crutcher [42] for a review).

2.4.1.4 Tracers of Astrophysical Environments

Because the chemical composition is highly dependent on the present and past physical conditions, it is possible to use some species, or some spectral lines as tracers of specific environment. As for the ionization fraction, the definition is somewhat subjective, hence the applicability of the tracers listed below to a new dataset must be carefully compared with both theoretical models and previous observations. Because the subject is evolving fast, the list of tracers presented below must be viewed as incomplete.

PDR Tracers

Photodissociation regions (PDRs) are defined as regions where the chemistry is dominated by photons. Hence the chemistry of PDR tracers must be dominated by photo-induced processes, at least indirectly. Reactive species, rapidly destroyed by reactions with H_2 or abundant neutrals are therefore good tracers of the illuminated outer layers of molecular clouds. The list includes the gas coolants [CII], [OI] and [CI], radicals like HCO, CCH or c-C_3H_2, as well as reactive ions like CO^+, HOC^+ or CF^+. The rotational and rovibrational lines of H_2 are also bright in PDRs.

Shock Tracers

Good shock tracers must show the highest abundance contrast between the ambient medium and the shocked gas. Because most of the silicon is locked up in grain cores, the SiO abundance is very low in the interstellar medium, except in shocked regions where sputtering of some silicated materials from the grains leads to the release of silicium and the formation of SiO. A similar story can be told for water vapour, which is locked up in grain mantles as solid ice and released in the gas phase in shocks. Other shock tracers can involve abundant species in grain mantles like methanol (CH_3OH) but their signature is not as clear as those of SiO and H_2O because the methanol abundance of the quiescent gas is not negligible. Finally, shocks are also bright sources of H_2 emission because of strong gas heating in the shocks.

Tracers of Turbulent Dissipation Regions

The presence of reactive ions like CH^+ in the diffuse interstellar medium has been a challenge for interstellar chemistry since CH^+ is easily destroyed by reactions with H_2 but slow to form under the known physical conditions of the diffuse interstellar medium. It now appears that a "warm chemistry" can develop in the tiny dissipative structures of the interstellar turbulence, enabling the formation of transient species like CH^+ and SH^+ [43]. The opening up of the sub-millimetre sky by the Herschel telescope has led to the discovery of several new reactive ions, enabling a better characterization of their chemistry. In the future, these tracers should bring interesting constraints on the properties of the interstellar turbulence.

Cold Cores

With their larger density compared to the background, and the cold grain temperatures, dense pre-stellar cores appear as bright sub-millimetre sources. A specific chemistry develops at the cold temperatures and high densities. While the main molecular species like CO freeze onto dust grains because of the very low grain temperatures ($\sim$10 K), the chemistry remains active. In particular, the cold temperatures are particularly favourable for enhancing the deuterium fractionation, with observed abundance ratios of the deuterated species compared to the main isotopologue larger than one tenth, while the elemental abundance of deuterium relative to hydrogen is 20 ppm. The presence of multiply deuterated species with up to three D atoms is another key characteristic of pre-stellar cores. This includes the deuterated forms of $H_3{}^+$ [18], Hence the presence of strong rotational lines from deuterated molecular ions (H_2D^+, DCO^+, N_2D^+) can be used to locate cold dense cores. The abundances of deuterated molecular ions is sensitive to the temperature because the fractionation is induced, in the gas phase, by the exothermic reaction between H^+ and HD:

$$H_3{}^+ + HD \rightarrow H_2D^+ + H_2 (\Delta E = 232K) \qquad \text{(R2.1)}$$

At the low temperatures of cold dense cores, the rate of the forward reaction is much larger than the rate of the reverse reaction, and the deuterium fractionation is favoured. The fractionation is limited by the H_2D^+ destruction processes, namely the dissociative recombination with electrons and reactions with neutral species (CO, N_2, etc.) that destroy H_2D^+ and $H_3{}^+$ faster than the reaction with HD. Similar reactions with D_2H^+, $D_3{}^+$, CH_2D^+ must also be considered in chemical networks, all contributing to deuterium enrichment.

In addition to deuterated species, some nitrogen bearing species can also be used to study the properties of cold dense cores, notably NH_3 and N_2H^+ [26] as emission maps show that they are spatially associated with the cold dense cores. Indeed, in time dependent models, these species need a rather long time to reach their steady state abundance which may explain this spatial association.

Hot Cores and Hot Corinos

The chemical composition of hot cores and hot corinos is much richer than at any other place in the interstellar medium. In particular, a huge variety of organic species can be detected in these warm and compact objects, such as CH_3OH, $HCOOCH_3$, CH_3CN, $(CH_3)_2O$. At the warm temperatures created by strong heating by the newly formed stars, ice mantles can evaporate, liberating in the gas phase the frozen molecules and simultaneously triggering a rich chemistry. Therefore many of the polyatomic species listed in Table 2.3 are only found in hot cores and hot corinos. As the chemical time scales associated with the processing of different abundant ices are different, high spatial resolution studies have shown that oxygen bearing species may show a different spatial pattern than nitrogen bearing species in some cases. Two sources stand out among the known hot cores, the hot core in the Orion KL region, close to the infrared source IRc2 because of its small distance (400 pc), and the "large molecule heimat" in the SgrB2 molecular complex because of the high abundance of organic species. The class 0 proto-star IRAS 16293-2422 (Fig. 2.2) is considered as a prototype of hot corino. We recall that the main difference between hot cores and hot corinos is the association with high mass stars for hot cores, and with low to intermediate mass stars for hot corinos. Furthermore, deuterated saturated molecules, like D_2CO are usually more abundant in hot corinos than in hot cores [21].

2.4.2 *Dust Grains*

2.4.2.1 Composition

The physics of dust grains is a vast topic that is only briefly introduced here as other chapters in the book present chemical aspects. The term "dust grains" refers to the solid particles present in the interstellar medium (Fig. 2.7) as well as in

Fig. 2.7 Image of a fraction of the Rosette nebula obtained at far infrared wavelengths with the Herschel telescope. The *colours* correspond to the three monochromatic images that have been combined, with *blue coding* for the shortest wave- length (60 μm) and *red* for the longest wavelength (500 μ*m*) [45]. Regions with the warmest dust appear in *blue* while the cold dust regions, further away from the heating sources, appear in *red*

circumstellar disks, proto-planetary and debris disks, as well as in evolved star envelopes. These sub-micronic solid particles are mainly formed in the last evolutionary phases (Red Giant phase and beyond) of stars, and possibly in other more energetic events (e.g. supernovae shocks). The diversity of progenitors leads to a diversity in the dust grain materials, grouped in two main categories, silicate grains and carbonaceous particles (Table 2.4).

In the interstellar medium, the mass of solid particles is about one hundredth of the total mass of the gas. The composition of interstellar dust grains can be established using several sources of information. By comparing the elemental abundances in the sun and in nearby stars with those in the interstellar medium, it is possible to identify the elements showing a deficit in the gas, because they are locked in solid particles. For instance gas phase abundances of silicon, magnesium and iron is lower by more than one order of magnitude in the diffuse interstellar medium than in the sun, because silicon is mostly locked up in solid silicate particles that also include some magnesium and iron. By contrast, gas phase abundances of sulfur and nitrogen are very similar in the sun and in the diffuse interstellar medium indicating that these elements do not contribute much to the building of solid particles in this environment.

Another source of information is the analysis of the preserved dust grains in meteorites. Indeed, some meteorites contain grains that can be identified through their isotopic compositions as being formed before the birth of the solar system. These "pre-solar grains" carry interesting information on the formation places of dust grains, and on their possible composition. In addition to silicates and oxides, one important component of pre-solar grains in primitive meteorites (carbonaceous chondrites) is nanodiamonds (size ∼2 nm).

Table 2.4 Composition and properties of interstellar dust grains

Category	Material	Size	Main bands
Amorphous silicates[a]	Pyroxene ($Mg_x\ Fe_{1-x}\ SiO_3$)	~0.1 μ*m*	9.7, 18 μ*m*
	Olivine ($Mg_x\ Fe_{1-x}\ SiO_4$)		
Oxides	SiO_2, MgO, Fe_3O_4, etc.		
Carbon material	Amorphous carbon	~10 *nm*	3.4 μ*m*
	Graphite[b]	~10 *nm*	2175 Å
	PAHs[c]	~1 *nm*	3.3, 6.2, 7.7, 8.6, 11.3 μ*m*
Ice mantles[d]	H_2O		3.07, 6.0 μ*m*
	CO		4.67 μ*m*
	CO_2		4.27, 15 μ*m*
	CH_3OH		8.5, 3.9, 8.9, 9.65 μ*m*

[a]Most silicates are amorphous. Crystalline silicates have been identified through their infrared spectral features in sources where they form (see text).
[b]Assignment debated between graphite and possibly PAHs and cluster of PAHS (see text).
[c]Polycyclic Aromatic Hydrocarbon (e.g. coronene $C_{24}H_{12}$).
[d]Ice mantles are mixtures of several components, only the most abundant ices are listed. See e.g. [44] for a more complete list of ice constituents.

Finally, information on the composition, size and quantity of solid material can be extracted from analysis of the spectra of interstellar sources. Dust grains are the main sources of reddening in the UV/visible and near infrared spectral regions, and contribute most of the sub-millimetre continuum radiation through their thermal emission. Several spectral bands due to solid materials have been found in the spectra, leading to important information on their size and composition (Table 2.4). The information listed in Table 2.4 must be viewed as a short summary of a complex topic. For instance the size distribution of dust grains extends from nanometre particles up to nearly micron size solids in the variety of interstellar environments, and could extend to even larger particle sizes in proto-planetary disks. The profile of the silicate absorption band at 9.7 μ*m* indicates that the grains responsible for this absorption are composed of amorphous silicates, in contrast with the nature of solar system grains that show both crystalline and amorphous components. Crystalline silicates are relatively easy to detect since they have a richer band spectrum than amorphous silicates, with bands specific of each material. The ISO satellite, followed by the Spitzer Space Telescope, discovered the presence of crystalline silicates in some circumstellar disks as well as in the ejecta of red giant stars where these particles form. Therefore, the physics of silicate grains is not as simple as initially thought, as it now appears that these particles continuously evolve, possibly through the combined effect of energetic radiation, cosmic ray particles, collisions and shocks.

The carbonaceous grain population, including polycyclic aromatic hydrocarbons (PAHs) and fullerenes is of particular interest for astrochemistry. They account for at least 10% of the total carbon content and can contribute up to ~20% of the total infrared luminosity of a star forming galaxy. The PAH family has been identified through strong vibrational bands in the infrared (notably at 3.3, 6.2, 7.7, 8.6 and 11.3 μm, see Fig. 1.4), that indicate that this population is ubiquitous, in the interstellar medium as well as in external galaxies up to very large redshifts (see the book by Joblin and Tielens [46] for a presentation of the recent results on PAHs). Although no single molecule has been identified so far, the combined work on observations, laboratory experiments and theory has enabled this population to be better defined. It is now believed that PAHs exist in the diffuse ISM and can be either neutral, positively or negatively charged. Their sizes range from a few tens to a few hundreds of carbon atoms. In dense gas, shielded from the FUV radiation, PAHs are thought to aggregate forming PAH clusters of several hundreds of carbon atoms. Fullerenes have been recently detected in the ISM and circumstellar envelopes but these symmetrical particles contain a smaller fraction of the carbon, at the percent level.

2.4.2.2 Physics

Dust grains play an important role in the physics of dense interstellar gas. When irradiated by far ultra violet photons, dust particles can get positively charged by the ejection of an energetic electron with typically 1 eV kinetic energy, through the photo-electric effect. This energetic electron loses its energy by collisions with gas species, and hence heats the gas. Since the effect is most efficient for small particles that have the largest surface area to volume ratio, the small carbonaceous dust particles and the PAHs are the most efficient dust particles for this heating mechanism. Although the efficiency is only a few percent, the photo-electric effect is one of the most important heating mechanisms of the gas since it allows transfer of energy from radiation to matter.

Dust grains can also contribute to the thermal balance as they efficiently cool through their thermal emission. At low densities the gas and grain temperatures are different since the gas cools less efficiently than the dust. However, for gas densities greater than about 10^5 cm^{-3}, collisions enable an efficient coupling between the gas and dust particles, leading to similar temperatures. Depending on the difference in temperatures, these collisions can be considered as either a cooling or a heating mechanism.

Because they are subject to photo-electric heating, dust grains usually carry charges, and therefore contribute to the charge balance. It is important to know the charge on dust grains since the efficiency of the photoelectric effect heavily depends on the grain charge, the ejection of an electron being more difficult from a positively charged grain than from a neutral particle. In addition grains can contribute to the neutralization of charged particles. Their role in the charge balance is especially important in the dense and well shielded regions where the ionization fraction is low since they can help to maintain a coupling of matter with the magnetic field.

Another key role of dust grains is their role in the formation of molecular hydrogen. Except in the very early universe, the formation of H_2 on dust grains is significantly more efficient than a pure gas phase route. First deduced from the comparison of measurements of H and H_2 column densities with a simple model [47], the formation of H_2 on solid surfaces is now the subject of laboratory and theoretical studies that are subsequently used in chemical models [48]. The newly formed H_2 is released with excess energy corresponding to a fraction of the binding energy of H_2 (4.5 eV). Therefore the formation of H_2 contributes to the heating of the gas.

2.4.2.3 Chemistry

Dust grains act as small chemical reactors. They also participate in the circulation of matter through the various phases of the interstellar medium. Ice mantles built up on silicate grains in dense cores, both through condensation of gas phase species and through additional processing in the solid phase. Because the molecules are closer to each other in ice mantles than in the gas phase, the chemistry can be more efficient in ice mantles. Started with difficult observations from the ground, the inventory of ice mantles has fully benefited from the ISO and Spitzer missions, with their sensitive spectrometers fully covering the infrared spectral domain [44]. This field has fully benefited from the synergy between astrophysical observations and laboratory experiments, where the formation and evolution of interstellar ices can be studied. The excellent agreement of the infrared spectra of material produced in the laboratory with astrophysical data indicates that the most important features are well understood. The effort is now put into understanding the physical and chemical processes leading to the complex ice composition when starting with relatively simple ice mixtures (e.g. [49]).

Indeed, one of the most important roles of dust grains is the production of complex organic molecules in grain mantles, which would not be possible through pure gas phase chemistry. The detection of organics such as those listed in Table 2.3 in hot cores and hot corinos with radio telescopes shows the chemical richness produced in ice mantles, both during the cold pre-stellar phase and during the warm-up phase following the birth of the proto-star(s). The energetic radiation associated with young proto-stars, with intense FUV and X-Ray emission, likely contributes to the processing of the ice mantles and helps in building complex organics. The detection through radio techniques is biased towards polar species, with radio spectra strong enough to be identified. It is likely that the inventory of organics in astrophysical ices is significantly richer than those identified so far.

The analysis of the soluble organic matter of primitive meteorites provides interesting clues on this issue, since this material has condensed in the primitive nebula in an analogous way as ices condense in dense interstellar clouds. The processing induced by the radiation of the young sun is also similar to that expected in the ISM. The composition of the soluble organic matter is very rich, with numerous amino-acids. The direct detection with high spectral resolution techniques of a particular amino-acid is extremely difficult since the expected spectrum in the astrophysical sources will be composed of a very large number of weak spectral features that may not be individually detectable. A better approach is probably to focus on laboratory experiments to learn more about the possible routes to chemical complexity in space. For instance, [50] have recently detected in ice residues a key intermediate species in the synthesis of peptides, hydantoin, confirming the important role of ice mixtures in the synthesis of complex organics.

2.5 Observations of the Dense Interstellar Medium

The properties of the interstellar medium have been deduced from analysis of extensive observations covering most of the electromagnetic spectrum (Fig. 2.8). Except for the analysis of pre-solar grain inclusions in meteorites, no in-situ measurement is possible because of the large distance of the objects. In the following, we describe the main observation methods and the information carried out by photons across the electromagnetic spectrum.

2.5.1 Photometry

For an extended object like the interstellar medium, imaging is a natural source of information, providing clues on the spatial structure, geometry and many other physical properties depending on the wavelength of interest.

2.5.1.1 Visible and Near Infrared

Images in the visible and UV are dominated by the combination of scattering and absorption of the radiation from surrounding and embedded stars. Strong visible lines, as those of hydrogen (Hα at 656.3 nm) can also contribute to the detected flux. Scattering and absorption are produced by the dust grains, and are therefore used to study the dust grain size distribution and composition. Both effects cause a dimming of stellar radiation which varies as a function of wavelength and is more pronounced at short wavelengths. The variation of the extinction as a function of wavelength is called the extinction curve. It decreases continuously from the FUV to IR wavelengths, except for the "extinction bump" at 217.5 nm. Dense regions of interstellar clouds have a large enough column density of dust grains that the light from the stars in their background is totally dimmed and they appear as black patches in the bright trail of the Milky Way. The first catalogues of so-called "dark clouds" were therefore established in the beginning of the twentieth century, by e.g. Barnard and co-workers. Barnard 68 (Fig. 2.1) is a good example of such dark clouds. Because the extinction decreases at near infrared wavelengths, background stars that are totally invisible in the optical can be detected in the near infrared J (1.2 μm), H (1.6 μm) and K (2.1 μm) bands.

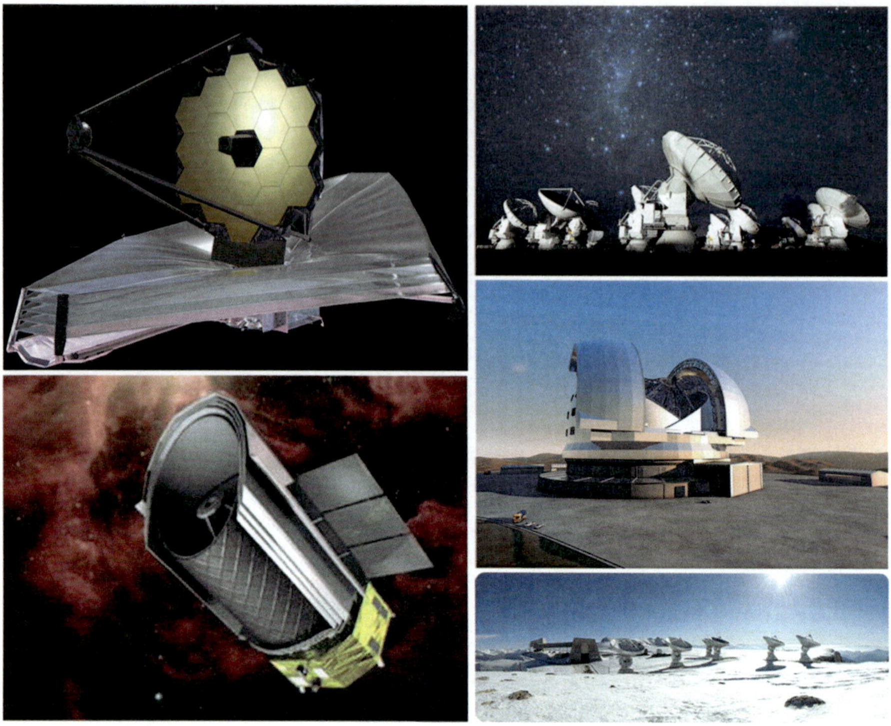

Fig. 2.8 Sketches of planned or proposed future telescopes. From left to right, *top row*: the space telescopes JWST (James Webb Space Telescope, currently in construction, image credit NASA) and SPICA (JAXA project, image credit JAXA); *bottom row*: the ground based telescopes ALMA (in construction, image credit ESO/NOAJ/NRAO), Extremely Large Telescope (ELT, ESO project), Plateau de Bure Interferometer (PdBI) (in operation, proposal for extension as NOEMA, image credit IRAM)

With the increase in the size and sensitivity of visible and near infrared cameras, it is now possible to map the extinction across large areas of the sky, covering the full extent of molecular complexes [51]. These studies make use of full sky surveys such as the Two Micron Sky Survey (2MASS [52]) and the associated stellar catalogue to derive extinction maps. To reach very large column densities, access to the infrared wavelength region has proven to be the most valuable tool. First images of the Galactic Plane at 8 μm revealed dark patches, reminiscent of the classical dark clouds, that were therefore named "infrared dark clouds". As explained above, some of these infra red dark clouds are now thought to be the birth place of massive stars. Current catalogues include 10,000 of objects [20].

The efficiency of scattering as a function of wavelength of the incoming radiation is a strong function of the particle size. Therefore a large fraction of current information on the particle size distribution has been extracted from analysis of the scattered light, in visible nebulae as well as at infrared wavelengths to probe the population of dust grains in dense cores. While the classical population of dust grains has sub-micron size (0.1 μm) in the diffuse clouds and in visible nebulae, the analysis of the deep images taken by the IRAC camera on board the Spitzer satellite has revealed the presence of significantly bigger grains, with micrometre sizes [53] likely resulting from the coagulation and aggregation of smaller particles.

2.5.1.2 Far Infrared

In the far infrared, the continuum radiation from the interstellar medium is dominated by the dust thermal emission. At long wavelengths in the far infrared and sub-millimetre domains, the emission is dominated by the sub-micrometre dust grains that reach a stable temperature. The resulting emission can be described by a modified black-body law, namely:

$$I(\nu) = \varepsilon(\nu)(2k_B\nu^2)/c^2 1/(e^{h\nu/k_BT} - 1) \tag{2.5}$$

where $\varepsilon(\nu)$ describes the emissivity of dust grains as a function of frequency, and T is the dust grain temperature. ε can be derived either empirically from the analysis of multi-wavelength observations, or from theoretical models of the dust grain composition and structure. To first order, ε scales as ν^{β}, with β ranging between 1.5 and 2. The assumption of a single dust grain temperature is a simplification given the spread in dust grain size and composition, but is usual in order to obtain a first insight into the mean properties of the source. The mean grain temperature is about 18 K [54, 55] in the diffuse interstellar medium, and decreases to ~12 K in dense clouds and even down to ~6 K in the coldest dense pre-stellar cores [26] (Fig. 2.7).

The spectra of interstellar nebulae deviate notably from a modified black-body at short and mid-infrared wavelengths (from about 3 to 60 μm), with a significantly stronger radiation than that expected from thermal produced by the smallest solid particles of nanometre size, that can fluctuate in temperature following the absorption of a far UV photon. This phenomenon can heat the small particles to temperatures of a few hundred Kelvin, sufficient to emit in the mid infrared [3].

Therefore, by combining information across the electromagnetic spectrum, it is possible to reveal the dust content, temperature and spatial distribution, as well as the direction of the heating sources.

2.5.1.3 Polarimetry

The analysis of light polarization properties is the best tool to probe the magnetic field in the interstellar medium. Indeed, there are two main mechanisms producing polarized radiation in the visible and near infrared: (1) through scattering of the incoming radiation from a nearby star, or (2), as a result of the dust grain partial alignment with the magnetic field. These two mechanisms can be easily identified with their different spatial and spectral properties. In the former case, the polarization is detected at the same wavelength as the incoming radiation and has a specific spatial pattern determined by the source geometry. In the latter case, the polarization of radiation is caused by asymmetries in the light absorption by dust grains, induced by the magnetic fields. Sub-micron dust grains are not perfectly spherical and can therefore induce an alignment of the long axis of the dust grains perpendicular to the local direction of the magnetic field. Such a geometry is indeed a stable configuration since spinning elongated particles tend to align their angular momentum with the local direction of the magnetic field. The most likely grain alignment mechanism is thought to be due to radiative torques (e.g. [56] and references therein) induced by asymmetry in the grain illumination and their irregular shapes. Aligned grains absorb with higher efficiency light polarized with the electric field parallel to the long axis of the dust grain. Therefore aligned dust grains create a polarization pattern in the extinction of background stars, with the direction of polarization parallel to the direction of the magnetic field. While the dust alignment is a local effect, the resulting polarization pattern is a global effect since the signal is integrated along the line of sight. The efficiency of polarization amounts to a few percent, but its dependency on the intensity of the field is not completely understood yet. The most useful information is the direction of polarization, which gives the direction of the component of the magnetic field parallel to the plane of the sky. Aligned dust grains also cause a polarization pattern in their thermal radiation at far infrared and sub-millimetre wavelength, because the efficiency of thermal radiation is larger in the direction parallel to their long axis. The resulting polarization pattern is perpendicular to the extinction polarization pattern, with the polarization angle being perpendicular to the local direction of the magnetic field.

In both cases, the direction of polarization gives access to the average direction of the magnetic field, projected on the plane of the sky. No information is available on the magnetic field component along the line of sight with this method. The polarization degree ranges from a few percent up to slightly more than 10%.

A third method for studying the magnetic field is the detection of the Zeeman effect using molecular lines. The presence of a magnetic field breaks the degeneracy of the energy levels of molecule. This effect induces a splitting of the molecular spectral lines, together with specific polarization pattern in the different spectral components. The current detection schemes make use of these properties. Usually, the Zeeman effect is detected in circular polarization, resulting from the difference between the left hand side and the right hand side polarizations. The magnitude of this effect scales with the intensity of the magnetic field along the line of sight. The number of suitable spectral lines for such studies is relatively limited, because it requires paramagnetic molecules that strongly interact with the magnetic field. Such species have spectral lines which can be significantly split in the presence of the magnetic field. In the interstellar medium, the hyperfine structure line of atomic hydrogen at 21 cm, the Λ doubling lines of OH at 18 cm, and the rotational lines of the CN radical near 113 GHz are the main spectral features used for measuring the magnetic field [42].

2.5.2 Spectroscopy

Spectroscopy is a key method of investigation for astrochemistry. It can be combined with imaging methods in so-called "spectro-imaging techniques", which provide three dimensional data cubes, having as first two dimensions the position on the sky, and as third dimension a spectral axis, labelled in wavelength, frequency or velocity.

Spectroscopy is a very important technique because it allows measurements of most of the constituents of the interstellar gas, thanks to their unique spectral signatures (see Chap. 1) (Fig. 2.9). Depending on the wavelength domain, molecules, atoms and ions in the gas phase can be detected through their electronic, vibrational, rotational and/or fine-structure transitions. Given the typical values of the Doppler broadening of interstellar lines of a few km s^{-1}, high spectral resolution is required for obtaining spectral information on the line profiles, namely $R = \lambda/\delta\lambda \geq 10^4$. This figure is a minimum, with higher spectral resolutions reaching 10^6 being favoured for studies requiring a detailed description of the line profiles. Given the low pressures of the interstellar medium, the sole broadening mechanism of spectral lines is the Doppler effect. It relates the shift in line frequency ν (or wavelength λ) relative to the rest frequency $\nu_0(\lambda_0)$ to the velocity of the molecule or atom v_z along the line of sight joining the source and the telescope. To first order, it can be written as,

$$\delta\nu/\nu_0 = (\nu - \nu_0)/\nu_0 = -\delta\lambda/\lambda_0 = -v_z/c \tag{2.6}$$

Therefore line profiles carry precious information on the gas dynamics along the line of sight, as the Doppler effect is only sensitive to the velocity component parallel to the direction of observation. The centroid of the line profile provides the mean velocity along the line of sight, the line width provides information on the velocity dispersion that can have several sources:

- Thermal broadening resulting from the random motions of molecules or atoms at a finite temperature, $\sigma_{th} = \sqrt{k_B \mathrm{T}/m}$
- Turbulent motions associated with the local fluid dynamics, σ_t.
- Systematic motions along the line of sight, such as ordered rotation, expansion or more complex motions.

The first two broadening mechanism produce Gaussian line profiles,

$$f(v) = (1/\sigma\sqrt{\pi})e^{-(v-v0)^2/\sigma^2} \tag{2.7}$$

with v_O the source velocity in the rest frame of interest, and σ the broadening parameter, while the last mechanism can produce any line profile, either symmetric or asymmetric. It is therefore customary to use Gaussian profiles to fit astronomical line profiles. For this purpose, astronomers often use the Full Width at Half

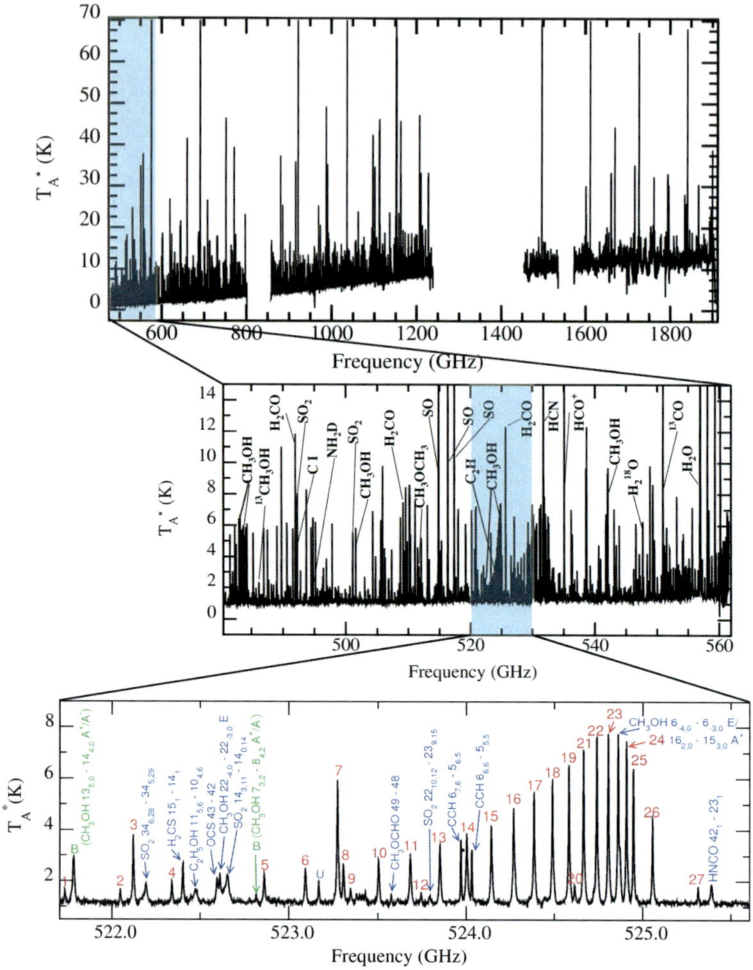

Fig. 2.9 (continued)

Maximum *FWHM* to characterize the line width. It is related to the broadening parameter σ with $FWHM = 2\sigma \sqrt{ln(2)} \sim 1.665\sigma$.

Although their spectral signatures are not as narrow as gas phase species, the composition of the dust particles can also be established through analysis of astronomical spectra, especially in the infrared spectral domain. The best tools are the vibrational bands of solid state materials. For instance, two deep absorption bands at 9.7 and 18 μm track the presence of silicate grains. The exact position and shape of the band carry important information on the silicate composition, amorphous versus crystalline character, as well as the grain sizes. The physical meaning is as follows: these bands correspond to the elongation and bending modes of the Si – O bond in silicates whose properties depend on the environment of these

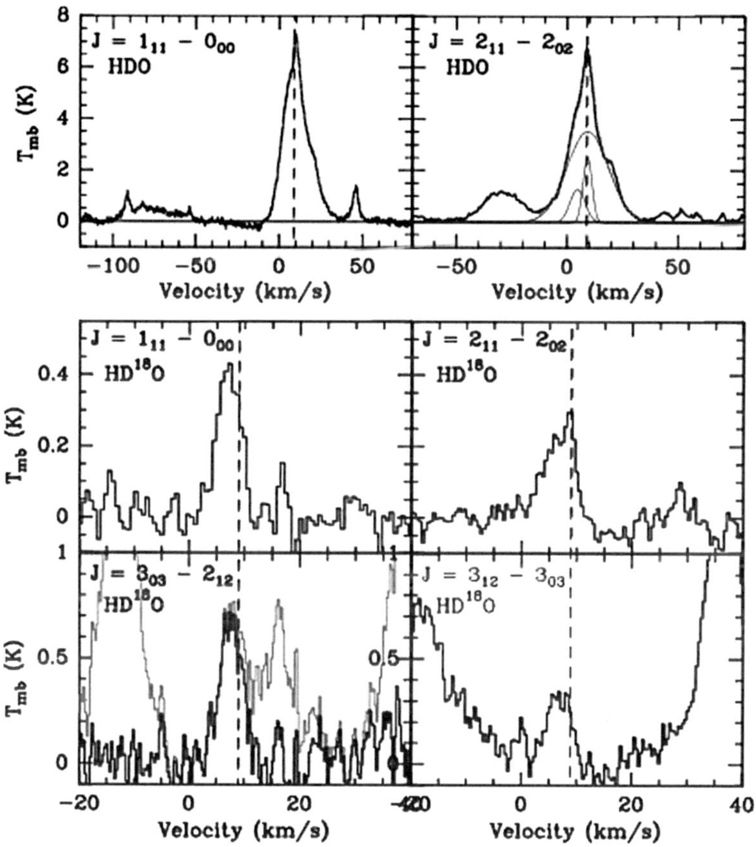

Fig. 2.9 Subset of the full spectral survey of the Orion-KL region performed with the Herschel space observatory. The *first panel* shows the frequency coverage with two successive zooms on a small subset of the data, where the carriers of the strongest lines are identified. In the *bottom panel*, all lines with numbers ranging from 1 to 27 are produced by CH_3OH. The *second panel* shows a collection of lines from HDO and $HD^{18}O$. The line profile is fitted with four separate spectral components, corresponding to four spatially distinct sources of emission, that are too close to be spatially separated at the moderate angular resolution allowed by Herschel [57]

atoms in the solid. The dependency with the grain size is introduced because a competition between absorption and scattering will be dimming the bands when grains reach a size comparable to the wavelength of the band [3].

The presence of mantles on silicate grains is revealed by absorption bands of water ice, superimposed with features from carbon monoxide, carbon dioxide, methanol and other less abundant constituents. As for the silicate bands, the spectral features associated with molecular ices are most prominent in the infrared spectral region, between 3 and 40 μm mainly. Because water ice is the main constituent of ice mantles, the spectral features due to water ice are the strongest, especially the 3.1 μm feature due to the stretching mode of the O-H bound in solid water. This band can be fully saturated for deeply embedded young stellar objects.

2.5.3 Instruments

Specific instruments have been developed to enable the detection of the faint and usually extended emission of interstellar nebulae, as well as the narrow spectral lines, that can appear either as emission or absorption features. The key parameters for such studies are the spatial and spectral resolution of the instruments, their wavelength coverage and sensitivity. Depending on the wavelength domain, specific detectors and instrument layout must be used that take advantage of the most recent advances in light collecting and detection.

For a telescope of diameter D, the scattering of the incoming radiation upon the dish provides the most stringent limit on the angular resolving power $\delta\theta$ defined as the angular size of the smallest object that can be individually resolved. This resolving power is related to the telescope diameter and the observing wavelength λ as

$$\delta\theta \sim \lambda/D \sim 1''(\lambda/1\mu m)(0.25m/D)$$

In this formula, the resolving power is expressed in arcsec ($1rd = 206265''$). The theoretical resolution is seldom achieved at visible wavelengths because the atmospheric turbulence limits the resolving capabilities of ground based telescopes to a fraction of an arcsec, unless they are equipped with adaptive optics correcting systems. In the far infrared and (sub)millimetre domain, the seeing induced by the atmosphere has no impact and telescopes can achieve their theoretical angular resolution. For the closest molecular clouds situated in the Taurus or Ophiuchus complexes, at distances d of about 120 parsecs, such a resolving power corresponds to a linear size l of $l = \delta\theta d = 1.8 \times 10^{13}$ m ~ 120 AU ~ 0.6×10^{-3} pc.

In the following, we give a short summary of the main observatories and instruments that have led to great advances in the field of astrochemistry.

Instruments operating at ultraviolet wavelengths must be space borne as the Earth's atmosphere efficiently shields the energetic radiation. Recent instruments include the Hubble Space Telescope (HST) with its long suite of instruments combining imaging cameras and spectrometers (see http://hubblesite.org/the_telescope/). The Far Ultraviolet Spectrum Explorer (FUSE) has explored the shortest wavelength, a key domain for the study of molecular hydrogen.

In the visible, ground based telescopes provide the necessary instruments, with imaging cameras with a large field of view and sensitive high resolution spectrometers. Similar instruments are developed in the near infrared spectral region, that is well accessible from the ground. (see e.g. the instrument available at the European Southern Observatory (ESO) http://www.eso.org).

The atmospheric transmission becomes very poor past ~10 μm, with the increased attenuation of the mid and far infrared radiation by molecules in the atmospheres, most notably water vapour H_2O and carbon dioxide CO_2. Therefore the information has been acquired using space borne or airborne telescopes such as the ESA mission Infrared Space Observatory (ISO http://sci.esa.int/iso) and the NASA Spitzer Space Telescope (http://www.spitzer.caltech.edu), or the Stratospheric Observatory for Infrared Astronomy (SOFIA), (http://www.sofia.usra.edu). The need for space borne systems is also true in the far infrared. This wavelength domain is currently accessible thanks to SOFIA and the joint ESA/NASA mission Herschel Space Observatory (http://herschel.esac.esa.int).

The sky becomes more transparent at longer wavelengths, allowing the use of large (sub)millimetre telescopes from ground based observatories. Because the presence of water vapour in the atmosphere is one of the major source of atmospheric opacity, these observatories are located in high altitude sites such as the top of the Mauna Kea extinct volcano or the high altitude Chajnantor plateau in the Atacama desert about 5,000 m above sea level in Chile. This frequency domain, from about 80 GHz to about 900 GHz is one of the main sources of information for studying interstellar molecules since it corresponds to the frequencies of rotational lines of molecules. However, the relatively long wavelength gives access to limited spatial resolution as the spatial resolution power of a given telescope scales linearly with the wavelength and inversely with its size, as λ/D. Therefore at (sub) millimetre wavelengths, even the largest monolithic telescopes like the IRAM 30 m telescope (http://www.iram-institute.org) have a modest spatial resolution of several tens of arcsec, significantly worse than the arcsec achievable with visible and IR telescopes because of the long wavelengths. Therefore, it is customary to build radio interferometers, in which the incoming radiation is coherently detected by several identical dishes spaced by a large distance, from a few hundred meters up to several kilometres. The resolving power is now given by the telescope separation, rather than by the telescope diameter, enabling an excellent spatial resolution. The price to pay is in terms of sensitivity, especially for extended emission since the sparse spacing of the telescopes composing the interferometer (typically six to nine element telescopes, giving access to $N(N-1)/2$ independent measurement points) introduces a spatial filtering in the incoming radiation.

The Plateau de Bure Interferometer (PdBI) located in the French Alps, and the Atacama Large (sub)Millimeter Array (ALMA) (http://www.almaobservatory.org/) located in Chile are among the most sensitive (sub)millimetre radio interferometers. When its construction is complete, ALMA will provide unprecedented sensitivity for detection and imaging, with its 50 antenna network and its complementary arrays increasing the sensitivity to extended emission.

References

1. Bertoldi F, Cox P, Neri R et al (2003) High-excitation CO in a quasar host galaxy at $z = 6.42$. Astron Astrophys 409:47
2. Noterdaeme P, Petitjean P, Srianand R, Ledoux C, López S (2011) The evolution of the cosmic microwave background temperature. Measurements of T_{CMB} at high redshift from carbon monoxide excitation. Astron Astrophys 526:L7
3. Draine B (2011) Physics of the interstellar and intergalactic medium, Princeton series in astrophysics. Princeton University Press, Princeton

4. Ostriker EC, McKee CF, Leroy AK (2010) Regulation of star formation rates in multiphase galactic disks: a thermal/dynamical equilibrium model. Astrophys J 721:975
5. Wood KH, Alex S, Joung MR, Mac Low MM, Benjamin RA, Haffner LM, Reynolds RJ, Madsen GJ (2010) Photoionization of high-altitude gas in a Supernova-driven turbulent interstellar medium. Astrophys J 721:1397
6. Field GB, Goldsmith DW, Habing HJ (1969) Cosmic-ray heating of the interstellar gas. Astrophys J 155:L149
7. Jenkins EB, Tripp TM (2011) The distribution of thermal pressures in the diffuse, cold neutral medium of our galaxy. II. An expanded survey of interstellar C I fine-structure excitations. Astrophys J 734:65
8. Heiles C, Troland TH (2005) The millennium Arecibo 21 centimeter absorption-line survey. IV. Statistics of magnetic field, column density, and turbulence. Astrophys J 624:773
9. Audit E, Hennebelle P (2010) On the structure of the turbulent interstellar clouds. Influence of the equation of state on the dynamics of 3D compressible flows. Astron Astrophys 511:A76
10. Rachford BL, Snow TP, Destree JD et al (2009) Molecular hydrogen in the far ultraviolet spectroscopic explorer translucent lines of sight: the full sample. Astrophys J Suppl 180:125
11. Miville-Deschênes M-A, Martin PG, Abergel A et al (2010) Herschel-SPIRE observations of the Polaris flare: structure of the diffuse interstellar medium at the sub-parsec scale. Astron Astrophys 518:L10
12. Roman-Duval J, Jackson JM, Heyer M, Rathborne J, Simon R (2010) Physical properties and galactic distribution of molecular clouds identified in the galactic ring survey. Astrophys J 723:492
13. Kolmogorov A (1941) The local structure of turbulence in incompressible viscous fluid for very large Reynolds' numbers. DoSSR 30:301
14. Falgarone E, Hily-Blant P, Pety J (2009) Intermittency of interstellar turbulence: extreme velocity-shears and CO emission on milliparsec scale. Astron Astrophys 507:355; Lis DC, Vaillancourt JE, Goldsmith PF, Bell TA, Scoville NZ, Zmuidzinas J (eds) (2009) Submillimeter astrophysics and technology: a symposium honoring Thomas G. Phillips. ASP conference series, vol 417. Astronomical Society of the Pacific, San Francisco, p 243
15. Lada CJ, Lombardi M, Alves J (2010) On the star formation rates in molecular clouds. Astrophys J 724:687
16. André P et al (2010) From filamentary clouds to prestellar cores to the stellar IMF: initial highlights from the Herschel Gould Belt Survey. Astron Astrophys 518:L102
17. Alves JF, Lada CJ, Lada EA (2001) Internal structure of a cold dark molecular cloud inferred from the extinction of background starlight. Nature 409:159
18. Caselli P, van der Tak FFS, Ceccarelli C, Bacmann A (2003) Abundant H_2D^+ in the pre-stellar core L1544. Astron Astrophys 403:L37
19. Pérault M, Omont A, Simon G (1996) First ISOCAM images of the Milky Way. Astron Astrophys 315:L165
20. Peretto N, Fuller GA (2010) A statistical study of the mass and density structure of infrared dark clouds. Astrophys J 723:555
21. Herbst E, van Dishoeck E (2009) Complex organic interstellar molecules. Annu Rev Astron Astrophys 47:427
22. Parise B, Castets A, Herbst E, Caux E, Ceccarelli C, Mukhopadhyay I, Tielens AGGM (2004) First detection of triply-deuterated methanol. Astron Astrophys 416:159
23. Stark R, Sandell G, Beck SC et al (2004) Probing the early stages of low-mass star formation in LDN 1689 N: dust and water in IRAS 16293-2422A, B, and E. Astrophys J 608:341
24. Commerçon B, Hennebelle P, Audit E, Chabrier G, Teyssier R (2010) Protostellar collapse: radiative and magnetic feedbacks on small-scale fragmentation. Astron Astrophys 510:L3
25. Mouschovias T, Spitzer L (1976) Note on the collapse of magnetic interstellar clouds. Astrophys J 210:326
26. Bergin E, Tafalla M (2007) Cold dark clouds: the initial conditions for star formation. Annu Rev Astron Astrophys 45:339

27. Pety J, Gueth F, Guilloteau S, Dutrey A (2006) Plateau de Bure interferometer observations of the disk and outflow of HH 30. Astron Astrophys 458:841
28. Burrows CJ, Stapelfeldt KR, Watson AM (1996) Hubble space telescope observations of the disk and jet of HH 30. Astrophys J 473:437
29. Levinson HF, Morbidelli A, Tsiganis K, Nesvorný D, Gomes R (2011) Late orbital instabilities in the outer planets induced by interaction with a self-gravitating planetesimal disk. Astron J 142:152
30. Lagrange AM, Bonnefoy M, Chauvin G et al (2010) A giant planet imaged in the disk of the young star β Pictoris. Science 329:57
31. Gerin M, Pety J, Goicoechea JR (2009) The horsehead nebula, a template source for interstellar physics and chemistry. In: Lis DC, Vaillancourt JE, Goldsmith PF, Bell TA, Scoville NZ, Zmuidzinas J (eds) Submillimeter astrophysics and technology: a symposium honoring Thomas G. Phillips. ASP conference series, vol 417. Astronomical Society of the Pacific, San Francisco, p 165
32. Gonzalez GM, Le Bourlot J, le Petit F, Roueff E (2008) Radiative transfer revisited for emission lines in photon dominated regions. Astron Astrophys 485:127
33. Glover SCO, Federrath C, Mac Low MM, Klessen RS (2010) Modelling CO formation in the turbulent interstellar medium. Mon Notices R Astron Soc 404:2
34. Draine BT, Mc Kee CF (1993) Theory of interstellar shocks. Annu Rev Astron Astrophys 31:373
35. Nisini B, Benedettini M, Codella C et al (2010) Water cooling of shocks in protostellar outflows: *Herschel*-PACS map of L1157. Astron Astrophys 518:L120
36. Troscompt N, Faure A, Wiesenfeld L, Ceccarelli C, Valiron P (2009) Rotational excitation of formaldehyde by hydrogen molecules: ortho-H_2CO at low temperature. Astron Astrophys 493:687
37. Maret S, Faure A, Scifoni E, Wiesenfeld L (2009) On the robustness of the ammonia thermometer. Mon Notices R Astron Soc 399:425
38. van der Tak FFS, Black JH, Schöier FL, Jansen DJ, van Dishoeck EF (2007) A computer program for fast non-LTE analysis of interstellar line spectra with diagnostic plots to interpret observed line intensity ratios. Astron Astrophys 468:627
39. Goicoechea JR, Pety J, Gerin M, Hily-Blant P, Le Bourlot J (2009) The ionization fraction gradient across the Horsehead edge: an archetype for molecular clouds. Astron Astrophys 498:771
40. Indriolo N, McCall BJ (2012) Investigating the cosmic-ray ionization rate in the galactic diffuse interstellar medium through observations of H_3^+. Astrophys J 745:91
41. Neufeld D, Goicoechea JR, Sonnentrucker P et al (2010) Herschel/HIFI observations of interstellar OH^+ and H_2O^+ towards W49N: a probe of diffuse clouds with a small molecular fraction. Astron Astrophys 521:L10
42. Crutcher RM (2009) OH and CN Zeeman observations of magnetic fields in molecular clouds. Rev Mex de Astron y Astrofisica (Serie de Conferencias) 36:107
43. Godard B, Falgarone E, Pineau Des Forets G (2009) Models of turbulent dissipation regions in the diffuse interstellar medium. Astron Astrophys 495:847
44. Gibb EL, Whittet DCB, Boogert ACA, Tielens AGGM (2004) Interstellar ice: the infrared space observatory legacy. Astrophys J 151:35
45. Motte F, Zavagno A, Bontemps S et al (2010) Initial highlights of the HOBYS key program, the Herschel imaging survey of OB young stellar objects. Astron Astrophys 518:L77
46. Joblin C, Tielens AGGM (eds) (2011) PAHs and the universe: a symposium to celebrate the 25th anniversary of the PAH hypothesis. EAS publications series, vol 46. EDP Sciences, Les Ulis
47. Jura M (1974) Formation and destruction rates of interstellar H_2. Astrophys J 191:375
48. Le Bourlot J, Le Petit F, Pinto C, Roueff E, Roy F (2012) Surface chemistry in the interstellar medium – I – H2 formation by Langmuir-Hinshelwood and Eley-Rideal mechanisms. Astron Astrophys 541:A76

49. Danger G, Bossa J-B, de Marcellus P, Borget F, Duvernay F, Theulé P, Chiavassa T, D'Hendecourt L (2011) Experimental investigation of nitrile formation from VUV photochemistry of interstellar ices analogs: acetonitrile and amino acetonitrile. Astron Astrophys 525:30
50. de Marcellus P, Bertrand M, Nuevo M, Westall F, Le Sergeant d'Hendecourt L (2011) Prebiotic significance of extraterrestrial ice photochemistry: detection of hydantoin in organic residues. Astrobiology 11:847
51. Lombardi M, Alves J, Lada CJ (2011) 2MASS wide field extinction maps. IV. The Orion, Monoceros R2, Rosette, and Canis Major star forming regions. Astron Astrophys 535:A16
52. Skrutskie MF, Cutri RM, Stiening R et al (2006) The two micron all sky survey (2MASS). Astron J 131:1163
53. Pagani L, Steinacker J, Bacmann A, Stutz A, Henning T (2010) The ubiquity of micrometer-sized dust grains in the dense interstellar medium. Science 329:1622
54. Planck Collaboration, Abergel A, Ade PAR, Aghanim N et al (2011) *Planck* early results. XXIV. Dust in the diffuse interstellar medium and the galactic halo. Astron Astrophys 536:24
55. Compiègne M, Verstraete L, Jones A, Bernard J-P, Boulanger F, Flagey N, Le Bourlot J, Paradis D, Ysard N (2011) The global dust SED: tracing the nature and evolution of dust with DustEM. Astron Astrophys 525:A103
56. Whittet DCB, Hough JH, Lazarian A, Hoang T (2008) The efficiency of grain alignment in dense interstellar clouds: a reassessment of constraints from near-infrared polarization. Astrophys J 764:304
57. Bergin E, Phillips TG, Comito C et al (2010) Herschel observations of EXtra-Ordinary Sources (HEXOS): the present and future of spectral surveys with Herschel/HIFI. Astron Astrophys 521:L20

Chapter 3
Chemical Processes in the Interstellar Medium

Michael J. Pilling

Abstract Models of the chemical composition of the interstellar medium incorporate networks of chemical reactions. The rate coefficients and the products of these reactions are important components of the model. In this chapter I review the determinants of these components and the methods used to measure them experimentally and calculate them using theory. The bulk of the chapter is devoted to ion + neutral molecule and neutral molecule + neutral molecule reactions. I also briefly discuss radiative association, dissociative recombination and reactions occurring on surfaces. The conditions of low pressure and low temperature in the interstellar medium place considerable demands on experiment and theory, which are particularly severe for reactions between neutral species. Many reactions can be estimated with tolerable accuracy. Others require a combination of high level electronic structure calculations, coupled with detailed theory and low temperature experimental measurements.

3.1 Introduction

The use of complex models, with large networks of chemical reactions, is discussed in Chap. 4. The models are directly linked to, and dependent on, observations and are designed to help us understand how interstellar molecules are created in their observed abundances [1]. Models are based on extensive theoretical and experimental research, both in the construction, evaluation and optimisation of the models themselves and of the networks on which they rely, and on the underlying science that provides quantitative information on the rates of those reactions. This chapter is primarily concerned with the underlying science.

M.J. Pilling (✉)
School of Chemistry, University of Leeds, Leeds LS2 9JT, UK
e-mail: m.j.pilling@leeds.ac.uk

I.W.M. Smith et al. (eds.), *Astrochemistry and Astrobiology*, Physical Chemistry in Action, DOI 10.1007/978-3-642-31730-9_3,

The study of the rates of chemical reactions has a long history as both a fundamental and applied field of research. Fundamental interest centres on the determinants of the reaction rates of elementary reactions, which are generally regarded as single step processes, although this limitation will be somewhat revised in the discussion below. From an experimental perspective, the development of direct methods of measuring rates of fast reactions in the 1950s and over subsequent decades played a central role. Advances in optical and mass spectrometries have been essential in these developments. Theories of such reactions also have a long history, largely, but not exclusively, through the conceptual framework provided by transition state theory, although dynamical theories have also played an important role. Advances in computing have revolutionised reaction kinetics theory by providing routes to calculating aspects of the potential energy surfaces on which reactions occur, as well as improving the level at which rate theory can be applied. The main areas of application of kinetics, especially in the gas phase, have been combustion and atmospheric chemistry, so that the emphasis has been on temperatures above ~200 K, with experimental techniques largely restricted to such conditions. Indeed, many measurements of key reactions have been limited to room temperature. This limitation has clear consequences for the provision of rate data for modelling chemical processes in the Interstellar Medium (ISM). The development of new techniques has been essential even to approach the lower temperatures needed, and relatively few reactions have been studied under such conditions. Progress has depended on combining theory with experiment.

This has been achieved at a variety of levels. In many cases, no experimental data are available and rate coefficients are simply estimated, often using theory at the most elementary of levels. Where measurements have been made at room temperature, rate data are again estimated at the temperatures appropriate to the model conditions, but with the benefit of some degree of model tuning using the available data. Measurements at very low temperatures have proved invaluable in reducing or even eliminating the extent of extrapolation needed, but also in providing a testing ground for the evaluation and refinement of theory. There are so many reactions in the major networks and experimental measurement of all of them is just not feasible, so estimation plays a key role. Recent years have seen substantial theoretical developments, both in the calculation of the potential energy surface and in the kinetic model itself, so that, in some cases, rate data of high accuracy can be achieved through theory alone. The ideal solution though is one of combined theory and experiment.

One aim of experimental and theoretical kinetics is the determination of the rate coefficient, $k(T)$ (Sect. 1.5), but it is not the only one. Some reactions form more than one set of products, as in the reaction between H_3^+ and $O(^3P)$:

$$H_3^+ + O(^3P) \rightarrow OH^+ + H_2 \tag{R3.1a}$$

$$H_3^+ + O(^3P) \rightarrow H_2O^+ + H \tag{R3.1b}$$

The rate coefficient for the overall reaction (R3.1) (= (R3.1a) + (R3.1b)) is k_1, which is equal to the sum of the rate coefficients for the two reaction channels, so $k_1 = k_{1a} + k_{1b}$. The *branching ratio* for the two channels is the ratio of their rate coefficients, so that in the example above the branching ratio is k_{1a}/k_{1b}, while the *channel efficiency* for channel (R3.1a) is k_{1a}/k_1.[1] The determination of branching ratios is as important as the determination of overall rate coefficients but presents greater challenges.

3.2 Reaction Networks

Reaction networks and their construction and use are discussed in much greater detail in Chap. 4. This section aims simply to provide a context for the discussion of chemical kinetics that follows.

Models of the ISM consist of a physical framework, that may involve dynamical processes, but certainly involves the timescales of the overall processes being modelled, a chemical context, e.g. the abundances of key species, and a set of ordinary differential equations (odes), that describe the kinetics of the component *elementary* reactions that make up the network [2]. These odes are of the form:

$$d[X]/dt = \text{Total rate of forming X} - \text{Total rate of removing X}. \qquad (3.1)$$

The rate terms contain the rates of each of the elementary reactions forming or removing species X. For example, for X = H_2O^+, the rate of forming X includes the rate of reaction (R3.1b), $k_{1b}[H_3^+][O]$, while for X = H_3^+, the rate of removing X includes the overall rate of reaction (R3.1), $k_1[H_3^+][O]$, where the square brackets denote concentration.

Reaction (R3.1) is an elementary reaction; it is a single step reaction in which the reactants approach, under the influence of, in this case, an ion-quadrupole interaction, and electronic rearrangement in the resulting collision complex leads to formation of the two sets of products. Quantitative information on the rate coefficient and branching ratios, including their temperature dependence, is needed to construct and use the reaction network.

Millar et al. [3] were the first to describe a database of reactions and rate coefficients for use in astrochemistry, and this formed the basis of the Manchester database, which is available on line (http://www.udfa.net). The other major network is that at Ohio State University (osu-09-2008 is the latest version) (http://www.physics.ohio-state.edu/~eric/research.html). Similar web-mounted networks are

[1] These definitions of '*branching ratio*' and '*channel efficiency*' are not universally agreed. For example, the KIDA data base uses 'branching ration' for the ratio of the rate coefficient for a particular channel to that for the overall reaction, so that the branching ratios sum to unity.

available for combustion (e.g. GRI-Mech, http://www.me.berkeley.edu/gri-mech/) and atmospheric chemistry (e.g. master chemical mechanism, http://mcm.leeds.ac.uk/MCM/)

Reaction networks are large: osu-09-2008 for example, includes reactions of 455 species and 4,457 reactions. The construction of such a network has taken place over many years and its current form reflects increasing knowledge of the elementary reactions and increasing understanding of the reactions that need to be included. The latter depends on evaluating simulations based on a network against observational data. Evaluation is easier for the equivalent networks used in combustion [4] and atmospheric chemistry [5, 6], where simulations are compared not only with observations of a flame or the atmosphere, but also with observations in relatively controlled environments in the laboratory, such as flow or stirred reactors in combustion and simulation chambers in atmospheric chemistry, where the experimentalist has some control over the system studied. This is not the case in the ISM, because the conditions of pressure, temperature and species concentration are difficult to achieve in the laboratory, and model and network evaluation are necessarily restricted to comparisons between model predictions and observations of the ISM itself. Tests on sub-systems – combinations of reactions, such as those occurring on surfaces – make an important contribution to the construction of realistic networks.

The effort required to determine rate coefficients is such that some focus is needed, so that key reactions are studied. This is increasingly achieved using uncertainty propagation [1, 7] and sensitivity analysis [8]. The former provides an estimate of the overall uncertainty of the model. The latter helps identify those reactions contributing most to this uncertainty, or which are the major reactions forming or removing a key species. The former activity requires knowledge of the parameters in the model; in the context of the present chapter, these are the rate coefficients and branching ratios of the component elementary reactions. This process is made more difficult because reaction kinetics is not as exact a science as one might wish, and measurements of the same rate coefficient in different laboratories can differ significantly. The provision of databases of evaluated rate coefficients has been an important activity in combustion [9] and atmospheric chemistry [10] for many years; the KIDA (Kinetic Database for Astrochemistry, http://kida.obs.u-bordeaux1.fr/) has recently been set up to provide the same service in astrochemistry [11]. Each datasheet reviews all available measurements and calculations of the rate coefficient for a specific reaction, and recommends rate parameters over a specified temperature range, together with an estimate of the uncertainty. The rate coefficients are usually expressed through the Kooij equation (1.29): $k = \alpha(T/300)^{\beta}\exp(-\gamma/T)$.

The emphasis in this chapter is on bimolecular reactions, i.e. reactions of the type A + B. Both the theoretical framework and the experimental techniques differ for reactions between an ion and a neutral molecule and between two neutral molecules, and they are discussed separately, in Sects. 3.3 and 3.4 respectively. In ion + neutral molecule reactions, the intermolecular forces are comparatively long range and theory concentrates on the calculation of the so-called *capture rate coefficient*.

For most reactions of this type, all that is necessary to ensure reaction is that the reactants approach and are held together by the strong intermolecular force. Any rearrangement of the electrons to form new chemical bonds is then assured and so is not rate determining. In neutral molecule + neutral molecule reactions, by contrast, the attractive intermolecular forces are shorter range and generally overlap the region in which the chemical forces operate. Calculation of the rate coefficient therefore must acknowledge both types of force and the approach is necessarily different.

Radiative association is examined in Sect. 3.5 and dissociative recombination and surface reactions are discussed briefly in Sects. 3.6 and 3.7. Photo-processes are not covered.

3.3 Ion-Neutral Reactions

The majority of reactions in the OSU kinetic database involve reaction of a cation with a neutral atom or molecule. They are initiated by the ionization of H_2 by cosmic rays [12, 13] followed by proton transfer to H_2 to form ${H_3}^+$:

$$ {H_2}^+ + H_2 \rightarrow {H_3}^+ + H \tag{R3.2} $$

The rate coefficient, k_2, is 1.7×10^{-9} cm^3 s^{-1} at 10 K [14], which is typical of many of the reactions of this type that are of importance in the ISM. The timescale for conversion of ${H_2}^+$ to ${H_3}^+$ is $(k_2[H_2])^{-1}$ which is less than a day for a molecular hydrogen density of 10^4 cm^{-3}. ${H_3}^+$ reacts rapidly with a wide range of neutral species. The key data needed for modelling purposes are the rate coefficients for these reactions and the branching ratios between competing product channels. For example, ${H_3}^+$ reacts with CO to form both HCO^+ and HOC^+ [15]:

$$ {H_3}^+ + CO \rightarrow HCO^+ + H_2 \tag{R3.3a} $$

$$ {H_3}^+ + CO \rightarrow HOC^+ + H_2 \tag{R3.3b} $$

Mass spectrometry, in principle, provides a means of determining the product branching ratio, but it is not always easy to do, especially for reaction (R3.3), where the product ions in channels a and b have the same mass. Most experimental studies simply provide the rate coefficient. The problem of providing appropriate rate data is further compounded by the difficulty of making measurements of k at temperatures appropriate to the ISM; most are determined at room temperature. It is often necessary, therefore, to extrapolate experimental data to low temperatures, or even, when measurements are unavailable, simply to make estimates to provide the necessary rate data.

3.3.1 Experimental Methods

Two general approaches have been used in the study of cation-neutral reactions, $A^+ + B$, based on trapping the ion in a field and observing its loss by reaction with the neutral [16] and flowing the ion down a flow-tube, in the presence of the neutral, and observing its loss at a point down the tube [17, 18].

Most of the earlier rate data for ion neutral reactions were determined using ion cyclotron resonance (ICR), where the ions, A^+, are trapped by a combination of magnetic and electric fields. The neutral reactant, B, is then introduced in large excess and A^+ decays exponentially with a time constant $(k[B])^{-1}$, where k is the rate coefficient. After a known time delay, the ions are transferred through an analysis region, where their concentration and hence the rate coefficient are determined. Varying the time delay allows a whole time profile to be developed, on a 10 ms timescale, for a selected ion.

The total gas density in an ICR is ~10^{11} cm^{-3}, while the extraction time is typically 10 ms. Since the collision frequency is ~100 ms, the reaction is little affected by collisions with the bath gas. This has some advantages, in that the conditions are closer to those found in the ISM than is the case for the flow methods discussed below, but observing the effects of collisions on the rate coefficient can provide useful quantitative information on the reaction mechanism.

Measurements made using ion cyclotron resonance are mainly restricted to room temperature. Low temperatures have been achieved using liquid helium cooled Penning ion-traps [19] and Gerlich and co-workers [20] have used inhomogeneous trapping fields, coupled with cooling by collision with a buffer gas to reach temperatures of ~10 K. They studied the reaction:

$$CH_4{}^+ + H_2 \rightarrow CH_5{}^+ + H \qquad \text{(R3.4)}$$

and showed that the rate coefficient varied as $T^{-1.1}$ over the range 300–30 K, and reached a roughly constant value of 4×10^{-10} cm^3 s^{-1} over the range 30–10 K.

Flow methods operate at gas densities of ~10^{16} cm^{-3}. In the flowing afterglow (FA) method, ions in a buffer gas, usually He, flow down a tube and are detected mass spectrometrically at a point downstream [17, 18]. The neutral co-reactant can be introduced at a number of injectors, so that the reaction time, between mixing of the ion and neutral and detection, can be varied. Many reactions have been studied using the selected ion flow tube (SIFT) method, in which the ions introduced into the flow tube are mass selected using a quadrupole mass filter [21]. Flow tubes can be cooled cryogenically and typical lower temperatures of ~80 K are achievable using liquid nitrogen. A major limitation is the condensation of lower volatility materials at the cooled flow tube walls, which can limit the molecules studied and the reactant concentrations. Snow and Bierbaum [22] recently reviewed the SIFT and other techniques for measuring rate coefficients for ion-neutral reactions. A selection of rate coefficients taken from their review is given in Table 3.1, for a few of reaction types.

Table 3.1 Examples of rate coefficients for ion-neutral reactions

Reactants	Products	Channel efficiency	k/cm^3 s^{-1}	T/K
$C_6H_6^+ + H$	$C_6H_7^+$	~0.65	2.1 ± 10^{-10}	298
	$C_6H_5^+ + H_2$	~0.35		
$CH_4^+ + H$	$CH_3^+ + H_2$	1.0	6×10^{-10}	50
$CO^+ + N$	$NO^+ + C$	1.0	8.2×10^{-11}	298

Significantly lower temperatures have been achieved by Rowe and co-workers [23, 24] using the CRESU (Cinétique de Réaction en Ecoulement Supersonique Uniforme) technique, in which the gas is expanded through a Laval nozzle and the flow parameters, density, temperature, pressure and velocity in the central 10–20 mm of the resulting uniform supersonic jet are invariant in both axial and radial directions, since the flow is isentropic over a flow distance of tens of centimetres. The jet crosses an electron beam to provide the ions and into a mass spectrometer, which is movable so that the reaction times can be varied. Measurements have been made down to 8 K. Measurements on $He^+ + N_2$ gave $k = 1.2 \times 10^{-9}$ cm s^{-1} at 8 K and 1.3×10^{-9} cm s^{-1} at 30 K, showing the temperature invariance predicted by the Langevin equation (see below), which gives a value of 1.7×10^{-9} cm^3 s^{-1}.

3.3.2 Theoretical Considerations

An understanding of the factors affecting the rate coefficient for reaction between an ion and a neutral atom or molecule centres on the calculation of the capture rate coefficient. Capture brings the reactants into sufficiently close proximity for chemical interaction to occur and reaction to take place. Intermolecular forces were discussed in Sect. 1.4 for the reaction A + B. The attractive potential varies as $R_{AB}{}^{-n}$, where R_{AB} is the distance between A and B. The effective potential energy, $V_{eff}(R_{AB})$, is obtained by adding the energy of orbital motion of A and B, giving:

$$V_{eff}(R_{AB}) = \left[E_{trans} b^2 / R_{AB}{}^2\right] - C/R_{AB}{}^n \tag{3.2}$$

where b is the impact parameter, which is the distance of closest approach of the centres of the two particles in the absence of intermolecular forces. C and n depend on the nature of the intermolecular potential. Table 3.2 gives values of n and expressions for C for a number of different interactions.

The requirement for reaction in a specific collision is that the reactants have sufficient energy to overcome the centrifugal barrier presented by their orbital motion. A strategy was presented in Sect. 1.4 for calculating the cross section for reaction with an ion-induced dipole interaction by (1) finding the value of R_{AB} ($R_{AB,max}$) at which $V_{eff}(R_{AB})$ has its maximum value (by differentiating the right-hand-side of (1.9) and setting the result to zero); (2) finding the corresponding

Table 3.2 Types of interaction between an ion and a neutral molecule (cgs units)[a]

Type	n	C
Ion-induced dipole	4	$\alpha q^2/2$
Ion-dipole	2	$q\mu_D\cos\theta$
Ion-quadrupole	3	$(q\mu_Q/4)(3\cos^2\theta - 1)$

[a]The form of the potential is $-C/R_{AB}{}^n$. q is the charge on the ion, α, μ_D and μ_Q are the polarizability, dipole moment and quadrupole moment of the neutral, θ is the angle formed between the dipole and R_{AB}

value of $V_{eff}(R_{AB})$, that is, $V_{eff}(R_{AB,max})$, and (3) using these results to find the maximum value of b at which collisions with relative energy E_{trans} can 'surmount' the 'centrifugal barrier', $V_{eff}(R_{AB,max})$. The rate coefficient is obtained by multiplying the expression for the cross-section by the Maxwell-Boltzmann expression for the distribution of relative velocities and integrating the result.

The A-B distance at the maximum, $R_{AB,max}$, moves to larger values as E_{trans} falls. Similarly, the average value of R_{AB} at temperature T increases as T is reduced. This behaviour applies to all types of interaction and also to reactions between neutral species, and has important mechanistic consequences as discussed in the following sections.

For reaction between an ion and a spherical molecule, $C = \alpha q^2/2$ and $n = 4$ (see Table 3.2 for definitions of α and q). The resulting capture rate coefficient is given by the Langevin expression:

$$k = 2\pi q(\alpha/\mu)^{1/2} \tag{3.3}$$

where μ is the reduced mass; for the reaction A + B, $\mu = m_A m_B/(m_A + m_B)$, where m_X is the mass of X. Typically, k_L is ~2×10^{-9} cm^3 s^{-1}, e.g. $k_L = 1.7 \times 10^{-9}$ cm^3 s^{-1} for $He^+ + N_2$ (treating N_2 as a spherical molecule) and $k_L = 2.4 \times 10^{-9}$ cm^3 s^{-1} for $H_3{}^+ + CH_4$. k_L is independent of temperature.

The Langevin expression was extended by Su, Chesnavich and Bowers [25, 26] to reaction between an ion and a dipolar molecule, using trajectory calculations and variational rate theory (see below). The reactants were treated as rigid bodies and only their translational and rotational energies were considered. This is a realistic assumption, because the vibrational energies are unlikely to change during approach and capture (see the discussion of adiabatic channels below). For reactions of this type, the potential in (3.3) includes not only the ion-induced dipole interaction but also the orientation-dependent ion-dipole interaction, $-q\mu_D\cos\theta/R_{AB}{}^2$ (Table 3.2). The effective potential energy for the system now depends not only on the orbital motion of the reactants and the interaction energy deriving from the charge on A and the induced and permanent dipole moments on B, but also on the orientation of B.

They showed that the ratio of the capture rate coefficient to the Langevin value, k_D/k_L, depends primarily on a reduced parameter, x, where

$$x = \frac{\mu_D}{(2\alpha k_B T)^{1/2}} \tag{3.4}$$

$x = 1/\sqrt{T_R}$, where T_R is a reduced temperature. They found that, under most conditions,

$$k_D/k_L = 0.4767x + 0.620 \quad \text{for } x \geq 2, \tag{3.5}$$

and

$$k_D/k_L = (x + 0.5090)^2/10.526 + 0.9754 \quad \text{for } x < 2 \tag{3.6}$$

As $x \to 0$, $k_D \to k_L$, as required, since at high temperatures the charge dipole interaction averages to zero. Wakelam et al. [1] give alternative expressions in terms of T and recommend that the Su-Chesnavich formulae are used where no experimental data are available and that the formulae are suitably scaled when room temperature experimental values of the capture rate coefficient have been measured. Applications of the formulae have been considerably aided by the publication by Woon and Herbst [27] of quantum chemical calculations of polarizabilities and dipole moments for 200 neutral species with up to 12 atoms that occur in astrochemical reaction networks. Where such data are available, the calculated values are compared with experimental results.

Maergoiz et al. [28–30] performed classical trajectory calculations for ion-dipole, ion-quadrupole and dipole-dipole collisions, deriving results for capture rate coefficients and expressing them in terms of two reduced parameters, the reduced temperature, θ, closely related to the T_R parameter in the work of Chesnavich and co-workers, and ξ, the Massey parameter, which is equal to the ratio of the collisional timescale to the rotational period of the neutral. $\xi \gg 1$ corresponds to the adiabatic limit (see below). They gave parameterized expressions for k_D/k_L, similar to those given by Su and Chesnavich [26], but extending the range of validity.

While the results of trajectory calculations provide an accurate testing ground for more approximate theories, and, in the parameterised form developed by Su, Chesnavich and Bowers [25, 26], a widely applied means of calculating capture rate coefficients for these more complex interactions, they provide less insight into reaction mechanisms and rate coefficient determinants than more analytic approaches. The simplest approach is provided by phase space theory (PST) which assumes an isotropic potential between the reactants [31]. The centrifugal term in the effective potential in (3.2) can be expressed in terms of the orbital angular momentum quantum number, ℓ , for the collision, so that the equation for $V_{eff}\,(R_{AB})$ becomes:

$$V_{eff}(R_{AB}) = \frac{\hbar^2\ell(\ell+1)}{2\mu R_{AB}^2} - \frac{C}{R_{AB}^n} \tag{3.7}$$

The capture rate for a given total energy, E, and total angular momentum, J, is proportional to the number of states $N(E,J)$ whose relative translational energy exceeds the maximum in the potential, V_{eff}. This maximum energy in turn depends on C,n, ℓ and μ. The total capture rate coefficient is obtained by integrating over a Boltzmann distribution, which gives:

$$k(T) = \frac{1}{hq}\int N(E,J)\exp\left(-\frac{E}{k_{\mathrm{B}}T}\right)dE\,dJ \tag{3.8}$$

where q is the partition function for the reactants (Sect. 1.3). $N(E,J)$ includes the vibrational and rotational states of each reactant, although the former change little in the capture process and can be omitted from the sum. This leaves the rotational states of each reactant. PST assumes that these do not interact so that the attractive potential remains isotropic. For an ion-induced dipole interaction, (3.8) leads to the Langevin expression.

When the interaction depends on the orientation of the neutral molecule, as is the case, for example, for ion-dipole reactions, the simple treatment outlined above is no longer appropriate. The adiabatic channel method is often used in this context [32]. The rotational energy levels in the separated reactants are coupled with the orbital energy levels to define a set of channels for the collision complex. The number of open states, $N(E,J)$, is the number of channels with an energy maximum below the energy E. Examples of this approach include the adiabatic channel centrifugal sudden approximation (ACCSA) of Clary [33] and the statistical adiabatic channel model (SACM) of Troe and co-workers [34].

Troe examined reactions of an ion with a dipolar molecule using SACM [35]. He found that $k_{\mathrm{D}}/k_{\mathrm{L}}$ increases with decreasing temperature in quantitative agreement with the expressions of Su and Chesnavich [26] for a number of ion + linear molecule reactions, including HCl + H_3^+ and HCN + H_3^+ although the results diverged at temperatures below the characteristic rotational temperature, $T^* = B/k_{\mathrm{B}}$, where B is the rotational constant ($= h/8\pi^2 I$, where I is the moment of inertia). For HCl + H_3^+, $T^* = 15.2$ K, while for HCN + H_3^+, $T^* = 2.12$ K. Below these temperatures the SACM values for $k_{\mathrm{D}}/k_{\mathrm{L}}$ reach a limiting value while those from (3.5) and (3.6) continue to increase. Troe also considered reactions in which the neutral, B, is a symmetric or an asymmetric top, where the rotational energy levels are described by more complex expressions than is the case for linear molecules.

Maergoiz et al. [28–30] found good agreement between their trajectory calculations and the Troe results in the adiabatic limit ($\xi \gg 1$). They also showed that the divergence of the SACM results from the expressions of Su and Chesnavich [26] at very low temperatures can be expressed in terms of $\alpha B/2\mu_{\mathrm{D}}^2$.

Georgevskii and Klippenstein [36] applied variational transition state theory (VTST) to the derivation of rate coefficients for reactions governed by long range interactions. VTST is an alternative approach to the adiabatic channel method and is discussed in greater detail in Sect. 3.4, They applied the microcanonical, rotationally resolved version of the theory (μJ-VTST, where J is the total angular momentum of the system), obtaining the final rate coefficient by integrating over the appropriate Boltzmann distributions of translational and rotational energies, as discussed above for PST. The result they obtained, for example, for reaction between an atomic ion and a dipolar molecule at low temperatures is very simple:

$$k = q\mu_{\mathrm{D}}\sqrt{\frac{2\pi}{\mu k_{\mathrm{B}}T}} \tag{3.9}$$

The result agrees well with their extensive classical trajectory calculations. Note that this expression is for a simple ion-dipole interaction – it does not include any effects of the induced dipole. The paper includes a large number of different interaction potentials, and much of the discussion relates to reactions between neutral molecules.

3.3.3 *Ion-Molecule Reactions: Specific Examples*

Two specific ion molecule reactions are examined in this section, $H_3^+ + O$ and $H_3^+ + CO$. Both experimental and theoretical results are discussed.

$H_3^+ + O(^3P)$

$O(^3P)$ is the ground electronic state of the oxygen atom. Its reaction with H_3^+ provides one of the main channels for removal of the ion in dense star forming regions. There are two reaction channels:

$$H_3^+ + O(^3P) \rightarrow OH^+ + H_2 \tag{R3.1a}$$

$$H_3^+ + O(^3P) \rightarrow H_2O^+ + H \tag{R3.1b}$$

There have been two experimental determinations of the overall rate coefficient. Fehsenfeld [37] used a flowing afterglow technique at 300 K, obtaining an overall rate coefficient of $(8 \pm 4) \times 10^{-10}$ cm^3 s^{-1}. Milligan and McEwan [38] used a SIFT method to generate H_3^+ and a microwave discharge to form $O(^3P)$. The experiments were conducted under conditions where $[H_3^+] \ll [O(^3P)]$, so that the atom concentration was needed to determine the rate coefficients; this was achieved by observing the decay of CH_3^+ in the presence of the same $[O(^3P)]$, and calculating $[O(^3P)]$ using the well established rate coefficient for $CH_3^+ + O(^3P)$. They obtained $k_1 = (1.2 \pm 0.5) \times 10^{-9}$ cm^3 s^{-1} and also determined the fractional yields as 0.7 and 0.3 for channels a and b respectively.

Theoretical analyses of the reaction have been performed by Bettens et al. [39] and by Klippenstein et al. [15]. The former used classical trajectories on a calculated potential energy surface. Klippenstein et al. used high level electronic structure methods to calculate the potential energy surface, coupled with the analytical solutions outlined above [36], detailed transition state theory and trajectory calculations. Their paper contains a great deal of insight into the mechanism and the temperature dependence of the overall rate coefficient.

Milligan and McEwan [38] compared their experimental result with that from the Langevin expression, obtaining agreement within their uncertainty range.

The reaction is, however, much more complex. It turns out that the ion-quadrupole interaction is the most important term in the long range potential, especially at low temperatures, where the variational transition state occurs at large internuclear separations [15]. Assuming that only the ion-quadrupole interaction is important, Klippenstein et al. [15] showed that the rate coefficient is given by:

$$k = 8.46\mu^{-1/2}(\mu_Q q)^{2/3}(k_B T)^{-1/6} \qquad (3.10)$$

where μ_Q is the quadrupole moment. This expression agrees well with their more exact calculations based on the electronic structure calculations, differing slightly at higher temperatures, where other interactions, in addition to the charge-quadrupole interaction, begin to contribute to the potential.

Complications arise from the differing reactivities of the spin orbit components of $O(^3P_J)$. There are three low lying energy levels, with J values of 2, 1 and 0, energies equivalent to 0, 228 and 326 K and degeneracies of 5, 3 and 1 respectively. These states are labelled $O(^3P_{2,1,0})$; $O(^3P_2)$ is lowest in energy. Only three components of the fivefold degenerate $O(^3P_2)$ state are reactive. The other two components and all the components of $O(^3P_{1,0})$ interact with H_3^+ on repulsive electronic surfaces, so that they are unreactive. There are two consequences. Firstly, the fraction of $O(^3P)$ that can react is temperature dependent and decreases with increasing temperature, as the fraction of unreactive states increases. Secondly, the interplay between the spin-orbit and charge-quadrupole interactions makes the potential energy curve slightly less attractive at low T, reducing the rate coefficient by a factor of ~2 in the low temperature limit. Klippenstein et al. [15] obtained the rate coefficient expression $k_1 = 1.14 \times 10^{-9}$ $(T/300\ \mathrm{K})^{-0.156}$ $\exp(-1.41\ \mathrm{K}/T)$ $\mathrm{cm^3\ s^{-1}}$, over the temperature range 5–400 K. This gives a room temperature value of $1.1 \times 10^{-9}\ \mathrm{cm^3\ s^{-1}}$, in very good agreement with the experimental result of Milligan and McEwan [38].

The reaction has been evaluated in the KIDA database (http://kida.obs.u-bordeaux1.fr/), and the expression of Klippenstein et al. adopted, with a reliability of ~40%. In the absence of a theoretical analysis of the temperature dependence of the channel yields, they recommended the values determined by Milligan and McEwan, independent of temperature, i.e. $k_{1a}/(k_{1a} + k_{1b}) = (0.7 \pm 0.2)$ and $k_{1b}/(k_{1a} + k_{1b}) = (0.3 \pm 0.2)$.

H_3^+ + CO

This reaction also has two channels forming HCO^+ and HOC^+.

$$H_3^+ + CO \rightarrow HCO^+ + H_2 \qquad (R3.3a)$$

$$H_3^+ + CO \rightarrow HOC^+ + H_2 \qquad (R3.3b)$$

It again occurs in dense clouds and is a major source of HCO^+, which is employed as a tracer for such environments.

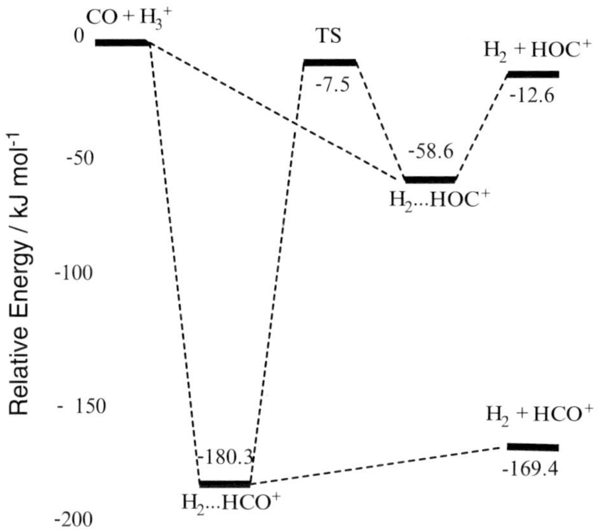

Fig. 3.1 Schematic diagram of the stationary points on the potential energy surface for CO reacting with H_3^+ [15]. Energies are in kJ mol^{-1} relative to CO + H_3^+ and include zero-point corrections

While CO has a dipole moment, it is very small (0.11 D), so that other interactions are also important, and the charge-quadrupole and charge-induced dipole interactions are of a comparable magnitude to the charge-dipole interaction. They have different distance dependences (Table 3.2), so that the relative contributions from the different potential energy components change with distance and hence their relative impact on the rate coefficient changes with temperature. Klippenstein et al. [15] have provided the most detailed analysis of the reaction, determining the rate coefficient with the potential energy calculated using both analytic (with the above three components) and high level electronic structure methods. The two approaches agree very well, except at the highest temperature studied (400 K), where a difference of ~5% indicates additional contributions to the potential. The channel yields depend primarily on the direction of approach of the reactants, with H_3^+ approaching from the C direction for channel (R3.3a); since the C atom is the negative end of the dipole, this direction is favoured and the fractional yield of channel a is greater than that of channel (R3.3b). An approach perpendicular to the CO bond is repulsive, so that it is straightforward to calculate the relative efficiencies of approach, using either trajectory or transition state calculations.

The reaction channels proceed through bound intermediates, $H_2 \ldots HCO^+$ and $H_2 \ldots HOC^+$, which have zero point energies 180.3 and 58.6 kJ mol^{-1} below the combined zero point energy of the reactants (Fig. 3.1). The relative efficiencies discussed in the previous paragraph refer to the formation of these two adducts. Calculation of the overall rate coefficients for channels (R3.3a) and (R3.3b) requires an assessment of the rates of the dissociation of the two adducts, which

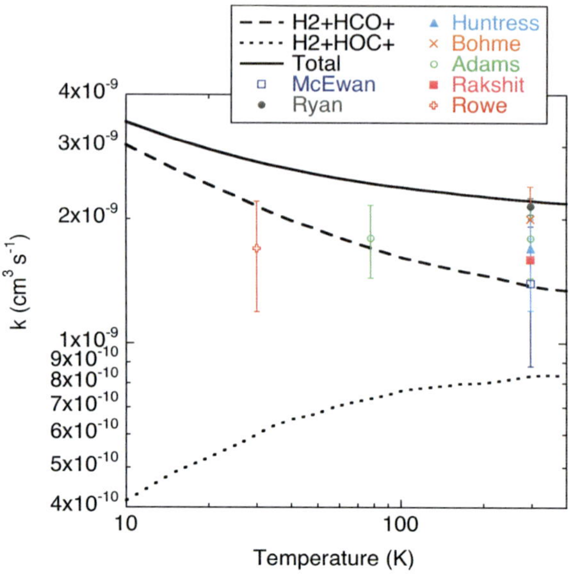

Fig. 3.2 Plot of the rate coefficient for CO + H_3^+ as a function of temperature. The *dash*, *dot*, and *solid lines* denote the theoretical predictions for formation of H_2 + HCO^+, H_2 + HOC^+, and the total products, respectively [15]. The *symbols* denote the experimental results [40–47]

can not only form the reaction products, but can also regenerate the reactants. In addition, as shown in Fig. 3.1, isomerisation between the adducts is also possible. This problem was solved using a master equation approach (discussed below in Sect. 3.4.4) at zero pressure, given the low pressure conditions pertaining in the ISM. The results show that forward dissociation of the adducts occurs much more rapidly than reverse dissociation to the reactants, even for $H_2 \ldots HOC^+$, and that isomerisation is not significant, so that the rates of formation of the adducts are equal to the rates of forming the products in channels a and b.

The rate coefficients are given by:

$$k_{3a} = 1.36 \times 10^{-9}\,(T/300\text{ K})^{-0.142}\,\exp(3.41\text{ K}/T)\text{ cm}^3\text{ s}^{-1}$$
$$k_{3b} = 8.49 \times 10^{-10}\,(T/300\text{ K})^{0.0661}\,\exp(-5.21\text{ K}/T)\text{ cm}^3\text{ s}^{-1}$$

and are compared with experimental results in Fig. 3.2. The fractional contribution from channel b to the overall rate coefficient increases from ~0.12 at 10 K to ~0.38 at 400 K, as increasing thermal energy washes out the orientational effects of the charge-dipole interaction. The experimental results lie below the calculated overall rate coefficient and close to the value calculated for channel a, leading Klippenstein et al. [15] to speculate that, since changing the energies of the adducts and transition states within likely limits has little effect, the discrepancy might be due to errors arising from mistreatment of quantum or dynamical effects in the analysis.

3.4 Neutral-Neutral Reactions

Reactions between neutrals include atom/radical + radical and atom/radical + molecule reactions. As discussed above, the intermolecular forces are shorter range than is the case with ion-molecule reactions, so that it is necessary to consider chemical interactions explicitly when modelling a reaction. After a section on experimental methods, the ideas behind transition state (TS) theory and its variational modification are discussed, together with theories of reactions where the TS switches, as the temperature increases, from A-B distances mainly controlled by the potential arising from electrostatic interaction to shorter distances where chemical forces are important. While the pressure in the ISM is too low for pressure dependent reactions, this topic is important in the conditions used to measure rate coefficients and in the chemistry of planetary atmospheres, including those of the exoplanets (see Chap. 5). This topic is discussed in Sect. 3.4.4, which also introduces the ideas that lie behind master equation models, which are widely used for such reactions. These models can also be used for reactions in which the adduct AB from an A + B reaction dissociates into several products, and these ideas are discussed in Sect. 3.4.5. Section 3.4 concludes with discussion of two examples of neutral + neutral reactions.

3.4.1 Experimental Methods

The technique which has been most widely used to study radical + radical and radical + molecule reactions is pulsed laser photolysis. A short laser pulse (~10 ns) is used to generate a radical reactant by photolysis. The photolysis laser is typically an excimer laser, which can generate a number of fixed wavelengths between 193 and 350 nm, or a Nd:YAG laser operating at 266 nm. For example, CN can be generated by photolysis of ICN at 248 nm [48] and CH by the multiphoton dissociation of $CHBr_3$ at 248 nm [49]. The radical is then detected, usually using some form of optical spectroscopy, a known time delay after the photolysis pulse. The most widely used detection technique is laser induced fluorescence (LIF) [50], in which a dye laser beam, at right angles to the photolysis beam, is tuned to an electronic transition of the radical and the resulting fluorescence detected at right angles to the two laser beams. LIF is restricted to species with suitably large fluorescence quantum yields. Techniques without this restriction include absorption spectroscopy, especially using the cavity ring down method [51] and frequency modulated infra-red laser spectroscopy [52]. Mass spectrometry can also be employed. One difficulty is interfering contributions to the reactant ion signal by ions formed by dissociative electron impact ionisation of the radical precursor. Soft ionisation using photons has been used to overcome this problem employing rare gas lamps, tuneable laser beams [53] or synchrotron radiation [54].

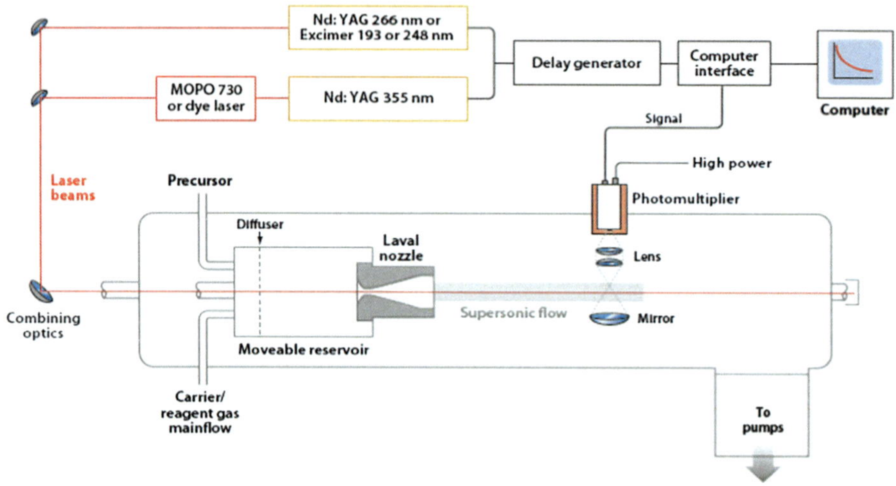

Fig. 3.3 Sketch of a CRESU (Cinétique de Réaction en Ecoulement Supersonique Uniforme) apparatus configured for the study of radical-neutral reactions. In this arrangement, radicals are generated by photolysis of a suitable precursor using radiation from a fixed-frequency pulsed laser operating at one of the three wavelengths, 226, 248, or 193 nm, and are detected by laser-induced fluorescence excited by tuneable radiation from a dye laser or a master oscillator parametric oscillator (*MOPO*) [56]

The limit of conventional, cryogenically cooled pulsed laser photolysis experiments is ~80 K, and the technique suffers from the problem noted for flow tube experiments on ion + neutral reactions, viz. freezing out of reactants or precursors on the cold walls of the reaction cell or the pipes leading into the cell. The CRESU technique has been applied to neutral + neutral reactions by Smith and co-workers to overcome this problem. A diagram of the apparatus is shown in Fig. 3.3. Temperatures as low as 13 K have been obtained. An alternative approach is to introduce the gas mixture into the nozzle via a pulsed valve. This is less demanding on the pumping capacity, but produces less stable flows. It is employed in a number of laboratories. Mullen and Smith [55], for example, have studied NH + NO at temperatures down to 53 K.

The rate coefficient in a pulsed photolysis experiment is determined by repeating the experiment a number of times with different delays between the photolysis and LIF lasers. An example of such a set of data is shown in Fig. 3.4 for CN + C_2H_2[48]. The concentrations are such that [CN] $\ll$ [C_2H_2], so that pseudo first order kinetics pertain and

$$[\mathrm{CN}](t) = [\mathrm{CN}](0)\exp(-k't) \tag{3.11}$$

where $k' = k[C_2H_2]$. Non-linear least squares fitting to the exponential decay gives k' as a fitting parameter and k is then determined by repeating the experiment at several [C_2H_2], as shown in Fig. 3.4. The rate coefficient is equal to the slope of the

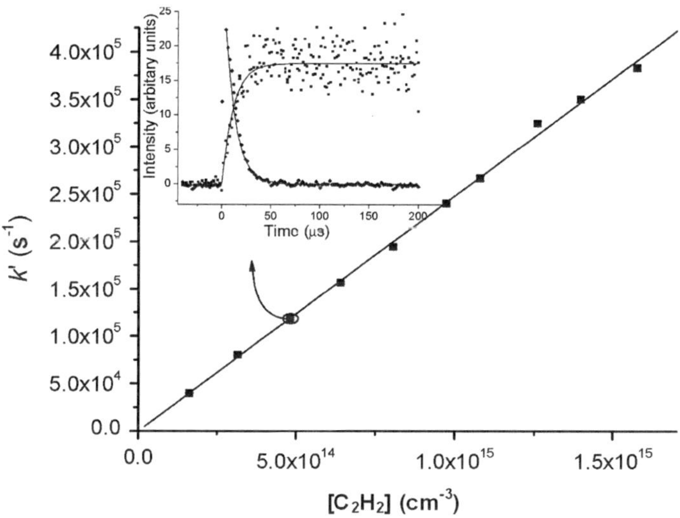

Fig. 3.4 Experimental data from a study of CN + C_2H_2 using pulsed laser photolysis/laser induced fluorescence [48]. The *inset* shows typical time profiles of CN and H in the reaction of CN + C_2H_2 at 60 Torr, total pressure, and 298 K; the signal decaying with time is that for CN and that increasing with time is for H. The *points* show the experimental data and the *solid lines* are nonlinear least-squares fits, using (3.11) and (3.12). The main graph shows a plot of the pseudo first order rate coefficient, k', determined from profiles like those shown in the inset, against $[C_2H_2]$. The slope gives the bimolecular rate coefficient, k

plot. Note that only the relative CN concentration is needed for an exponential decay of this sort, so that absolute calibration of the signal is unnecessary.

Figure 3.4 also shows the laser induced fluorescence signal for H, obtained using the Lyman α transition at 121.6 nm. Analysis of the exponential growth:

$$[\mathrm{H}](t) = [\mathrm{H}](\infty)\{1 - \exp(-k't)\} \tag{3.12}$$

returns an essentially identical value for k', showing that H is a product of the reaction. It is not possible to determine the yield of H simply from the time dependence and it is necessary to calibrate the H atom signal, as is discussed below for CN + NH_3.

Discharge flow provides an alternative technique, which is quite similar to the flowing afterglow method, and has been widely employed to study reactions of atoms and small radicals such as OH. The transient species is produced in a microwave discharge, or by secondary chemistry following production of a precursor in the discharge. It is mixed with excess co-reactant, introduced via a movable injector, and the transient measured a certain distance downstream. The reaction time is determined from the distance from the mixing zone to the detector and the flow velocity. A range of detectors, similar to those used for pulsed photolysis, can be employed. Early versions operated at total gas densities of ~10^{17} cm^{-3}, usually in a helium diluents, to ensure laminar flow [57]. More recently, turbulent flow reactors have been used, allowing an increase of two orders of magnitude in the density [58].

3.4.2 Theoretical Considerations: Variational Transition State Theory (VTST)

In the simplest sort of reaction, in which an atom or group of atoms is exchanged between the reactants to form the products, e.g.

$$OH + CH_4 \rightarrow CH_3 + H_2O \tag{R3.5}$$

the reactants collide and have to overcome an energy barrier, resulting from the requirements of rearranging the electrons in the molecular orbitals of the system, known as a *chemical* interaction. The rate coefficient depends on the passage of the reactants over the energy barrier. Only the more energetic collisions have sufficient energy to overcome the energy barrier, so that the rate coefficient falls as the temperature is reduced and this is reflected in a large value for γ in the Kooij equation. For reaction (R3.5), $\gamma = 1{,}350$ K, [9] so that this reaction is of no consequence in the gas phase in the ISM. To be significant, reactions must have very small barriers. Since thermal energies are low in the ISM, weaker interactions, resulting from intermolecular forces (Sect. 1.4) are important in the reactions and have a close interplay with the chemical interactions. This interplay generally determines the magnitude and temperature dependence of the rate coefficient.

Transition state theory (TST) is an established method for calculating the rate coefficient for a reaction that takes place over a well defined potential energy barrier [2]. It also provides a means of understanding the factors that determine the magnitude of the rate coefficient. The top of the barrier (Fig. 1.5) is termed the transition state or activated complex. The rate coefficient for the schematic reaction

$$A + B \rightarrow AB^{\ddagger} \rightarrow \text{Products},$$

where $AB^{\ddagger}$ is the transition state, is given by:

$$k = \frac{k_B T}{h} \frac{q_{AB^{\ddagger}}}{q_A q_B} \exp\left(-\frac{E_0}{k_B T}\right) \tag{3.13}$$

The threshold energy, E_0 is the difference between the zero point energies of the reactants and the transition state and q is the partition function:

$$q = \sum\nolimits_i g_i \exp\left(-\frac{E_i}{k_B T}\right) \tag{3.14}$$

where g_i is the degeneracy of the ith energy level and E_i its energy (Sect. 1.3). The summation is over all the translational, rotational, vibrational and electronic levels for A, B and $AB^{\ddagger}$. The partition functions depend on the masses, the moments of inertia and the vibrational frequencies of the three species and the energies of any low-lying electronic states. The moments of inertia and vibrational frequencies of A and B are determined by spectroscopy; those for $AB^{\ddagger}$, and the energy of the transition state, E_0, are generally determined by electronic structure calculations which, if carried out at an appropriately high level, can provide reliable values, with uncertainties of 1–2 kJ mol^{-1}. E_0 is the most difficult and sensitive parameter to determine; its sensitivity is particularly high at low T because of its appearance in the exponential term.

The partition functions depend on the spacings of the energy levels – the more closely spaced they are, the higher is q at a particular temperature. The molar entropy, S, of the reactants and transition state are also related to the partition functions, while the change in molar enthalpy, ΔH, is related to E_0. Since $\Delta G = \Delta H - T\,\Delta S$ (1.20 and 1.21), where G is the molar Gibbs energy, (3.13) leads to:

$$k = \frac{k_B T}{h} \exp\left(-\frac{\Delta G^{\ddagger}}{RT}\right) = \frac{k_B T}{h} \exp\left(-\frac{(\Delta H^{\ddagger} - T\Delta S^{\ddagger})}{RT}\right) \tag{3.15}$$

This equation is the basis of Fig. 1.5. The transition state is located at the maximum of the Gibbs energy along the so-called reaction coordinate, which traces out the minimum energy path between the reactants and the transition state. If the maximum potential energy along the reaction coordinate is well-defined, and E_0/k_B is much greater than the ambient temperature, then the position of the maximum Gibbs energy is very close to the maximum in the potential energy – we say that the transition state is constrained and, in order to calculate k using (3.13), it is necessary only to locate the position of the maximum and properties of the transition state at that location.

The energy barriers are necessarily much smaller for reactions occurring in the ISM than is the case for reaction (R3.5). In addition, the attractive intermolecular forces are comparable in magnitude to these small chemical interactions, and vary along the reaction coordinate. As a result, the transition states in reactions of interest in the ISM are rarely constrained, so that a *variational* approach is needed. This could be achieved by calculating the variation in the energy and entropy of the reaction system along the reaction coordinate and locating the transition state at the maximum in the Gibbs energy; this uses so-called *canonical* variational transition state theory (CVTST) and can provide useful insights. A more accurate approach, though, is to use microcanonical variational transition state theory, μJVTST, which determines the rate coefficient for a specific energy, E, and overall rotational quantum number, J, which are constants of motion in a reaction that is not subject to external influences during the course of a reactive collision. The microcanonical rate coefficient is given by:

$$k(E,J) = \frac{N^{\ddagger}(E,J)}{h\rho(E,J)} \tag{3.16}$$

where $N^{\ddagger}(E,J)$ is the sum of states in the transition state, at E,J, lying above the threshold energy and with energy less than E. $\rho(E,J)$ is the density of reactant states, including relative translational energy, i.e. $\partial N(E,J)/\partial E$. The transition state is located at the minimum in $N^{\ddagger}(E,J)$, which acts as a bottleneck in the reaction flux from reactants to products. The overall canonical rate coefficient is then obtained by integrating over a Boltzmann distribution, resulting in (3.17):

$$k(T) = \frac{1}{hq}\int N^{\ddagger}(E,J)\exp(-E/k_{\mathrm{B}}T)dEdJ \tag{3.17}$$

where q is the partition function of the reactants. If the transition state is constrained, (3.17) reduces to (3.13).

The microcanonical variational approach is essential when there is no energy barrier, as is the case for radical + radical reactions. The ideas were elaborated by Wardlaw and Marcus [59], who used the reaction between CH_3 + CH_3 as one of their first examples. The two unpaired electrons enter a bonding σ orbital as the reaction proceeds and the potential energy decreases without going through a maximum. As a result, the amount of energy that can be distributed between the modes of motion increases as the reactants approach. The minimum in the sum of states arises because of the changes in some of those modes of motion. Some, such as the C-H stretching vibrations, change little as C_2H_6 is formed; these are termed the *conserved* modes. As was the case in our discussion of ion-neutral reactions, the orbital motion of the two radicals gives rise to a centrifugal barrier. There are substantial changes in the rotational motion of the radicals; for example, taking the forming C-C bond as the z-axis, rotations about the x and y coordinates in the radicals correlate with rocking motions of the CH_3 groups in the new molecule – a change from a free rotational motion in the separated reactants into vibrations in C_2H_6, with an accompanying large increase in the spacing of the energy levels. This change is much more marked than those taking place in ion-molecule reactions, because the transition state is located at much shorter distances, so that the angular motions are much more strongly hindered. Similar changes occur for other angular modes; they are termed the *transitional* modes and the changes must be treated explicitly in a realistic model. The effect of the increase in the energy level spacing in the transitional modes is to *decrease* $N(E,J)$. Combined with the increasing energy available for distribution amongst the modes, resulting from the decrease in the potential energy, this leads to a minimum in $N(E,J)$,which is the location of the microcanonical transition state. As E and J increase, the transition state moves to smaller C-C distances.

In canonical terms, the location of the transition state depends on the balance between the decreasing energy and the decreasing entropy as the angular motion becomes more constrained as the CH_3 groups hinder one another. The transition state occurs at the maximum in the Gibbs energy.

A key problem is that of accurately calculating the potential energy along the reaction coordinate, to determine the available energy for redistribution amongst the internal modes, and the angular potential, to determine the changes in the transitional modes. The accuracy of the calculation, depends sensitively on the quality of the electronic structure calculations of the potential energy, as discussed by Harding et al. [60]. The transition states occur at quite small distances between the reactants (<0.5 nm), so that chemical interactions are dominant. For ion molecule reactions, the intermolecular forces are longer range and stronger, so that the transition states occur at longer distances and the angular motion remains close to that of the separate reactants. It is still necessary, in ion-dipole reactions, for example, to describe the rotational motion of the dipolar molecule, but the chemical interactions that are important in $CH_3 + CH_3$ can largely be neglected.

$CN + O_2$ provides an example of a radical + radical reaction of importance in the ISM. It involves initial formation of NCOO, which rapidly dissociates in two channels:

$$CN + O_2 \rightarrow NCOO^* \rightarrow NCO + O \quad \text{(R3.6a)}$$

$$\rightarrow CO + NO \quad \text{(R3.6b)}$$

where the * indicates excess energy. The rate determining step in the overall reaction is the formation of the adduct, NCOO.

The reaction has been studied experimentally at temperatures up to 3500 K because of its importance in combustion. Sims et al. have measured the overall rate coefficient using conventional pulsed laser photolysis [61] and the CRESU technique [62, 63], both with LIF detection of CN, over the temperature range 13–761 K, obtaining $k = 2.5 \times 10^{-11}\ (T/298\ \mathrm{K})^{-0.63}$ The channel branching ratio has been determined at 296–475 K by Feng and Hershberger [64], using pulsed laser photolysis and detecting the NO product by infra-red diode laser spectroscopy to determine the yield of channel b. They obtained a yield of 0.20 ± 0.02 for channel b and, by difference, a yield of 0.80 ± 0.02 for channel a. An evaluation for combustion applications by Baulch et al. [9] shows that the rate coefficient continues to decrease at higher temperatures; they recommend $k = 1.2 \times 10^{-11} \exp(210\ \mathrm{K}/T)$ over the temperature range 290–4,500 K.

There has been no recent theoretical analysis of the reaction, but Klippenstein and Kim [65] used a canonical version of the variational transition state model discussed above. The transitional modes in the nascent molecule, NCOO, which derive from the rotational motions of CN and O_2 are the in-plane C-O-O and N-C-O bends and the out-of-plane N-C-O-O bend. The wavenumbers of the vibrations increase, respectively, from 93, 45 and 64 cm^{-1} at a C-O bond length of 0.3 nm to 477, 142 and 224 cm^{-1} at 0.19 nm, leading, in canonical terms, to a significant decrease in the entropy as the C-O bond shortens. This bond length has values of 0.30 and 0.17 nm at the canonical transition state at temperatures of 50 and 3,000 K respectively.

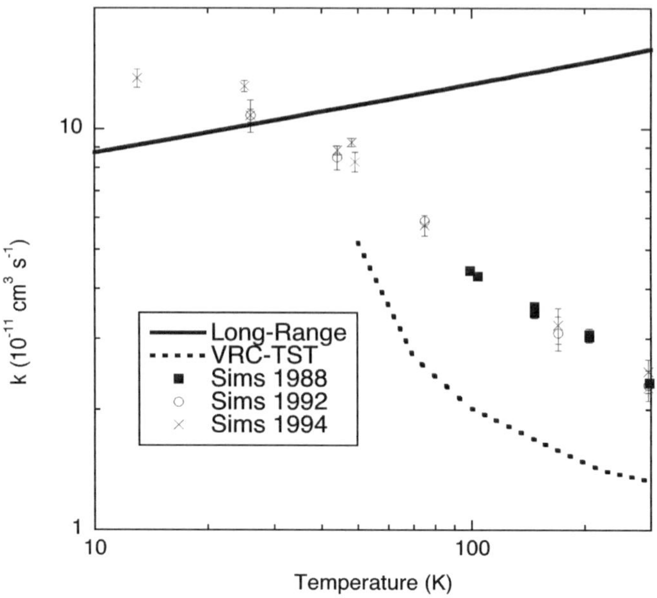

Fig. 3.5 Plot of experimental [61–63], long range ACCSA [33] and short range canonical variational TST (VRC-TST) [65] calculations of the temperature dependence of the rate coefficient for CN + O_2

Figure 3.5 shows a plot of the lower temperature data of Sims et al. and the values calculated by Klippenstein and Kim [65]. The model reproduces the strong increase in the rate coefficient at low temperatures, but underestimates the values, most markedly in the region of ~100 K, where the difference is a factor of ~2.5. Klippenstein has commented that the discrepancies are mainly due to inadequacies in the treatment of the potential energy surface, reflecting the importance of the level of the ab initio method employed for radical + radical reactions, as discussed by Harding et al. [60]. Calculations using today's computing resources would probably lead to a significant improvement in the agreement. Figure 3.5 also shows the results of long range ACCSA capture calculations by Stoecklin et al. [66], which include only the long range terms in the potential. The agreement is, in this case, best at low temperatures, where the variational transition state is located at comparatively large distances, so that the chemical interactions are less important and the long range contributions to the potential significant.

3.4.3 Transition State Switching

The interplay between longer range attractive forces and shorter range chemical forces can lead to interesting effects of considerable relevance to the ISM, especially in radical + molecule reactions, and especially where there is a relatively strong

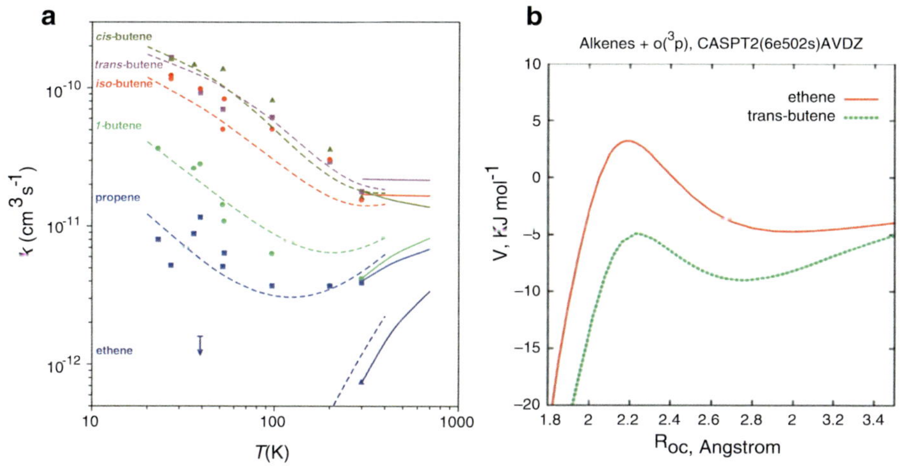

Fig. 3.6 (**a**) Measured rate coefficients (*symbols*) and calculated values (*dashed lines*). (**b**) Ab initio potential energy curves along the minimum energy paths for $O(^3P)$ + ethane and *trans*-butene [68]

long range interaction. This can lead to a van der Waals well which lies outside an energy barrier caused by the chemical interaction. There are now two transition states. The outer transition state lies at the minimum in the sum of states as the reactants approach and experience the decreasing potential energy in the van der Waals well; the location of the outer TS can be determined by the methods discussed above. The inner transition state lies at or close to the energy barrier caused by the rearrangement of the electrons in the formation of the new chemical bond. Figure 3.6b shows potentials of this type.

A good example is provided by the reaction

$$OH + C_2H_4 \rightarrow HO{-}C_2H_4 \tag{R3.7}$$

Although this is an association reaction, like CH_3 + CH_3, considerably more electronic rearrangement is involved, leading to a well defined potential energy maximum at short distances. The non-bonding interaction between OH and the double bond in ethene gives a van der Waals well at longer distances. The inner transition state is located at a shorter distance and lies below the energy of the separate reactants: it is *submerged*. The inner transition state is much tighter than the outer one, so that its energy levels are more widely spaced. The reaction has been discussed in detail by Greenwald et al. [67]. The sum of states in (3.16) is replaced by an effective sum of states $N^{\ddagger}(E,J)_{\mathrm{eff}}$ where

$$\frac{1}{N^{\ddagger}(E,J)_{\mathrm{eff}}} = \frac{1}{N^{\ddagger}(E,J)_{\mathrm{inner}}} + \frac{1}{N^{\ddagger}(E,J)_{\mathrm{outer}}} \tag{3.18}$$

At low energies, just above the long range asymptote, there is much more energy available for redistribution among the vibrations at the inner TS, so that the sum of states at the inner TS is much greater than that at the outer TS, the rate coefficient is mainly determined by $N^{\ddagger}(E,J)_{\text{outer}}$ and the TS lies at long distances. As the energy increases, however, the wider spacing in the energy levels at the inner TS becomes more significant and both transition states make a contribution to the effective sum of states, until, at still higher energies, the inner transition state is the main determinant of k.

This behaviour is termed transition state switching; its importance in the ISM can be illustrated by reference to $O(^3P)$ + alkene reactions. The reactions involve the formation of an adduct, which can either be collisionally stabilised or can dissociate to form products. In the case of the reaction with ethene, the latter include $CH_2CHO + H$ and $CH_3 + HCO$. The overall rate determining step is adduct formation.

Figure 3.6a shows a plot of the rate coefficients for $O(^3P)$ + ethene, propene and four butene isomers, measured by Sabbah et al. [68] using the CRESU technique over the temperature range 23–298 K. $O(^3P)$ was generated by photolysis of NO_2 and detected by monitoring the chemiluminescence from the reaction between $O(^3P)$ and a small concentration of added NO. The rate coefficients for reaction with the *trans-, cis-* and *iso*-butenes increase as the temperature decreases across the whole of the experimental range. That for reaction with ethene decreases as the temperature decreases and soon lies outside the range that can be studied with the CRESU technique. The rate coefficient for reaction with propene shows interesting intermediate behaviour, at first decreasing as the temperature falls and then increasing. The figure clearly shows the difficulties facing modellers, who may only have experimental data available to them close to room temperature – extrapolation of the propene data from this region leads to an underestimate of the rate coefficient by over an order of magnitude. The reaction moves from being one of little interest in the ISM to one of potential significance.

Figure 3.6b shows a plot of the potential energy vs C-O bond distance for ethene and for *trans*-butene and shows clearly the van der Waals well, resulting from the interaction of $O(^3P)$ with the double bond, and the inner transition state. Crucially, for ethene, the energy of the inner transition state lies above that of the reactants, while for *trans*-butene it is submerged. Figure 3.6a shows the rate coefficients calculated using μJVTST. At low temperatures, the outer transition state dominates for the butenes and is becoming increasingly important for propene as the temperature falls. As the temperature increases, the inner transition state becomes more important in determining the rate, especially for propene. The inner transition state dominates the kinetics for ethene at all temperatures.

Detailed calculations of this sort are not feasible for all of the radical + molecule reactions that might be significant in the ISM. Measurements at the critical temperatures are even more time-consuming. Smith et al. [69] examined ways of predicting whether or not the inner transition state has an energy compatible with a sufficiently high rate coefficient at low temperatures for the reaction to be potentially significant in the ISM. The reaction may be thought to involve a virtual transfer of an electron from the molecule to the radical. A measure of the ease of this process is the difference between the ionization energy (E_i) of the molecule and the electron affinity (E_{ea}) of the radical [70].

Smith et al. examined this difference for a large number of reactions studied at low temperatures and proposed that if ($E_i - E_{ea}$) >8.75 eV, then the reaction is likely to possess an inner barrier above that of the reactants, so that it will be negligibly slow at 20 K. For the O(^{3}P) + alkene reactions, they estimated that ($E_i - E_{ea}$) = 9.05 eV for ethene, 8.27 for propene, and smaller still for the other alkenes studied. The observed behaviour is in excellent agreement with the proposed criterion, thus lending weight to the original proposal to use the value of ($E_i - E_{ea}$) to assess the importance or otherwise of radical + molecule reactions at low temperatures.

3.4.4 Pressure Dependent Reactions

Rate coefficients for reactions that form an adduct can show a dependence on pressure [2]. Examples include reaction (R3.7) and the association reactions of two methyl radicals to form ethane and of CH_3^+ and H_2 to form CH_5^+. In the absence of external influences, the adduct molecule will dissociate to regenerate the reactants, since the energy of the reaction system is conserved. Collisions with a third body, M, serve to stabilise the initially formed adduct by energy transfer.

At the simplest level, reactions of this sort can be described by the Lindemann model [2]:

$$A + B \rightleftharpoons AB^* \qquad k_a, k_{-a} \qquad \text{(RA, -A)}$$

$$AB^* + M \rightarrow AB + M \qquad k_s \qquad \text{(RS)}$$

Reaction A is the association or capture process that forms the energised adduct, AB*. Reaction -A is the dissociation of this adduct to regenerate the reactants and reaction S is the stabilisation of AB* by collision with a third body, M. The overall rate of forming the product, AB, is equal to k_s[AB*][M]. At intermediate pressures, it is easy to show that the overall rate coefficient is given by [2]:

$$k = \frac{k_a k_s[M]}{k_{-a} + k_s[M]} \qquad (3.19)$$

At high pressures, $k_{-a} \ll k_s[M]$: all the AB* formed is stabilised and the rate coefficient is equal to k_a, the capture value. At low pressures, $k_{-a} \gg k_s[M]$: stabilisation is the rate determining step and $k = (k_a/k_{-a})k_s[M]$ where (k_a/k_{-a}) is equal to the equilibrium constant for the (RA, -A) reaction pair.

The Lindemann treatment is oversimplified. It assumes that k_a and k_{-a} are independent of energy and our earlier discussion shows that this is not the case. In addition it assumes that collisional stabilisation is a single step process – the so-called strong collision assumption. In reality, relaxation is a multistep process. These problems are overcome using a *master equation* approach, by reformulating the problem to recognise the energy dependence of the various processes, and then setting up a set of coupled, energy resolved differential equations of the sort [71]:

$$dn(E)/dt = k_{a,}g(E)[A][B] + k_s[M]\int dE'[P(E,E')n(E') - P(E',E)n(E)] - k_{-a}(E)_n(E) \qquad (3.20)$$

where $n(E)$ is the population density of the adduct at energy E, $g(E)$ is the fraction of the association reaction that forms adducts at energy E, $P(E,E')$ is the probability of transfer from states at E' to states at E on collision and $k_{-a}(E)$ is the energy dependent rate coefficient for dissociation of the adduct at energy E. Pseudo first order conditions are applied, i.e. $[A] \ll [B]$. The master equation is solved by casting it in a discretized form describing the rates for a population in a set of *grains,* each spanning a small range of energies, typically ~0.5 kJ mol^{-1}. The coupled differential equations are then expressed in matrix form, symmetrised and solved to obtain the eigenpairs. There are as many eigenvalue and eigenvector pairs as grains. For the present association problem, it turns out that the absolute value of the eigenvalue of smallest magnitude is the pseudo first order rate coefficient, k[B]. The methodologies have been discussed in detail by Miller and Klippenstein [71] and by Robertson et al. [72].

Pressure dependent reactions are not significant in the ISM, but many experimental measurements of reactions of importance there have been determined under pressure dependent conditions, and a master equation analysis is needed to extrapolate the rate data to the conditions required for ISM applications. The reaction CH_2 + H, which is discussed below, is an example. The master equation formulation, outlined above also provides a framework for reactions discussed in Sects. 3.4.5, 3.4.6 and 3.5. Finally pressure dependent reactions are important at the higher pressures found in planetary atmospheres (Chap. 5).

3.4.5 Multiple Product Channels

In many of the reactions occurring in the ISM, the adduct can dissociate in one or more ways to form products. Examples discussed above include CN + O_2 and $O(^3P)$ + alkenes. For reactions of this sort, the following reactions of AB* are added to the scheme (RA, -A, RS):

$$AB^* \rightarrow C + D \qquad (RD1)$$

$$AB^* \rightarrow E + F \tag{RD2}$$

where C + D and E + F are product molecules. This problem can be solved straightforwardly using a master equation approach, by subtracting the term ($k_{d1}(E)n(E)$ + $k_{d2}(E)n(E)$), from the right hand side of (3.20). A simpler approach can be used at very low pressures, which are our main concern. The lifetime of AB* at energy E is equal to $(k_{-a}(E) + k_{d1}(E) + k_{d2}(E))^{-1}$ which depends on E and on the number of atoms in AB, but typically is in the range 10^{-6}–10^{-9} s. If we neglect collisional relaxation, AB* retains the energy it was formed with. The probability, $P_{d1}(E,J)$, of forming products C + D is given by:

$$P_{d1}(E) = \frac{k_{d1}(E)}{k_{-a}(E) + k_{d1}(E) + k_{d2}(E)} \tag{3.21}$$

The relative magnitudes of the microcanonical rate coefficients depend, in particular, on the energy of the transition state leading to those products. If the transition states for channels D1 and D2 lie well below the reactant energy, then $k_{-a}(E) \ll k_{d1}(E)$, $k_{d2}(E)$ and reverse dissociation to regenerate the reactants is unimportant. The overall rate coefficient is determined by the capture value, and the partitioning of the products between channels D1 and D2 depends on the relative magnitudes of these rate coefficients. A related example, H_3^+ + CO, was discussed in Sect. 3.3.

3.4.6 Examples of Neutral + Neutral Reactions

$CH_2 + H \rightarrow CH + H_2$

Wakelam et al. [1] carried out a sensitivity analysis to identify the key reactions in chemical modelling of cold cores. They identified those reactions that most influenced the concentrations of important species. Reaction R3.8 was found to be the most important source of CH:

$$CH_2 + H \rightarrow CH + H_2. \tag{R3.8}$$

The rate coefficient has been measured at room temperature in a discharge flow reactor by Böhland et al. [73, 74], using a discharge through H_2 to generate H and through ketene (CH_2CO) to form CH_2; they detected CH_2 using laser magnetic resonance and H by absorption spectroscopy on the Lyman-α transition. Boullart and Peeters [75] also used discharge flow and mass spectrometry, but generated the reactants less directly, from the reaction between O and C_2H_2; they measured the rate coefficient for reaction (R3.8) relative to the known value for CH_2 + O. Devriendt et al. [76] used a similar technique, at somewhat higher temperatures and observed a negative temperature dependence in k_9. There have been several measurements at higher temperatures, using shock tubes and complex reaction schemes; the reported values cover over an order of magnitude at 1,500–3,000 K, but each measurement reported no temperature dependence in k_9 [9]. All of these measurements are difficult, because of the nature of the reaction, which involves two transient species.

Baulch et al., in their evaluation for combustion applications recommended a temperature invariant value of 2×10^{-10} cm^3 s^{-1}, based on the room temperature measurements, with an uncertainty of a factor of 2 [9]. The KIDA database recommends a value of 2.2×10^{-10} cm^3 s^{-1}, on the same basis, with an uncertainty at 298 K of a factor of 2 and at 10 K of a factor of 4. CH_2 is a biradical, with two electrons with parallel spins, so that it has a spin quantum number of 1. Combining this with the spin of ½ on H means that the collision occurs on doublet and quartet potential energy surfaces with 1/3 of collisions occurring on the former; the quartet surface is repulsive. As discussed below, the reaction occurs by initial association on an attractive surface to form CH_3 (spin = ½), so that only 1/3 of collisions can lead to reaction.

The reverse reaction, which is one component of a multichannel reaction,

$$CH + H_2 \rightarrow CH_2 + H \tag{R3.9a}$$

$$CH + H_2 \rightarrow CH_3 \tag{R3.9b}$$

has been more widely studied, both experimentally and theoretically. Brownsword et al. [77] used heated and cryogenically cooled cells over the temperature range 86–744 K and the CRESU apparatus over the range 13–295 K. They generated CH by pulsed multiple photon dissociation of $CHBr_3$ and detected it by laser induced fluorescence. Their experimental pressures covered the range 4–400 Torr in conventional cells and 0.28–4.5 Torr at 53 K in the CRESU apparatus. Fulle and Hippler [78] studied the reaction at much higher pressures (1–160 bar) and at 185–800 K, again using pulsed laser photolysis of $CHBr_3$. The reaction is pressure dependent and the major channel, especially at higher pressures and lower temperatures, is CH_3 formation, via the mechanism discussed in Sect. 3.4.4. The pressure dependent rate coefficient does not tend to zero at zero pressure, but reaches a limit, corresponding to reaction R3.9a. The limiting value increases with increasing temperature, reflecting the endothermic nature of the reaction ($\Delta_r H_0{}^0 = 14.15 \pm 0.18$ kJ mol^{-1} [79–81]). The analysis of the rate data to obtain reliable values for k_{9a} is facilitated by accurate determination of the limiting high pressure rate coefficient, which gives, in effect, the overall capture rate coefficient. Extrapolation to this limit was difficult for Brownsword et al., so instead they measured rate coefficients for CH($v = 1$) + H_2, D_2 and CH($v = 0$) + D_2. The reaction of the vibrationally excited CH leads to formation of CH_3* and, provided rapid intramolecular vibrational relaxation takes place in the adduct, it dissociates to generate CH($v = 0$) much more rapidly than it does to regenerate CH($v = 1$), resulting in loss of the LIF signal for the vibrationally excited state. The alternative processes, reaction to form CH_2 + H and collisional stabilisation to give CH_3, also lead to loss of CH($v = 1$), so that all collisions leading to formation of CH_3* result in loss of the reactant: the technique provides a measurement of the capture rate coefficient, which is the limiting high pressure value. Measurement of CH($v = 0$) + D_2 also provides this limiting value, since dissociation of CHD_2* results in facile isotope exchange. In their measurements of all of these reactions, Brownsword et al. found the rate coefficients to be independent of pressure, as required. Correction for mass effects in the reactions with D_2 allowed all of the measurements to be used to determine $k_{9,\infty} = 1.6 \times 10^{-10}\,(T/298\ \text{K})^{-0.08}\ \text{cm}^3\ \text{s}^{-1}$. The measurements of Fulle and Hippler approached the high pressure limit more closely and they were able to extrapolate their data to obtain $k_{9,\infty} = 2.0 \times 10^{-10}$ $(T/300\ \text{K})^{0.15}\ \text{cm}^3\ \text{s}^{-1}$. Given the differences in the techniques, the agreement is extremely good, although the expressions differ by a factor of 2 at 10 K.

Brownsword et al. [77] used an empirical method to extract the rate coefficient for reaction R3.9a from the limit of the experimental data at zero pressure, obtaining a value of $k = 3.1 \times 10^{-10} \exp(-1650\ \text{K}/T)\ \text{cm}^3\ \text{s}^{-1}$. In principle, this expression could be used to refine the temperature dependence of k_8 using the equilibrium constant, K. There are two problems. The endothermic channel (R3.9a) makes only a small contribution to reaction (R3.9) at low temperatures, so that the limiting zero pressure rate coefficient is too small to measure directly and can only be obtained by extrapolation. The other is that, although the thermochemistry of the reaction is very well defined, with an uncertainty of $\pm 0.18\ \text{kJ mol}^{-1}$, this gives an uncertainty of a factor of nearly 10 in K at 10 K, because of its exponential dependence on $\Delta_r H/RT$.

There have been a number of theoretical studies of the reaction system, and the potential energy surface has been calculated at high level [82]. These calculations, though, have not contributed to a reduction in the uncertainty in k_8 at the temperatures found in the ISM.

CN + NH_3

The overall rate coefficient for this radical + molecule reaction has been studied over the temperature range 25–716 K by Sims et al. [63] using the CRESU apparatus and by Sims and Smith [61] using a conventional heated and cryogenically cooled cell. The rate coefficient shows a strong, negative temperature dependence, with $k = 2.77 \times 10^{-11} (T/298\ \text{K})^{-1.14}\ \text{cm}^3\ \text{s}^{-1}$. There are two exothermic branching channels

$$CN + NH_3 \rightarrow HCN + NH_2 \qquad \text{(R3.10a)}$$

$$\rightarrow NCNH_2 + H \qquad \text{(R3.10b)}$$

Channel (R3.10a) has been confirmed experimentally by Meads et al. [83], using pulsed photolysis and infra-red diode laser absorption, in a study of CN + ND_3. Channel b has been suggested as a source of cyanamide, $NCNH_2$, by Herbst et al. [84, 85], to explain the observed presence of low densities of this species in a molecular cloud.

Talbi and Smith [86] carried out high level ab initio calculations on the surface and showed that there is an NC-NH_3 complex, with a binding energy of

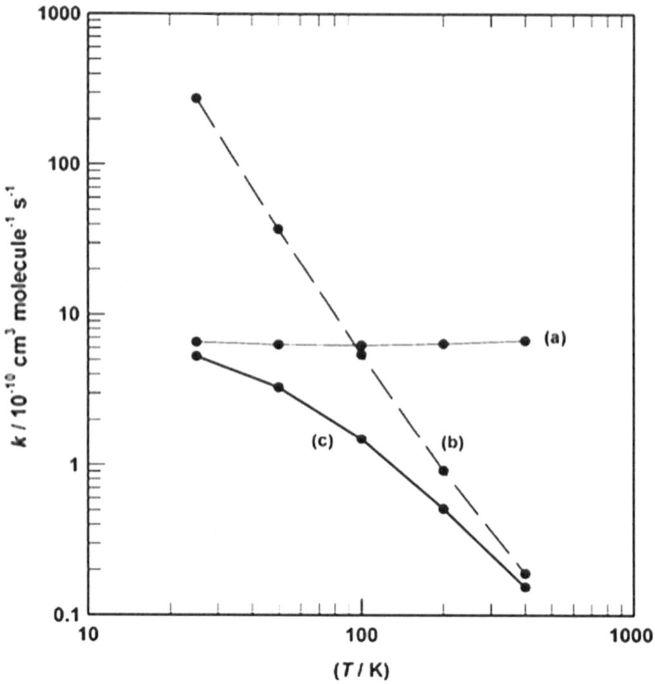

Fig. 3.7 Calculated components of the rate coefficient for CN + NH_3: (**a**) rate determined by the outer transition state, (—●—); (**b**) rate determined by the 'inner' transition state (– –●– –); (**c**) taking into account both transition states (—●—)

39.3 kJ mol^{-1} followed by a submerged transition state lying 10.9 kJ mol^{-1} below the zero point energy of the reactants. The geometry of the complex is such that the carbon atom on CN and the N atom on NH_3 are adjacent; reaction involves an internal rotation so that one of the equivalent H atoms presents itself to the carbon atom. The route to the products of channel a is then strongly exothermic, with no further potential maxima: the calculations show HCN + NH_2 lying 87 kJ mol^{-1} below the reactants, compared with the experimental exothermicity of 78.9 kJ mol^{-1}. By contrast, the transition state leading from NC-NH_3 to $NCNH_2$ + H lies 50 kJ mol^{-1} above the reactants; another route is also available, via an $HNCNH_2$ intermediate, but the barrier to the formation of this species is even higher.

Talbi and Smith [86] calculated the rate coefficient as a function of temperature using the inner and outer transition state model employed in the calculations on $O(^3P)$ + alkenes. They calculated both the axial and the angular components of the long range potential using analytic expressions, including six interaction terms: dipole-dipole, two dipole-quadrupole, two dipole-induced dipole and the dispersion interactions. The rate coefficients they calculated are shown in Fig. 3.7; the reaction fluxes through both transition states are significant across the experimental temperature range, but the inner transition state is the main determinant of the rate at high T and the outer TS at low T. Their calculated rate coefficients agree well with the experimental values, with minor tuning of the energy of the inner TS within the likely uncertainty limits of the calculation.

Blitz et al. [87] confirmed experimentally that the yield of channel b is insignificant using pulsed laser photolysis coupled with LIF detection of H at 121.6 nm. As discussed in Sect. 3.4.1, simply observing the relative LIF signal for H and its time dependence does not allow the yield of H to be determined: it is necessary to calibrate the signal in some way. This was achieved by comparing the H signal from $CN + C_2H_2$ in back-to back experiments. The yield of H in the latter reaction had previously been shown to be (100 ± 20) %, using a similar comparison against $CN + H_2$, which has a well established unit yield of H [48]. The experiments gave a yield of H and therefore of channel b of 0.024 ± 0.011 in a series of experiments at room temperature; the authors set a conservative upper estimate of the yield of channel b at 5%.

3.5 Radiative Association

Three body association, the low pressure limit of an association reaction, as discussed in Sect. 3.4.4, is not an efficient reaction in the ISM because of the very low pressures that pertain. An alternative process is radiative association, in which the adduct is stabilised by emission of radiation rather than by collision. An important radiative association identified by Wakelam et al. [1] is the reaction:

$$C + H_2 \rightarrow CH_2 + h\nu \qquad \text{(R3.11)}$$

The potentially competing reaction, forming CH + H, is strongly endothermic and uncompetitive. An analysis similar to that employed in Sect. 3.4.4, replacing $k_s[M]$ by k_r, the rate coefficient for radiative stabilisation, gives the following expression for the rate coefficient for radiative association, k_{RA}:

$$k_{RA} = \frac{k_a k_r}{k_{-a} + k_r} \approx \frac{k_a k_r}{k_{-a}} \text{ since } k_{-a} \gg k_r \qquad (3.22)$$

Reaction (R3.11) has not been studied experimentally, despite its importance [1]. Radiative association involving ions can be studied more easily using ion traps, where long trapping times can be used. The equivalent reaction of C^+:

$$C^+ + H_2 \rightarrow CH_2{}^+ + h\nu \qquad \text{(R3.12)}$$

has been measured by Gerlich [88] using radiofrequency ion traps. He measured $k_{12} = 1.7 \times 10^{-15}$ cm^3 s^{-1} for p-H_2 and $k_{13} = 6.8 \times 10^{-16}$ cm^3 s^{-1}, for n-H_2, both at 10 K. The measurements covered a hydrogen density from $<10^{12}$ to $>10^{14}$ cm^{-3}, so that rate coefficients for both three body and radiative association reactions could be determined.

The normal mode of radiative stabilisation is the emission of infra-red radiation in the ground electronic state of the adduct, for which k_r ~10^2–10^3 s^{-1}. k_r is larger for ions than for neutral species. The measured rate coefficient for (R3.12) is much faster than expected on this basis. A more efficient radiative route is available, however, because reaction can lead to formation of an electronically excited state of the molecular ion, which is radiatively connected to the ground state. k_r depends on the Einstein *A* coefficient for the transition, which is proportional to ν^3, where ν is frequency of the transition (1.4). Since electronic transitions have higher frequencies than infra-red transitions, they also have larger Einstein *A* coefficients and hence higher k_r values. In the present case, k_r is estimated to be ~10^5 s^{-1}. Both Herbst [89] and Smith [90] have discussed this process and Smith calculated an overall rate coefficient, which agrees remarkably well with the experimental values.

The method used by Smith was based on an earlier paper on radiative association for neutral-neutral reactions [91]. This approach used a microcanonical, *J*-resolved version of (3.22), in the limit that $k_{-a}(E,J) \gg k_r(E,J)$:

$$k_{RA}(E,J) = k_r(E,J)\left[\frac{k_a(E,J)}{k_{-a}(E,J)}\right] \tag{3.23}$$

where the term in square brackets can be replaced by $f(E,J)$, the equilibrium ratio of the energised molecule and the reactants which depends on density of states at (E,J),. The overall rate coefficient, k_{RA}, is obtained by integrating over energies and summing over *J*.

$$k_{RA} = \sum\nolimits_J \int_{E_{0,J}}^{\infty} dE\ k_r(E,J)\ f(E,J) \tag{3.24}$$

Smith approximated the collision as one between atoms, so that *J* is the orbital angular momentum and the maximum on the potential energy curve corresponds to the centrifugal barrier. This defines, in one model he used, the lower limit of the integral, $E_{0,J}$. He assumed that the long range potential varies as R^{-6} to obtain this maximum and then used unimolecular rate theory to calculate $f(E,J)$. The radiative rate coefficient was calculated from the Einstein *A* coefficients of fundamental infra-red transition probabilities for the various vibrational modes in the adduct with the assumption that the radiative rate coefficient for emission from the *n*th vibrational level is *n* times that for the fundamental, an exact result for harmonic vibrations. $k_r(E,J)$ was then calculated by summing $P_{i,n}nA_i$ over all the vibrational modes, *i*, and numbers of quanta, *n*, where $P_{i,n}$ is the fractional contribution of states with *n* quanta in the *i*th mode to $f(E,J)$. In a second model, Smith extended the lower limit of the integral in (3.24) by allowing tunnelling through the centrifugal barrier. He obtained rate coefficients for radiative association for nine neutral + neutral reactions, including H + OH and H + CN. The values obtained at 10 K for these two reactions, using the tunnelling model, were 8.5×10^{-18} and 6.6×10^{-17} cm^3 s^{-1}, respectively. Tunnelling increased the rate coefficient by factors of 6.5 and 1.8 respectively, for these two reactions, over the values obtained with the non-tunnelling model.

3.6 Dissociative Recombination (DR)

Dissociative recombination occurs following the association between a molecular cation and an electron followed by the dissociation of the adduct to form neutral fragments. The dissociation process is similar to that discussed in Sect. 3.4.5 except that the capture rate coefficient is very large (~10^{-7} cm^3 s^{-1}) and the adduct AB* is highly energetic. DR removes molecular cations from the ISM and can lead to formation of a stable molecular product, as in the reaction $CH_5^+ + e \rightarrow CH_4 + H$, thus terminating a specific branch in a chemical network. Rate coefficients can be measured in a modified flowing afterglow apparatus, incorporating a Langmuir probe (FALP) to determine the electron density along the flow tube. The cation is measured by mass spectrometry at the end of the tube. Often the cation is vibrationally and rotationally excited, compromising the applicability of the rate coefficient measurement. This problem has been overcome using storage rings, where the ions are introduced into the ring from a supersonic nozzle source and allowed to cool radiatively before being merged with an electron beam. The technique measures the cross section as a function of collision energy, and it is feasible to detect the neutral products by mass discrimination.

Examples of dissociative recombination include $H_3^+ + e$, which has two channels, $H_2 + H$ and $H + H + H$. Measuring the channel efficiencies presents significant problems. The high energy of the adduct can result in several channels; for example, there are four product channels in $H_3O^+ + e$. Estimating the channel yields theoretically is difficult because of the high adduct energies. DR, and its importance in the ISM, have been reviewed by Geppert and Larsson [92].

3.7 Surface Reactions

The density of H_2 in diffuse interstellar clouds, where the destruction rate by uv photolysis is very high, demands a high rate of H_2 formation. It is accepted that gas phase processes cannot account for this high rate and that formation on grains is essential. Grains consist of small silicate or carbonaceous particles with dimensions comparable with the wavelength of visible light. In diffuse clouds, at temperatures of 10–20 K, they are largely bare, but in denser clouds they are coated with molecules such as H_2O, CO_2, CO and CH_3OH.

The mechanism of forming H_2 on grain mimics has received a good deal of attention over the last few years. Vidali and co-workers [93, 94] studied the formation of HD after exposing a number of surfaces (olivine, amorphous carbon and water ice at 5–20 K) to beams of H and D generated in microwave discharges.

They then used temperature programmed desorption to detect HD, in addition to detection of 'prompt' HD during the exposure process. They observed formation of both prompt and delayed HD, the efficiency of the latter falling from 0.5 to 0.1 as the experimental temperature was increased. They found that the HD had a low translational temperature and they inferred significant energy transfer to the surface. Similar experiments were conducted by Hornekaer et al. [95] on both amorphous and non-porous water ice surfaces. They found that the product was fully thermalised on the surface of the porous amorphous ice but was rapidly desorbed from the non-porous surface suggesting that desorption occurred before thermalisation was complete. Zecho et al. [96] studied the effect of increasing the energy of the incident beams and showed that chemisorption on graphite is an activated process, with a barrier of ~0.2 eV and that the lower temperature H,D beam experiments led to physisorbed atoms.

Price and co-workers [97–99] looked at the internal energy of the desorbed HD (and H_2 in experiments with only H atom beams) and used resonance enhanced multiphoton ionisation (REMPI) to probe the vibrational and rotational states of HD. They found considerable internal excitation, with a marked maximum in the vibrational population at $v = 4$ and with coupled rotational excitation corresponding to a rotational temperature of ~300 K; nearly half of the vibrationally excited HD detected was in $v = 4$. Their results are compatible with both Eley-Rideal (ER) and Langmuir-Hinshelwood (LH) mechanisms (Sect. 1.6). In the former, a gas phase atom reacts directly with an adsorbed atom. In the latter, both species are adsorbed, migrate together and react. The experimental results indicate more excitation than suggested by calculations based on the ER mechanism and somewhat less than expected for LH. Given the generally low density of atoms on grain surfaces in diffuse clouds, the LH mechanism is more likely to operate there. The mechanism in the lab is closely linked to the degree of coverage associated with the experiment. The diffusion of H atoms on surfaces has been studied experimentally at low temperatures, and it is agreed that a thermal mechanism operates, with quite low barriers. Tunnelling of H is not thought to be important, because the barriers are quite wide. The formation of H_2 releases ~440 kJ mol^{-1} and the disposal of this energy is important. Energy left behind on the grain could assist desorption of other adsorbates. It is clear from the experiments of Price and co-workers that a significant fraction of the energy released goes into internal excitation in H_2 (or HD). Detection of this excitation could provide a means of probing the surface chemistry in the ISM. It has also been suggested that this degree of vibrational excitation could facilitate endothermic chemical reactions, such as $C^+ + H_2$, but relaxation of the vibrations by IR emission could be comparatively rapid.

Surface chemistry can be a source of other species. Gas phase ion molecule reactions predominantly lead to the formation of unsaturated organic compounds. Reactions of H on surfaces, on the other hand, lead to the formation of saturated compounds. Formation of CH_3OH is an important example, since its concentration in cold cores cannot be accounted for by gas phase reactions. The surface process involves a sequence of reactions starting with CO:

$$H + CO \rightarrow HCO\ (1); HCO + H \rightarrow H_2CO\ (2);$$
$$H_2CO + H \rightarrow CH_3O\ (3); CH_3O + H \rightarrow CH_3OH (4)$$

The sequence has been studied by Watanabe and co-workers [100–102] using beams of H at temperatures of 30 and 300 K incident on water ice at 10 K with sub-monolayer coverage of CO or H_2CO. The loss of the reactant and the formation of products were observed by infra-red spectroscopy. The rate of reaction was independent of the thermal energy of the H beam. Reaction (1) is the faster of the two atom + molecule steps, with k_1 ~$2k_3$ at 15 K; k_1 changed little on reducing the temperature to 10 K, but k_3 fell, giving an even larger rate coefficient ratio at the lower temperature. Experiments with deuterium showed some addition reaction involving D, but also the presence of abstraction reactions, giving a mechanism for D/H exchange.

Ward and Price [103] studied the reaction of $O(^3P)$ + ethene and propene. They dosed the atom and the alkene on to graphite at 12–90 K and detected the products by temperature programmed desorption and mass spectrometry. They used tunable-laser 2 photon ionisation, allowing them to distinguish, in the case of ethene, between the three potential products of formula C_2H_4O: vinyl alcohol, acetaldehyde and ethylene oxide, through their differing ionization energies. The results showed that the main product was ethylene oxide. The experiments demonstrated that the activation barrier is much reduced compared with the gas phase reaction (discussed in Sect. 3.4.3). The activation energy for ethene is equivalent to (70 ± 15) K, compared with a gas phase value of 974 ± 48 K.

It can be problematic to incorporate surface processes in astrochemical models. The availability of quantitative rate data on the rates of adsorption and desorption, diffusion and surface reaction is very limited. A further substantial problem derives from the small numbers of species adsorbed on grains – in diffuse clouds, the average number for a given species is probably less than unity. As a result, the deterministic form of conventional rate equations is inappropriate and a stochastic model is needed [104]. Approaches to developing a physically realistic description include Monte Carlo methods [105] and a master equation approach [106, 107]. It is difficult to incorporate these methods into the sets of ordinary differential equations, which constitute the conventional approach to modelling, without an enormous computational overhead. Some progress has been made, though, by Caselli et al. [108] using a semi-empirical approach to incorporate some of the stochastic effects. This topic is discussed in greater detail in Chap. 4.

3.8 Concluding Remarks

It is clear that progress in understanding and quantifying the rates of elementary chemical reactions requires a close interaction between experiment and theory. Until relatively recently, theory provided a framework for experiments, but in applications such as combustion and atmospheric chemistry, experimental values for rate coefficients held sway and dominated the recommendations of evaluation panels. Advances in theoretical understanding, coupled with increases in computer power, are changing significantly the ways in which rate parameters are determined.

The low temperatures in the interstellar medium provide a unique challenge to both experiment and theory. From an experimental perspective, the prevailing temperatures are difficult to achieve and the CRESU technique has made a unique and groundbreaking contribution by providing rate coefficients at appropriate temperatures. Theory faces problems in this area because precise energies of transition states are difficult to calculate – a two standard deviation uncertainty of ~1 kJ mol^{-1} is probably the best that can be achieved for high level calculations. When incorporated in the Kooij equation for a rate coefficient, this gives an uncertainty of a factor of over 10^5 at 10 K. The problems are less marked for ion molecule reactions, where relatively well understood long range potentials are the main determinants. Radical + radical reactions are also better described than this, but the uncertainties are still significant. For the two transition state radical + molecule reactions discussed in Sect. 3.4.3, the important question is whether or not the inner transition state is submerged, or at least lies only slightly above the reactant energies. To resolve these issues, even high quality theory should, where possible, be linked to experiment for applications to reactions in the ISM. The last few years have seen impressive collaborations across the theory/experiment 'divide', in recognition of the difficulties faced by both theory and experiment, for applications in the ISM in particular, but also in combustion and atmospheric chemistry.

The next chapter discusses, *inter alia*, chemical networks and the context in which research on the kinetics of elementary reactions is applied in the ISM. In particular it provides a rationale for the provision of rate coefficients, channel efficiencies and uncertainties in rate parameters. It provides the justification for many of the topics discussed in Chap. 3.

Acknowledgements I thank Dr. Stephen Klippenstein for helpful discussion of uncertainties in transition state energies in electronic structure calculations and Dr. Branko Ruscic for the provision of recent data for the Active Thermochemical Tables (ATcT).

References

1. Wakelam V, Smith IWM, Herbst E, Troe J, Geppert W, Linnartz H, Oberg K, Roueff E, Agundez M, Pernot P, Cuppen HM, Loison JC, Talbi D (2010) Reaction networks for interstellar chemical modelling: improvements and challenges. Space Sci Rev 156:13–72
2. Pilling MJ, Seakins PW (1995) Reaction kinetics. Oxford University Press, Oxford
3. Millar TJ, Rawlings JMC, Bennett A, Brown PD, Charnley SB (1991) Gas-phase reactions and rate coefficients for use in astrochemistry – the UMIST ratefile. Astron Astrophys Suppl Ser 87:585–619

4. Burke MP, Dryer FL, Ju YG (2011) Assessment of kinetic modeling for lean H(2)/CH(4)/O(2)/diluent flames at high pressures. Proc Combust Inst 33:905–912
5. Bloss C, Wagner V, Bonzanini A, Jenkin ME, Wirtz K, Martin-Reviejo M, Pilling MJ (2005) Evaluation of detailed aromatic mechanisms (MCMv3 and MCMv3.1) against environmental chamber data. Atmos Chem Phys 5:623–639
6. Bloss C, Wagner V, Jenkin ME, Volkamer R, Bloss WJ, Lee JD, Heard DE, Wirtz K, Martin-Reviejo M, Rea G, Wenger JC, Pilling MJ (2005) Development of a detailed chemical mechanism (MCMv3.1) for the atmospheric oxidation of aromatic hydrocarbons. Atmos Chem Phys 5:641–664
7. Wakelam V, Herbst E, Selsis F (2006) The effect of uncertainties on chemical models of dark clouds. Astron Astrophys 451:551–562
8. Wakelam V, Loison JC, Herbst E, Talbi D, Quan D, Caralp F (2009) A sensitivity study of the neutral-neutral reactions C + C(3) and C + C(5) in cold dense interstellar clouds. Astron Astrophys 495:513–521
9. Baulch DL, Bowman CT, Cobos CJ, Cox RA, Just T, Kerr JA, Pilling MJ, Stocker D, Troe J, Tsang W, Walker RW, Warnatz J (2005) Evaluated kinetic data for combustion modeling: supplement II. J Phys Chem Ref Data 34:757–1397
10. Crowley JN, Ammann M, Cox RA, Hynes RG, Jenkin ME, Mellouki A, Rossi MJ, Troe J, Wallington TJ (2010) Evaluated kinetic and photochemical data for atmospheric chemistry: Volume V – heterogeneous reactions on solid substrates. Atmos Chem Phys 10:9059–9223
11. Wakelam V et al (2012) A kinetic database for astrochemistry (KIDA). Astrophys J Suppl Ser 199:21. doi:10.1088/0067-0049/199/1/21
12. Solomon PM, Werner MW (1971) Low-energy cosmic rays and abundance of atomic hydrogen in dark clouds. Astrophys J 165:41–49
13. Herbst E, Klemperer W (1973) Formation and depletion of molecules in dense interstellar clouds. Astrophys J 185:505–533
14. Barlow SE, Luine JA, Dunn GH (1986) Measurement of ion molecule reactions between 10 K and 20 K. Int J Mass Spectrom 74:97–128
15. Klippenstein SJ, Georgievskii Y, McCall BJ (2010) Temperature dependence of two key interstellar reactions of H_3^+: $O(^3P) + H_3^+$ and $CO + H_3^+$. J Phys Chem A 114:278–290
16. McMahon TB, Beaucham JI (1972) Versatile trapped ion cell for ion-cyclotron resonance spectroscopy. Rev Sci Instrum 43:509–512
17. Fehsenfeld FC, Schmeltekopf AL, Goldan PD, Schiff HI, Ferguson EE (1966) Thermal energy ion-neutral reaction rates. I. Some reactions of helium ions. J Chem Phys 44:4087–4094
18. Dunkin DB, Fehsenfeld FC, Schmeltekopf AL, Ferguson EE (1968) Ion-molecule reaction studies from 300 to 600 K in a temperature-controlled flowing afterglow system. J Chem Phys 49:1365–1371
19. Barlow SE, Dunn GH, Schauer M (1984) Radiative association of CH_3^+ and H_2 at 13 K. Phys Rev Lett 52:902–905
20. Asvany O, Savic I, Schlemmer S, Gerlich D (2004) Variable temperature ion trap studies of $CH4^+$+H_2, HD and D_2: negative temperature dependence and significant isotope effect. Chem Phys 298:97–105
21. Adams NG, Smith D (1976) Selected ion flow tube (sift) – technique for studying ion-neutral reactions. Int J Mass Spectrom 21:349–359
22. Snow TP, Bierbaum VM (2008) Ion chemistry in the interstellar medium. Annu Rev Anal Chem 1:229–259
23. Rowe BR, Dupeyrat G, Marquette JB, Gaucherel P (1984) Study of the reactions $N_2^+ + 2\,N_2 \rightarrow N_4^+ + N_2$ and $O_2^+ + 2O_2 \rightarrow O_4^+ + O_2$ from 20 to 160 K by the CRESU technique. J Chem Phys 80:4915–4921
24. Rowe BR, Marquette JB (1987) CRESU studies of ion molecule reactions. Int J Mass Spectrom 80:239–254
25. Chesnavich WJ, Su T, Bowers MT (1980) Collisions in a non-central field – variational and trajectory investigation of ion-dipole capture. J Chem Phys 72:2641–2655

26. Su T, Chesnavich WJ (1982) Parametrization of the ion-polar molecule collision rate-constant by trajectory calculations. J Chem Phys 76:5183–5185
27. Woon DE, Herbst E (2009) Quantum chemical predictions of the properties of known and postulated neutral interstellar molecules. Astrophys J Suppl Ser 185:273–288
28. Maergoiz AI, Nikitin EE, Troe J, Ushakov VG (1996) Classical trajectory and adiabatic channel study of the transition from adiabatic to sudden capture dynamics. 1. Ion-dipole capture. J Chem Phys 105:6263–6269
29. Maergoiz AI, Nikitin EE, Troe J, Ushakov VG (1996) Classical trajectory and adiabatic channel study of the transition from adiabatic to sudden capture dynamics. 2. Ion-quadrupole capture. J Chem Phys 105:6270–6276
30. Maergoiz AI, Nikitin EE, Troe J, Ushakov VG (1996) Classical trajectory and adiabatic channel study of the transition from adiabatic to sudden capture dynamics. 3. Dipole-dipole capture. J Chem Phys 105:6277–6284
31. Pechukas P, Light JC (1965) On detailed balancing and statistical theories of chemical kinetics. J Chem Phys 42:3281–3291
32. Quack M, Troe J (1975) Complex-formation in reactive and inelastic-scattering – statistical adiabatic channel model of unimolecular processes III. Ber Bunsenges Phys Chem Chem Phys 79:170–183
33. Clary DC (1984) Rates of chemical-reactions dominated by long-range intermolecular forces. Mol Phys 53:3–21
34. Troe J (1987) Statistical adiabatic channel model for ion molecule capture processes. J Chem Phys 87:2773–2780
35. Troe J (1996) Statistical adiabatic channel model for ion-molecule capture processes. 2. Analytical treatment of ion-dipole capture. J Chem Phys 105:6249–6262
36. Georgievskii Y, Klippenstein SJ (2005) Long-range transition state theory. J Chem Phys 122(194103):1–17
37. Fehsenfeld FC (1976) Ion reactions with atomic oxygen and atomic nitrogen of astrophysical importance. Astrophys J 209:638–639
38. Milligan DB, McEwan MJ (2000) H_3^+ + O: an experimental study. Chem Phys Lett 319:482–485
39. Bettens RPA, Hansen TA, Collins MA (1999) Interpolated potential energy surface and reaction dynamics for $O(^3P) + H_3^+(^1A_1')$ and $OH^+(^3\Sigma^-) + H_2(^1\Sigma_g^+)$. J Chem Phys 111:6322–6332
40. Tanner SD, Mackay GI, Hopkinson AC, Bohme DK (1979) Proton-transfer reactions of HCO^+ at 298 K. Int J Mass Spectrom 29:153–169
41. Kim JK, Theard LP, Huntress WT (1975) Proton-transfer reactions from H_3^+ ions to N_2, O_2, and CO molecules. Chem Phys Lett 32:610–614
42. Burt JA, Dunn JL, McEwan MJ, Sutton MM, Roche AE, Schiff HI (1970) Some ion-molecule reactions of H_3^+ and proton affinity of H_2. J Chem Phys 52:6062–6075
43. Ryan KR (1974) Ionic collision processes in simple gas mixtures containing hydrogen. J Chem Phys 61:1559–1570
44. Bohme DK, Mackay GI, Schiff HI (1980) Determination of proton affinities from the kinetics of proton-transfer reactions. 7. The proton affinities of O_2, H_2, Kr, O, N_2, Xe, CO_2, CH_4, N_2O, and CO. J Chem Phys 73:4976–4986
45. Adams NG, Smith D (1981) A laboratory study of the reaction $H_3^+ + HD \leftrightarrow H_2D^+ + H_2$ – the electron-densities and the temperatures in inter-stellar clouds. Astrophys J 248:373–379
46. Rakshit AB (1982) A drift-chamber mass-spectrometric study of the interaction of H_3^+ ions with neutral molecules at 300 K. Int J Mass Spectrom 41:185–197
47. Marquette JB, Rebrion C, Rowe BR (1989) Proton-transfer reactions of H_3^+ with molecular neutrals at 30 K. Astron Astrophys 213:L29–L32
48. Gannon KL, Glowacki DR, Blitz MA, Hughes KJ, Pilling MJ, Seakins PW (2007) H atom yields from the reactions of CN radicals with C_2H_2, C_2H_4, C_3H_6, *trans*-2-C_4H_8, and *iso*-C4H8. J Phys Chem A 111:6679–6692

49. Blitz MA, Pesa M, Pilling MJ, Seakins PW (1999) Reaction of CH with H_2O: temperature dependence and isotope effect. J Phys Chem A 103:5699–5704
50. Wollenhaupt M, Carl SA, Horowitz A, Crowley JN (2000) Rate coefficients for reaction of OH with acetone between 202 and 395 K. J Phys Chem A 104:2695–2705
51. Brown SS, Ravishankara AR, Stark H (2000) Simultaneous kinetics and ring-down: rate coefficients from single cavity loss temporal profiles. J Phys Chem A 104:7044–7052
52. DeSain JD, Clifford EP, Taatjes CA (2001) Infrared frequency-modulation probing of product formation in alkyl plus O_2 reactions: II. The reaction of C_3H_7 with O_2 between 296 and 683 K. J Phys Chem A 105:3205–3213
53. Blitz MA, Goddard A, Ingham T, Pilling MJ (2007) Time-of-flight mass spectrometry for time-resolved measurements. Rev Sci Instrum 78:034103.1–034103.9
54. Meloni G, Selby TM, Osborn DL, Taatjes CA (2008) Enol formation and ring-opening in OH-initiated oxidation of cycloalkenes. J Phys Chem A 112:13444–13451
55. Mullen C, Smith MA (2005) Low temperature NH(X $^3\Sigma^-$) radical reactions with NO, saturated, and unsaturated hydrocarbons studied in a pulsed supersonic laval nozzle flow reactor between 53 and 188 K. J Phys Chem A 109:1391–1399
56. Smith IWM (2011) Laboratory astrochemistry: gas-phase processes. Annu Rev Astron Astrophys 49(49):29–66
57. Tyndall GS, Staffelbach TA, Orlando JJ, Calvert JG (1995) Rate coefficients for the reactions of OH radicals with methylglyoxal and acetaldehyde. Int J Chem Kinet 27:1009–1020
58. Dransfield TJ, Donahue NM, Anderson JG (2001) High-pressure flow reactor product study of the reactions of HO(X) + NO_2: the role of vibrationally excited intermediates. J Phys Chem A 105:1507–1514
59. Wardlaw DM, Marcus RA (1986) Unimolecular reaction-rate theory for transition-states of any looseness. 3. Application to methyl radical recombination. J Phys Chem 90:5383–5393
60. Harding LB, Klippenstein SJ, Jasper AW (2007) Ab initio methods for reactive potential surfaces. Phys Chem Chem Phys 9:4055–4070
61. Sims IR, Smith IWM (1988) Pulsed laser photolysis laser-induced fluorescence measurements on the kinetics of CN(v = 0) and CN(v = 1) with O_2, NH_3 and NO between 294 and 761 K. J Chem Soc Faraday Trans II 84:527–539
62. Sims IR, Queffelec JL, Defrance A, Rebrionrowe C, Travers D, Rowe BR, Smith IWM (1992) Ultra-low temperature kinetics of neutral-neutral reactions – the reaction CN + O_2 down to 26 K. J Chem Phys 97:8798–8800
63. Sims IR, Queffelec JL, Defrance A, Rebrionrowe C, Travers D, Bocherel P, Rowe BR, Smith IWM (1994) Ultralow temperature kinetics of neutral-neutral reactions – the technique and results for the reactions CN + O_2 down to 13 K and CN + NH_3 down to 25 K. J Chem Phys 100:4229–4241
64. Feng WH, Hershberger JF (2009) Reinvestigation of the branching ratio of the CN + O_2 reaction. J Phys Chem A 113:3523–3527
65. Klippenstein SJ, Kim YW (1993) Variational statistical study of the CN + O_2 reaction employing ab-initio determined properties for the transition-state. J Chem Phys 99:5790–5799
66. Stoecklin T, Dateo CE, Clary DC (1991) Rate-constant calculations on fast diatom-diatom reactions. J Chem Soc Faraday Trans 87:1667–1679
67. Greenwald EE, North SW, Georgievskii Y, Klippenstein SJ (2005) A two transition state model for radical-molecule reactions: a case study of the addition of OH to C_2H_4. J Phys Chem A 109:6031–6044
68. Sabbah H, Biennier L, Sims IR, Georgievskii Y, Klippenstein SJ, Smith IWM (2007) Understanding reactivity at very low temperatures: the reactions of oxygen atoms with alkenes. Science 317:102–105
69. Smith IWM, Sage AM, Donahue NM, Herbst E, Quan D (2006) The temperature-dependence of rapid low temperature reactions: experiment, understanding and prediction. Faraday Discuss 133:137–156

70. Clarke JS, Kroll JH, Donahue NM, Anderson JG (1998) Testing frontier orbital control: kinetics of OH with ethane, propane, and cyclopropane from 180 to 360 K. J Phys Chem A 102:9847–9857
71. Miller JA, Klippenstein SJ (2006) Master equation methods in gas phase chemical kinetics. J Phys Chem A 110:10528–10544
72. Robertson SH, Pilling MJ, Jitariu LC, Hillier IH (2007) Master equation methods for multiple well systems: application to the 1-,2-pentyl system. Phys Chem Chem Phys 9:4085–4097
73. Bohland T, Temps F (1984) Direct determination of the rate-constant for the reaction $CH_2 + H \rightarrow CH + H_2$. Ber Bunsenges Phys Chem 88:459–461
74. Bohland T, Temps F, Wagner HG (1987) A direct study of the reactions of $CH_2(X^3B_1)$ radicals with H and D atoms. J Phys Chem 91:1205–1209
75. Boullart W, Peeters J (1992) Product distributions of the $C_2H_2 + O$ and $HCCO + H$ reactions – rate-constant of $CH_2(X^3B_1) + H$. J Phys Chem 96:9810–9816
76. Devriendt K, Vanpoppel M, Boullart W, Peeters J (1995) Kinetic investigation of the CH2 (X3B1) + H → CH(X2Π) + H2 reaction in the temperature-range 400 K > T >1,000 K. J Phys Chem 99:16953–16959
77. Brownsword RA, Canosa A, Rowe BR, Sims IR, Smith IWM, Stewart DWA, Symonds AC, Travers D (1997) Kinetics over a wide range of temperature (13–744 K): rate constants for the reactions of $CH(\nu = 0)$ with H_2 and D_2 and for the removal of $CH(\nu = 1)$ by H_2 and D_2. J Chem Phys 106:7662–7677
78. Fulle D, Hippler H (1997) The temperature and pressure dependence of the reaction $CH + H_2 \rightarrow CH_3 \rightarrow CH_2 + H$. J Chem Phys 106:8691–8698
79. Ruscic B (2012) Private communication of unpublished ATcT datum for ver. 1.112 od ATcT TN
80. Ruscic B, Pinzon RE, Morton ML, von Laszevski G, Bittner SJ, Nijsure SG, Amin KA, Minkoff M, Wagner AF (2004) Introduction to active thermochemical tables: several "key" enthalpies of formation revisited. J Phys Chem A 108:9979–9997
81. Ruscic B, Pinzon RE, von Laszewski G, Kodeboyina D, Burcat A, Leahy D, Montoya D, Wagner AF (2005) Active thermochemical tables: thermochemistry for the 21st century. In: Conference on scientific discovery through advanced computing (SciDAC 2005), vol 16. San Francisco, pp 561–570
82. Medvedev DM, Harding LB, Gray SK (2006) Methyl radical: ab initio global potential surface, vibrational levels and partition function. Mol Phys 104:73–81
83. Meads RF, MacLagan RGAR, Phillips LF (1993) Kinetics, energetics, and dynamics of the reactions of CN with NH_3 and ND_3. J Phys Chem 97:3257–3265
84. Herbst E, Lee HH, Howe DA, Millar TJ (1994) The effect of rapid neutral-neutral reactions on chemical-models of dense interstellar clouds. Mon Not R Astron Soc 268:335–344
85. Bettens RPA, Lee HH, Herbst E (1995) The importance of classes of neutral-neutral reactions in the production of complex interstellar-molecules. Astrophys J 443:664–674
86. Talbi D, Smith IWM (2009) A theoretical analysis of the reaction between CN radicals and NH_3. Phys Chem Chem Phys 11:8477–8483
87. Blitz MA, Seakins PW, Smith IWM (2009) An experimental confirmation of the products of the reaction between CN radicals and NH_3. Phys Chem Chem Phys 11:10824–10826
88. Gerlich D (2008) In: Smith IWM (ed) Low temperatures and cold molecules. Imperial College Press, London, pp 121–174
89. Herbst E (1982) A reinvestigation of the rate of the $C^+ + H_2$ radiative association reaction. Astrophys J 252:810–813
90. Smith IWM (1989) Effects of quantum-mechanical tunneling on rates of radiative association. Astrophys J 347:282–288
91. Smith IWM (1989) Radiative association in collisions between neutral free-radicals. Chem Phys 131:391–401
92. Geppert WD, Larsson M (2008) Dissociative recombination in the interstellar medium and planetary ionospheres. Mol Phys 106:2199–2226

93. Vidali G, Pirronello V, Liu C, Shen LY (1998) Experimental studies of chemical reactions on surfaces of astrophysical interest. Astrophys Lett Commun 35:423–447
94. Vidali G, Roser JE, Manico G, Pirronello V (2004) Laboratory studies of formation of molecules on dust grain analogues under ISM conditions. J Geophys Res Planets 109: E07S14, doi:10.1029/2003JE002189
95. Hornekaer L, Baurichter A, Petrunin VV, Luntz AC, Kay BD, Al-Halabi A (2005) Influence of surface morphology on D_2 desorption kinetics from amorphous solid water. J Chem Phys 122:124701
96. Zecho T, Guttler A, Sha XW, Lemoine D, Jackson B, Kuppers J (2002) Abstraction of D chemisorbed on graphite(0001) with gaseous H atoms. Chem Phys Lett 366:188–195
97. Islam F, Latimer ER, Price SD (2007) The formation of vibrationally excited HD from atomic recombination on cold graphite surfaces. J Chem Phys 127:064701
98. Williams DA, Brown WA, Price SD, Rawlings JMC, Viti S (2007) Molecules, ices and astronomy. Astron Geophys 48:25–34
99. Latimer ER, Islam F, Price SD (2008) Studies of HD formed in excited vibrational states from atomic recombination on cold graphite surfaces. Chem Phys Lett 455:174–177
100. Hidaka H, Watanabe M, Kouchi A, Watanabe N (2009) Reaction routes in the CO-H_2CO-CH_3OH system, clarified from H(D) exposure on solid formaldehyde at low temperatures. Astrophys J 702:291–300
101. Hidaka H, Watanabe N, Shiraki T, Nagaoka A, Kouchi A (2004) Conversion of H_2CO to CH_3OH by reactions of cold atomic hydrogen on ice surfaces below 20 K. Astrophys J 614:1124–1131
102. Watanabe N, Nagaoka A, Shiraki T, Kouchi A (2004) Hydrogenation of CO on pure solid CO and CO-H_2O mixed ice. Astrophys J 616:638–642
103. Ward MD, Price SD (2011) Thermal reactions of oxygen atoms with alkenes at low temperatures on interstellar dust. Astrophys J 741:121
104. Tielens AGGM, Hagen W (1982) Model-calculations of the molecular composition of interstellar grain mantles. Astron Astrophys 114:245–260
105. Charnley SB (2001) Stochastic theory of molecule formation on dust. Astrophys J 562: L99–L102
106. Green NJB, Toniazzo T, Pilling MJ, Ruffle DP, Bell N, Hartquist TW (2001) A stochastic approach to grain surface chemical kinetics. Astron Astrophys 375:1111–1119
107. Biham O, Furman I, Pirronello V, Vidali G (2001) Master equation for hydrogen recombination on grain surfaces. Astrophys J 553:595–603
108. Caselli P, Hasegawa TI, Herbst E (1998) A proposed modification of the rate equations for reactions on grain surfaces. Astrophys J 495:309–316

Chapter 4
Astrochemistry: Synthesis and Modelling

Valentine Wakelam, Herma M. Cuppen, and Eric Herbst

Abstract We discuss models that astrochemists have developed to study the chemical composition of the interstellar medium. These models aim at computing the evolution of the chemical composition of a mixture of gas and dust under astrophysical conditions. These conditions, as well as the geometry and the physical dynamics, have to be adapted to the objects being studied because different classes of objects have very different characteristics (temperatures, densities, UV radiation fields, geometry, history etc); e.g., proto-planetary disks do not have the same characteristics as proto-stellar envelopes. Chemical models are being improved continually thanks to comparisons with observations but also thanks to laboratory and theoretical work in which the individual processes are studied.

4.1 Introduction

A large number of molecules have now been observed in the interstellar medium (ISM) and many more are expected to be discovered considering the unidentified lines in existing spectral surveys and new surveys to come [1]. The promise of

V. Wakelam (✉)
LAB, Univ. Bordeaux, UMR 5804, Floirac F-33270, France

LAB, CNRS, UMR 5804, Floirac F-33270, France
e-mail: wakelam@obs.u-bordeaux1.fr

H.M. Cuppen
Theoretical Chemistry, Institute for Molecules and Materials, Radboud University Nijmegen, Heyendaalseweg 135, Nijmegen 6525 AJ, The Netherlands

E. Herbst
Departments of Chemistry, Astronomy, and Physics, University of Virginia, Charlottesville, VA 22904, USA

I.W.M. Smith et al. (eds.), *Astrochemistry and Astrobiology*, Physical Chemistry in Action,
DOI 10.1007/978-3-642-31730-9_4,

ALMA,[1] a powerful new interferometric telescope, on this matter is endless. On a very basic level, simple molecules such as CO and CS are used to understand the physical properties of astrophysical objects such as dark clouds, star forming regions, proto-planetary disks, and galaxies through the excitation of their observed spectral lines [2] (see also Chaps. 1 and 2). Observers need, however, to know the distribution of these and other species in terms of abundances[2] in these sources. Chemical models are used for this purpose; e.g., to understand the chemical composition of astrophysical objects and their evolution. Combining model predictions with radiative transfer, astrochemists can make predictions on the detectability of some species. In addition, the abundance of "key" species that cannot be directly observed (molecules without a dipole moment for instance) can be computed with these models. These computations can be important for numerical simulations of protostellar collapse, for instance, since the energy budget (cooling and heating) depends on the abundances of some cooling species such as atomic oxygen that cannot be directly observed except in exceptional regions.

The gas and dust in the ISM are constantly being refreshed and modified, a process that is related to the life cycle of stars (see Chap. 2). After the death of a star, the elements that are formed inside the star are spread throughout the diffuse interstellar medium in the form of atoms and refractory dust particles, often known as "grains." This material is modified somewhat by chemical processes and by the strong interstellar UV field at play in such environments, before gravity collapses the material to form colder dense clouds, where external UV photons cannot penetrate. In the interior of these dense clouds, a rich gas-phase chemistry takes place and the gas interacts with grain surfaces where catalytic surface reactions can occur as well. Species such as molecular ions, radicals, isomers, and large linear unsaturated molecules such as $HC_{11}N$ can then be formed in the gas, while more saturated species such as water, methane, ammonia, and methanol can be formed on the grains along with CO_2. When stars and planets form from these clouds, this molecular complexity mutates into a more terrestrial-like organic chemistry including molecules such as basic esters, alcohols, and more saturated nitriles, which can survive and participate in the very long path towards life.

To study how the astrophysical environment modifies the composition of the gas and the dust, astrochemists have built increasingly complex chemical models over the years. In this chapter, we will describe those models (Sect. 4.2), discuss the chemical and physical processes that occur in the ISM (Sect. 4.3), and consider the synthesis of O_2 as an example of how these processes operate (Sect. 4.4).

---

[1] ALMA, for Atacama Large Millimetre Array: https://science.nrao.edu/facilities/alma

[2] Abundances are defined by the ratio of the density of a species to the total density (in cm^{-3}) of hydrogen atoms. Because hydrogen is mostly in the form of atomic H or molecular H_2, the proton density is $n_H = n(H) + 2n(H_2)$.

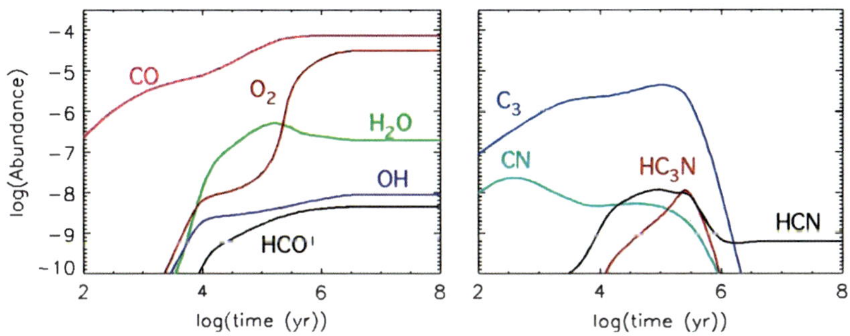

Fig. 4.1 Abundances (compared to total H) of species computed by a pure gas-phase chemical model for dense cloud conditions as a function of time

4.2 Astrochemical Models

From a numerical point of view, chemical models solve a system of differential equations of the type:

$$\frac{dn_i}{dt} = \sum production - \sum destruction \tag{4.1}$$

where n_i is the density of species i (in cm^{-3}). The production and destruction terms refer to all chemical and physical processes that produce and destroy this species. Numerical solvers using the Gear method [3] are typically employed in double precision to solve such systems as a function of time. In the end, modellers compute the evolution of the densities of species (or abundances) for a set of parameters and from an initial composition. For large systems, the number of equations (one for each species) and processes can be quite important and lead to hypersensitivity to some model parameters and even bi-stability[3] [4, 5].

In Fig. 4.1, we show the results of a model for dense cloud conditions (a temperature of 10 K, a hydrogen atom density of $2 \times 10^4\,cm^{-3}$ and a visual extinction of 10), which includes only gas-phase processes, except for the production of molecular hydrogen, which occurs on granular surfaces. The chemical processes that are involved in the chemistry of the interstellar medium are described in Sect. 4.3 of this chapter. In addition to the parameters described in Sect. 4.2.1, the geometry of the object, the presence of mixing, and physical dynamics can influence the chemical composition as well (Sects. 4.2.2 and 4.2.3).

[3] Bi-stability refers to two different steady-state results with the same set of parameters but different initial conditions.

4.2.1 Elemental Abundances and Initial Conditions

Material consisting of various elements, related to the life cycle of stars, is constantly modifying the existing material in the ISM. The elements are eventually stored in three forms [6]:

- *Phase 1*: in the gas-phase (in atomic or molecular form),
- *Phase 2*: in refractory cores of interstellar grains (formed in circumstellar envelopes),
- *Phase 3*: in a mantle of volatile species on top of grain cores.

The formation of grain mantles is efficient in cold regions shielded from UV photons. Depletion of the elements from purely gas-phase systems first occurs in those parts of stellar atmospheres where grains are first formed (see Fig. 4.3). Additional depletion from the gas-phase has been observed in diffuse clouds depending on the overall density of the clouds (from a few atoms and molecules per cm^3 to 100 per cm^3) [7, 8]. Although it seems logical to assume that the missing elements stick on grains, the mechanisms of depletion (considering the very low densities of these clouds) and the form of the depleted species remain a puzzle [9, 10]. Observations of unambiguous gas-phase elemental abundances are limited to purely atomic diffuse clouds with densities smaller than 10 cm^{-3}, where the UV field is still very efficient in preventing the formation of molecules in the gas-phase and at the surface of the grains. As a consequence, we do not have direct measurements of the fraction of the elements that are available for the formation of the molecules observed in dense clouds (in the gas and in the grain mantles). Most (if not all) chemical models of dense sources (molecular clouds, proto-stellar envelopes, proto-planetary disks) include as initial input an additional depletion of those elements heavier than oxygen compared with the gas-phase abundances observed in diffuse clouds, which are already depleted in these elements compared with stellar abundances, except for isolated exceptions such as sulphur. Many studies even consider the elemental abundances as free parameters that can be varied in order to reproduce the observations of gas-phase molecules in dense regions, even for elements lighter than oxygen such as nitrogen. Nitrogen (for example [11, 12]) and sulphur (for example [13, 14]) are good examples in this respect. Even when they are not varied intentionally, uncertainties in their exact values remain, depending on the reference used [15], and the results of chemical models strongly depend on these choices [16].

Another type of parameter, to which the model can be sensitive, is the choice in the initial composition of species (initial conditions). Since the chemistry is not at steady-state in most objects, the chemical composition predicted by the model will depend on the assumed initial conditions. There does not exist yet any model able to follow the chemical composition of the gas and dust during a complete cycle of evolution starting with material ejected from stars and ending with the collapse of clouds to form new stars, mainly because the evolution between different stages of star formation (e.g., diffuse to dense clouds, proto-stellar envelopes to proto-planetary disks) is not fully understood. For these reasons, assumptions about previous steps are usually made. For dense clouds for instance, it is typically assumed that they are formed from diffuse medium conditions in which elements are mainly in the atomic form and that the time scale of the "contraction" up to the densities of dense clouds is faster than the chemical evolution. Some models do explicitly include the physical transition from one stage to another as the chemistry progresses. As an example, Hassel et al. [17] followed the chemistry as a shock wave passes through diffuse material and a cold dense cloud begins to form behind the shock. Unfortunately, the abundances of gas-phase and grain-surface molecules synthesised were only followed up to a visual extinction of 3, at densities not as high as are found in cold dense clouds, because of limitations to the physical model. For proto-planetary disks, the chemical composition of the parent cloud is usually assumed to define the initial conditions [18]. In pursuing such an approximation, we of course ignore the transition between the cloud and the disk itself and so assume that the chemical composition is not modified during this transition. In the absence of a full physical model describing those transitions, it is probably the best that can be done.

4.2.2 Geometry

In astrophysical sources, where turbulence, or any kind of mixing, is not efficient (or is not constrained), the chemistry is usually treated in a zero-dimensional approximation, which consists of single spatial point. In such models, the temperature and densities are kept constant in space and time and the chemical composition is computed at each time step from an initial composition up to steady-state (or earlier if the time to reach steady-state is unrealistic). Many (if not all) astrophysical objects cannot be characterised by a single temperature and density, but present spatial gradients. Envelopes of gas and dust around proto-stars are characterised by an increase of the temperature and density towards the centre of the envelope in a spherical symmetry. Figure 4.2 shows the temperature and density in the envelope of the low mass proto-star IRAS16293-2422 as determined by Crimier et al. [19] based on the analysis of observations of the dust and the H_2O molecule at several wavelengths. In such objects, the chemistry will strongly depend on the radius. At high temperatures (larger than the specific sublimation temperature of the species), many molecules from the grain mantles will be evaporated in the gas-phase and will participate in the gas-phase chemistry (see for instance [20, 21]). Such an object (if considered as static) can be treated in a so-called pseudo 1D approach; i.e. the temperature and density will depend on the radius but the cells are independent. There is observational evidence from non-thermal broadening of the molecular lines that turbulent mixing exists in these objects although the exact nature of the mixing is not understood. Such mixing processes can be included in chemical composition calculations, as was done by Wakelam et al. [22] for massive proto-stellar envelopes. In proto-planetary disks, where radial and vertical mixing may exist, 2D chemical models with mixing have been developed [23].

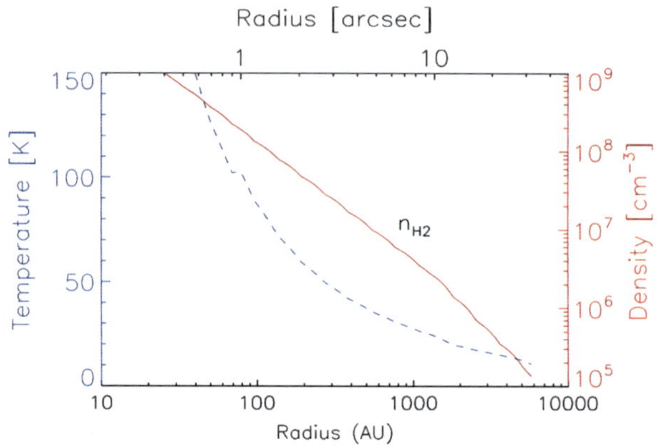

Fig. 4.2 Temperature and density of H_2 profiles in the envelope of the low mass proto-star IRAS 16293–2422 based on multi-wavelength observations from Crimier et al. [19]

The geometry is also important for the treatment of the interaction with UV photons. Borders of dense molecular clouds are exposed to the interstellar UV field produced by massive stars. The penetration of these photons into the cloud has to be computed as a function of depth since they are absorbed and scattered by the dust (see Sect. 4.3.1.1). In proto-planetary disks, the young central star usually presents a strong emission of UV (and X-ray) photons so that the UV penetration has to be computed in two dimensions to take into account the interstellar UV radiation field and the one coming from the star.

4.2.3 Physical Conditions Evolving with Time (Dynamics)

Most, if not all, astrophysical objects are not static over a period of time long enough for the chemistry to reach steady state, except perhaps in the diffuse medium as long as the initial H_2/H abundance ratio is assumed to be non-zero. As a consequence, the modifications of the physical conditions with time have to be considered. A good example is the modelling of circumstellar envelopes. The cells of material pushed away from the star encounter lower temperatures and densities. Close to the star, the temperatures are so high that species are only in atomic form. As the material moves away from the star, molecules will be formed and survive until they encounter a much thinner medium where the interstellar UV field will destroy them. A schematic view of the physical structure, which is far more complex than discussed here, and the chemical composition of a circumstellar envelope are given by Fig. 4.3.

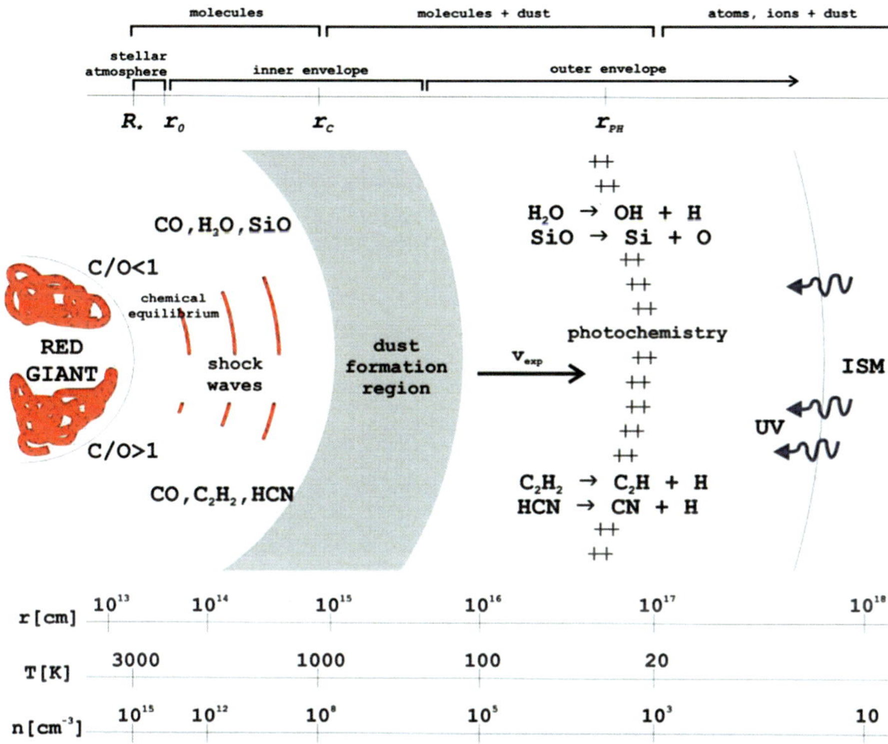

Fig. 4.3 Representation of a circumstellar envelope (Courtesy of Marcelino Agùndez)

As another example, consider the formation of stars, which begins with the collapse of a molecular cloud first into a pre-stellar core and then into a proto-star. This collapse occurs in a few times 10^5 year, during which the temperature and the density in the proto-stellar envelope will increase towards the centre. The environment is often referred to as a hot core or, for low-mass stars, a hot corino, when the temperature reaches 100–300 K [24]. During this time, the material of the envelope can fall towards the newly forming star in a rotating motion. The gas and dust will first encounter a small temperature and large density allowing most of the molecules to stick onto the grains, then a medium temperature during which grain-surface reactions will take place, and then a warmer temperature allowing the sublimation of the molecules formed on the grains. Although simplified models exist in which the time-dependent temperature is homogeneous throughout the warming region [25], a more accurate approach is to couple the chemistry to a 1D hydrodynamic collapse model, which computes the evolution of the physical conditions in a Lagrangian approach [21]. An extension to a full 3D hydrodynamic model is underway [26].

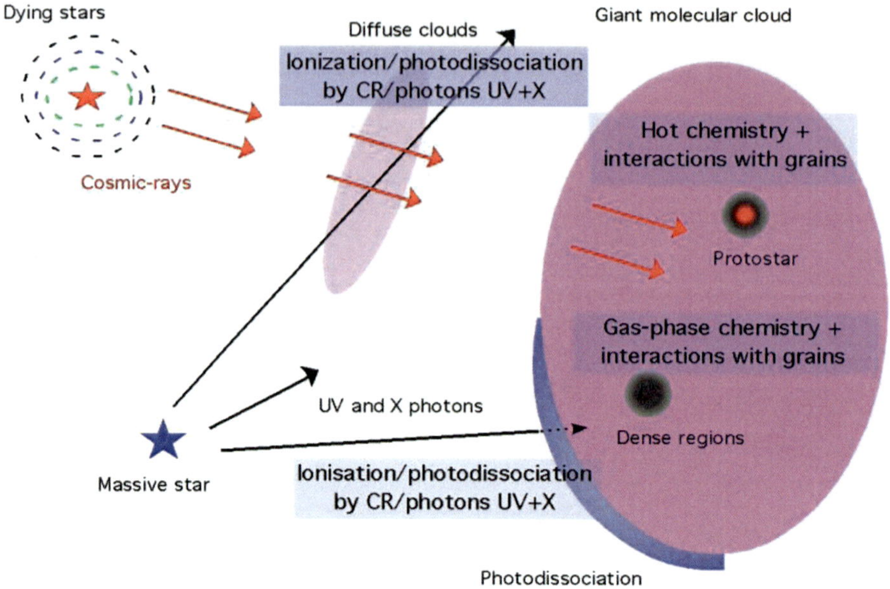

Fig. 4.4 Schematic view of the dominant physico-chemical processes in the interstellar medium depending on the physical conditions

4.3 Physico-chemical Processes

The chemistry of the ISM can be separated into four groups of processes. The first is the interaction with high-energy cosmic-ray particles, while the second consists of photo-processes induced by UV photons.[4] The third concerns bimolecular gas-phase reactions, and the last one concerns interactions with grains. A summary of the dominant processes in a selection of astrophysical objects is given in Fig. 4.4. We discuss the first three classes in the Sect. 4.3.1 and the last one in Sect. 4.3.2.

4.3.1 Gas-Phase Chemistry

4.3.1.1 Processes and Parameters

Cosmic-ray particles (with energies in the MeV to GeV range) produce the ionisation of molecular and atomic hydrogen and helium. The high-energy electrons from this process can in turn excite H_2, which then emits UV photons. This process, first proposed by Prasad and Tarafdar [27], is an efficient source for the

[4] X-rays are also important in some sources.

photodissociation of molecules in the dense internal part of molecular clouds. Although these first-order processes are not standard bimolecular chemical processes, their rate can be expressed in term of rate coefficients. The rates of direct and indirect dissociation and ionisation by cosmic rays are proportional to the total H_2 cosmic-ray ionisation rate ζ [28, 29]. As an example, the helium direct cosmic-ray ionisation rate is one-half of ζ. The value of ζ for any species is a function of depth into an interstellar cloud, although this dependence is most frequently ignored because it is difficult to calculate [30].

The rate coefficients for photodissociation caused by external UV photons depend on the visual extinction A_V:

$$k_{phot} = a \exp(-\gamma A_V) \tag{4.2}$$

with α and γ parameters specific to each species. The visual extinction A_V is proportional to the total hydrogen column density N_H ($A_V = N_H/1.6 \times 10^{21}$, [31]) where column density (cm^{-2}) is defined as the volume density integrated over a path through the cloud towards the observer.

Bimolecular reactions have been described in Chaps. 1 and 3 of this book. The efficiency of these processes can depend on the temperature of the gas. All production and destruction terms (in units of $cm^{-3}\ s^{-1}$) in equation (4.1) can be written as $k_{ij}n_in_j$ for bimolecular (second-order) reactions and k_in_i for first-order reactions, with k_{ij} the rate coefficient of the reaction between species i and j and n the density of the reactant(s). More details on these expressions can be found in Wakelam et al. [22].

Current networks for astrochemical models contain more than 4,000 gas-phase reactions for more than 400 atomic and molecular species, which comprise neutral, positively, and negatively charged species [21]. Most of these reactions have not been studied under the conditions of the cold ISM so that the rate coefficients can be quite uncertain if not completely wrong [16]. An example of the latter concerns the formation of OCS in the gas-phase by the radiative association reaction CO + S $\rightarrow$ OCS + hv. A rate coefficient of $1.6 \times 10^{-17}\ (T/300)^{-1.5}\ cm^3\ s^{-1}$ was included in models following a crude estimate by Prasad and Huntress [28], which gives a rate coefficient of approximately $2 \times 10^{-15}\ cm^3\ s^{-1}$ at 10 K. Recent calculations by Loison et al. [32] show that the rate was greatly overestimated at 10 K. The consequence for the OCS gas-phase production is then dramatic and the predicted abundance of OCS is much smaller than the observed one. In general, radiative association reactions, despite their importance, have not been studied in more than a few cases in the laboratory at any temperature.

4.3.1.2 Uncertainties and Sensitivity to the Parameters

Model predictions can be very sensitive to model parameters. Uncertainties in the parameters can therefore lead to strong differences in the predicted abundances of species. Two aspects of the problem can be considered. First, the model parameters are

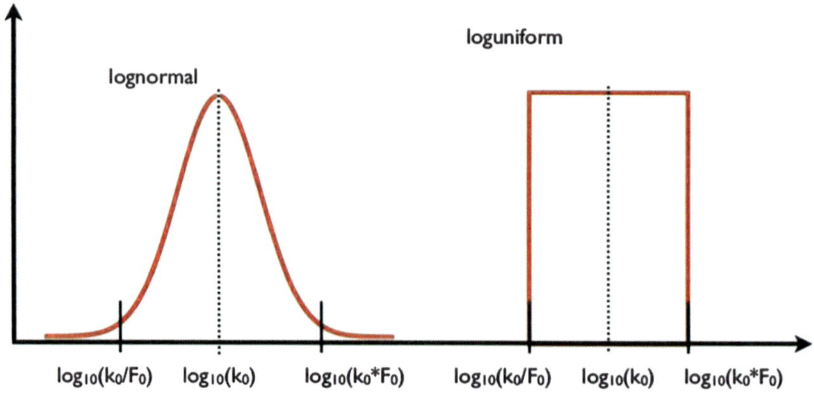

Fig. 4.5 Possible distributions (lognormal or loguniform) of the rate coefficients randomly varied within a factor F_0

more or less known within a range of uncertainty. These uncertainties propagate through the time-dependent calculations and lead to error bars for the model results. If those error bars have been derived, quantitative comparisons with observations (for which error bars are more frequently determined) can be undertaken. Secondly, it is possible through sensitivity analyses to identify key parameters for which a better estimate would reduce the model uncertainties. A number of such studies have been done recently with the aim of (1) computing model error bars, (2) understanding the sensitivity of these models to the parameters (sensitivity analysis), and (3) identifying key parameters in a variety of sources.

Estimation of model error bars and sensitivity analyses are based on the same principle. All rate coefficients (or other model parameters) of a system are randomly varied within a certain range. The chemical evolution is then computed for each set of rate coefficients. For a network containing 4,000 reactions, the model is typically run 2,000 times with different sets of rate coefficients. The distribution of the rate coefficients can be either log-normal or log-uniform (see Fig. 4.5). The first choice implies that the mean value k_0 is a preferred value. This is usually the case for rate coefficients, which are measured with an uncertainty defined by statistical errors. The factor F_0, which defines the range of variation, can be a fixed factor for all reactions for a sensitivity analysis or specific to each reaction for an uncertainty propagation study. Use of the same F_0 for all reactions, in the case of a sensitivity analysis, assures the modeller that an underestimated uncertainty factor will not bias the analysis. The results of thousands of runs are used differently to identify important reactions and to estimate model error bars.

For sensitivity analyses, one of the methods to identify "key"[5] reactions is to compute Pearson correlation coefficients for each species and each reaction. Such

[5] By "key" reactions, we mean reactions with rate coefficients that are quantitatively important for the computed abundance of species.

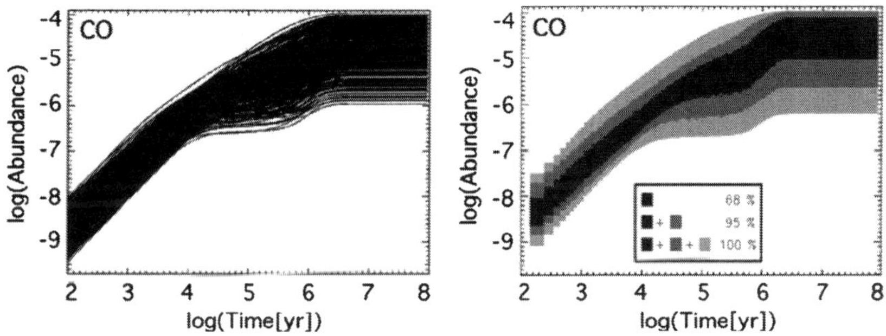

Fig. 4.6 CO abundance computed by a chemical model under dense cloud conditions (from [16]). Each curve on the *left* represents the result of one model in which the rate coefficients have been randomly modified. The *grey* levels on the *right* show the density of the curves (in percentage)

coefficients quantify how modified abundances of species correlate with the modifications of rate coefficients. For uncertain propagation, model error bars, or uncertainties in the computed abundances of species, are defined by the minimum interval containing 95 % of the curves (see Fig. 4.6 and [33]). When the abundances follow a log-normal distribution, these error bars are equivalent to twice the standard deviation of the curves. More details on these methods and their applications can be found in a review paper by Wakelam et al. [34].

4.3.1.3 Reduction of Chemical Networks

Large chemical networks are often needed to synthesise large molecules. For some applications, for example, in which the chemistry is calculated with a physical model that leads to lengthy computational times, it can be convenient or even necessary to reduce the size of a network of reactions, if only limited information is required. The simpler technique to accomplish this goal is to remove reactions one-at-a-time and check that the abundances of the species of interest remain within a certain range of tolerance. This of course can only be correct for one physical condition and one initial network. For the reduced network to be used over a range of physical conditions, the analysis has to be redone several times. In addition, if the rate coefficients of some reactions have to be changed, the outcome may be changed. More sophisticated methods based on correlations can be used, as they are probably less time consuming [35].

Such reduction methods have been applied by a number of groups for different cases. To the best of our knowledge, Ruffle et al. [35] were the first to apply such a technique (in a more complex version) to determine the minimum network for the computation of the gas-phase CO abundance in interstellar clouds of different density, temperature and visual extinction. Their network still contains more than two hundred reactions and more than 60 species. Reduced networks to compute the ionisation fraction in dense clouds [36] and in proto-planetary disks [37], and reduced networks for dense clouds [38] have also been proposed.

4.3.2 Gas-Grain Interactions and Surface Reactions

In the ISM, molecules from the gas phase interact with grains by sticking to their surface, possibly undergoing reactions with other adsorbates, and desorbing again through different thermal and non-thermal processes. The formation of molecular hydrogen, the most abundant molecule in the ISM, is the most obvious example for the importance of grain surface chemistry, since it is almost entirely formed through surface chemistry in cold environments. As explained in Chap. 1, species can either be physisorbed to the surface and then general reactions occur through the diffusive Langmuir-Hinshelwood mechanism, or be chemisorbed to the surface so that chemistry can also proceed through the Eley-Rideal mechanism, in which the adsorbate is struck by a gas-phase reactant. Molecular hydrogen is formed through both mechanisms; the Langmuir-Hinshelwood mechanism dominates at low temperature for both silicate and carbonaceous grains and the Eley-Rideal mechanism at high gas and/or grain temperatures for carbonaceous grains.

In the present section, we will present the different methods to model grain surface chemistry, the input parameters that go into these models, the experimental and theoretical methods to obtain these parameters, and the limitations of these models. Most gas-grain networks are predominantly designed for the low temperature regime and, as we will see in the following sections, the expressions that are used almost exclusively describe the Langmuir-Hinshelwood type of reaction.

4.3.2.1 Rate Equation Approximation

The most obvious way to introduce grain-surface chemistry into gas-phase astrochemical models is to use rate equations for both the surface chemistry and for gas-phase chemistry. The main advantages of this method are that the gas phase and grain surface can be easily coupled and that the rate equation method is computationally light, which allows for a large number of grain-surface species and reactions. Moreover, a long timescale can be simulated and simulating a large number of points to cover a grid of different conditions is feasible as well.

Here we follow the rate equation approach by Hasegawa et al. [39]. Consider a dust particle with N surface sites, on which there are $N(A)$ molecules of species A. The surface abundance, $N(A)$, changes in time according to

$$\begin{aligned}\frac{dN(A)}{dt} = k_{acc}n(A) - k_{des}N(A) - \sum (k_{hop,A} + k_{hop,i})N(A)N(i)/N \\ + \sum (k_{hop,i_1} + k_{hop,i_2})N(i_1)N(i_2)/N \qquad (4.3)\end{aligned}$$

where $n(A)$ is the gas phase abundance of species A and the different k's represent different rate coefficients. The first term accounts for accretion of A from the gas phase to the surface; the accretion rate coefficient k_{acc} is determined by the product of the velocity of A in the gas phase, the cross section of the grain, and the sticking fraction of A to the grain, which depends on the gas and surface temperatures. The second term represents the loss of A from the surface due to desorption. This can occur through a non-thermal desorption mechanism or through thermal desorption, for which the rate coefficient is given by

$$k_{des,therm} = \nu \exp\left(-E_{des}/k_B T_{grain}\right) \tag{4.4}$$

with ν the attempt frequency, E_{des} the desorption energy, k_B the Boltzmann constant, and T_{grain} the grain temperature. Non-thermal mechanisms include photodesorption [40, 41], cosmic-ray desorption due to sputtering or flash heating of the grain [42, 43], and desorption upon reaction [44].

The third term in (4.3) refers to the loss of species A due to Langmuir-Hinshelwood reactions with species i. The rate coefficient $k_{hop,i}$, defined as the rate of hopping of species i over the potential barrier between two adjacent sites, is itself given by the equation:

$$k_{hop,i} = \nu_0 \exp\left(-E_{b,i}/k_B T_{grain}\right) \tag{4.5}$$

Similarly the last term in (4.3) represents the gain in species A by formation mechanisms.

Apart from the clear advantages of this method, which lie mostly in its user-friendliness and the fact that it is computationally inexpensive, there are several disadvantages to the rate equation method, which become apparent in different physical environments. In conditions where the number of species on the surface is small ($<<1$), one can run into the so-called "accretion" limit [45] in which the formation rate of molecules can be overestimated by several orders of magnitude. The quantities *N(i)* are in fact expectation values of *N(i)*: $\langle N(i) \rangle$. The Langmuir-Hinshelwood term in (4.3) then becomes $\langle N(A) \rangle \langle N(i) \rangle$, where in fact the rate of reaction is determined by $\langle N(A)N(i) \rangle$. For small values of *N(A)*, this approximation is not valid and $\langle N(A) \rangle \langle N(i) \rangle$ overestimates $\langle N(A)N(i) \rangle$. Different methods have been developed to overcome this problem.

Another disadvantage of the rate equation method is that the actual surface and the positions of the atoms and molecules are not considered. Especially in conditions where the grain mantle is beyond the sub-monolayer regime, this can have severe consequences: species are allowed to interact with all other species, regardless of their relative position, species are assumed to hop and desorb with the same rate, regardless of their position (bulk vs. surface), and the grain mantle is assumed to be homogeneous, whereas observations show that the grain mantle consists of several layers of different composition. But even in the limit where the surface species add up to less than a monolayer, the rate equation method can overestimate formation rates due to the missing back diffusion term. When an atom scans the grain surface, it can visit a site more than once, which is called "back diffusion" and can lower the rate by as much as a factor of 3 [46].

4.3.2.2 Methods to Circumvent the Accretion Limit

Different stochastic methods have been considered to overcome the accretion limit problem. The main focus has been on the master equation method [47–49] and macroscopic Monte Carlo simulations [49–51]. Of the two, the master equation method can be more easily coupled to rate equations, which handle the gas phase chemistry in the most straightforward way, since it treats the gas and surface kinetics with equations of similar form.

In the Monte Carlo methods, the evolution of (discrete) number densities is followed in time by randomly selecting a sequence of processes. The probability of selecting a process is proportional to its rate, which is determined in a similar manner to the rate equation method. Because Monte Carlo methods use random numbers and probabilities instead of analytical expressions, coupling between the two methods (rate equations for the gas and a Monte Carlo procedure for the grain) is harder to achieve. Different (kinetic) Monte Carlo implementations are used and usually a distinction between macroscopic and microscopic Monte Carlo is made. In the macroscopic simulations, only the number density is followed in time; in the microscopic simulations the exact positions of the species are also considered. Recently, macroscopic Monte Carlo simulations of both the gas phase and grain surface chemistry have been carried out for a proto-planetary disk [52].

The master equation method specifically considers each possible configuration of species. For a system where only H and H_2 are considered on the grain, possible configurations would be (1, 0) with one H atom and no H_2 molecules on the grain, (0, 0), (0, 1), (1, 1), (2, 0), etc. For each configuration a separate rate equation is constructed. The number of possibilities is clearly infinite, but selective cut-offs can be used to exclude the higher order terms if their probability of occurrence becomes very small. However, if the number of species expands, either by a change in physical conditions or by increasing the number of considered species in the model, the number of equations blows up rapidly. Moment equations with a cut-off up to the second-order have been suggested to make the effect less dramatic and to extend the range in which the rate equation method is applicable [53, 54] and codes are available which allow this method to be used for large chemical networks in a hybrid sense in which the method is only used when the accretion limit is reached for a species [55].

Apart from stochastic methods, some modifications to the rate equations have been put forward to mimic the stochastic behaviour of the surface chemistry. Caselli et al. [56] made semi-empirical adjustments to the rates of a selection of reactions, for the case where the surface migration of atomic hydrogen is significantly faster than its accretion rate onto grains. While this method showed only limited success, the more recent modified-rate approach suggested by Garrod et al. [57] has been much more successful. In this approach, the expressions for the reactions are adjusted if the modified reaction rate is smaller than the classical rate. The modified reaction rate is determined by the accretion rate of one of the reactants multiplied by the product of the surface abundance of the other reactant and an efficiency term that takes into account the competition between diffusion across the surface for the reactants and the desorption of the reactants. Its main advantages are that it is computationally inexpensive compared with stochastic methods and that it automatically switches to the normal rate equation in the regime where those are still valid. Garrod et al. [58] have shown that the new method produces an excellent match to macroscopic Monte Carlo solutions to full gas-grain chemical systems for a range of physical conditions.

4.3.2.3 Methods that Include Layering

Species that land on an ice mantle can only react with species in the top few monolayers. Reactions deeper in the mantle either occur through bulk diffusion, which is much slower than surface diffusion, or through energetic processing such as UV photodissociation, or bombardment by energetic particles such as cosmic rays. Cuppen et al. [59] showed with a model that takes the positions of all individual species into account that the more common rate equation methods overestimate the effect of a changing gas phase composition on the grain composition by several orders of magnitude. Due to the layering of the ice, only the top layers are available for reaction and changes in the gas phase abundances are only reflected in these top layers. For simulations at constant temperature, grain models with layering show therefore many fewer fluctuations in the abundance of surface species.

The first gas-grain model that accounted for the distinction between mantle and surface was the three-phase model by Hasegawa and Herbst [60]. Here the mantle and the surface are treated as separate phases and all three phases are described by rate equations. Species can only react and desorb from the surface phase. Upon desorption the surface is replenished by species from the mantle phase and *vice versa* upon accretion. With only two solid phases (surface and mantle), they observed that the abundances at early times are preserved for a much longer period. Fayolle et al. [61] made some modifications to this model to account for bulk diffusion (or segregation), which leads to interaction between the two phases. This adjusted model is able to reproduce laboratory desorption experiments of mixed ice layers. Unfortunately, the three-phase model is not widely used, even though it has the clear advantage that it returns the correct desorption behaviour. Recently, a macroscopic Monte Carlo code has been developed that uses multiple phases for the ice mantle [62].

As mentioned in Sect. 4.3.2.2, the microscopic implementation of the Monte Carlo method follows the position of each species on the grain. It therefore automatically accounts for layering, since only species in the direct vicinity of each other are allowed to react. It can furthermore use environmentally dependent binding energies. The main disadvantage of this method is, however, that it is computationally very expensive and it can therefore not be as easily applied for the simulation of long timescales and large chemical networks. Efforts in this direction are however on the way, including the use of massively parallel processors (Q. Chang and E. Herbst, in preparation).

4.3.2.4 Input Parameters for Surface Models

All different surface models have similar input parameters. These result in (temperature dependent) hopping, desorption and reaction rates. Usually energy barriers are given from which a thermal or tunnelling rate is determined. The energy barriers, as for the gas phase, are often taken from experiments or (quantum) chemical calculations. It must be mentioned that the results of surface reaction experiments in the laboratory are not as easily converted to rates under interstellar conditions as for gas phase reactions for a variety of reasons. Firstly, the formation of molecules on the surface is the result of a combination of diffusion and reaction processes in competition with the desorption of the reactants. It is hard to disentangle the separate contributions and give barriers for the individual processes. Since all processes scale differently according to temperature and coverage, they should be disentangled before using them in an astrochemical model. Secondly, the experiments are not carried out on small particles but on large surfaces. Thirdly, the substrate, in monolayer regime experiments, is not always a good representative for a dust grain and in multilayer experiments the ice is usually much more homogeneous than an interstellar ice.

Binding Energies

Desorption, which is controlled by the binding energy, is a one-step process and therefore relatively straightforward to study. Binding energies are usually determined through Temperature Programmed Desorption (TPD). In this technique, a known quantity of the species is first deposited and then the substrate is linearly heated with time while the desorption is recorded using mass spectrometry. Several of these experiments are performed with either different initial deposition quantities, deposition temperature, or heating ramps to obtain the order of the desorption process, the desorption energy, and the pre-exponential factor ν in (4.4). The desorption energies of a wide collection of stable species have been determined in this way. Examples are N_2 [63], CO [63], O_2 [64], H_2O [65, 66], and CH_3OH [67]. The desorption energies have been mostly determined for the desorption of pure ices from different substrates. The differences between the different substrates are rather small and become negligible in the multilayer regime. In this regime, the molecules desorb with a (near) zeroth order rate whereas they desorb with a (near) first order rate in the monolayer regime. Since interstellar ices are not homogeneous, the desorption of mixed layers is more relevant for astrochemical modelling. However, the introduction of more species in the ice makes the desorption process immediately much more complex. First, the desorption energy can change depending on its surrounding material. Second, the dominant mantle species can prevent other species from desorbing. Collings et al. [68] showed, for instance, that molecules like CO and CO_2 can become trapped in an ice mantle that consists predominantly of water ice as the desorption of water occurs at much higher temperatures than for CO and CO_2. However, at the long timescales available in the ISM, some of these trapped species might be able to escape because of a segregation process where the two main fractions of the mantle slowly separate [69]. The model by Fayolle et al. [61], which was described earlier, is especially designed to handle this behaviour with the addition of only a few extra parameters.

Diffusion Barriers

Diffusion barriers are very hard to measure and are mostly determined by quantum chemical calculations. The quality of these data relies heavily on the potential and substrate used. Experimentally, diffusion barriers can be inferred by detecting reaction products between the mobile species of interest and some immobile, sparse other reactant. Matar et al. [70] were able to determine the diffusion of H on amorphous solid water by reaction of H and O_2. In chemical models, often the approximation is made that the diffusion barrier is a constant fraction of the desorption energy; values of 0.3, 0.5 or 0.78 are usually taken.

Reaction Barriers

Experimental studies of ice reactions can be roughly divided into two groups depending on the analysis technique and the thickness regime (sub-monolayer versus multilayer). The reactants and products can be probed mass spectrometrically. The surface is initially exposed to a small quantity of the reactants, after which the surface is heated until the products and reactants desorb and are detected via TPD. Different initial exposures and temperatures can be probed to obtain information on reaction order, etc. The main advantage of this technique is the sensitivity, which allows for sub-monolayer exposures and which is able to detect all species (masses). The method has, however, four major disadvantages: the products cannot be probed in-situ, i.e., during the atom bombardment, additional reactions during the heat-up to desorption cannot be excluded, quantifying the desorbing species is not straightforward, and some of the interesting species have equal masses.

A second method is to initially grow an ice of several monolayers and expose this ice to the atomic beam while recording reflection adsorption infrared spectra (RAIRS). In this way, the reactants and products are probed in-situ at the time and temperature that one is interested in, which is the main advantage of this technique. Quantifying the formed product is relatively simple, provided that the RAIRS is calibrated with an independent method. The main disadvantages are that not all species can be detected in this way and that the sensitivity is less than with the previous technique. Most systems have an additional quadrupole mass spectrometer (QMS) installed. So far this technique has been applied to unravel the formation of the main components of interstellar ices, *i.e.*, water, methanol, carbon dioxide, formaldehyde, and formic acid, mainly through H-atom additions to CO- and/or

O_2-ices under conditions relevant to the interstellar medium (e.g., [70–77]). The emphasis has been on the smaller species, with ethanol being the largest product formed [78], and mostly with H-atom beams. An important reason for the latter is that purely atomic beams can never be achieved and at these low temperatures pollutants like O_2 or N_2, will stick to the surface as well; the majority of H_2, on the other hand, will desorb. Another important reason is that H atoms diffuse quite rapidly even at low temperatures.

As explained above, the formation of new molecules on the surface is a combination of several independent processes, especially when the stable, detectable products are the result of multiple reactions. The barriers for the individual processes are very hard to disentangle and this is only possible with additional information from, for instance, TPD experiments. For this reason, usually only a qualitative measure for the efficiency of the reaction is given. Common assessments are: (1) effectively barrierless, (2) with a barrier that allows the reaction to proceed at 10 K, (3) with a larger barrier so that reaction does not proceed at this temperature, (4) thermally activated, or proceeds through quantum chemical tunnelling.

Heavier species become increasingly difficult to form by simple atom-addition reactions. H-atom addition to acetaldehyde leads not only to ethanol, but also to the smaller organic species formaldehyde and methanol [78]. Furthermore, there is no experimental evidence that one can form longer carbon chains on the grains by simply adding carbon atoms. A different type of mechanism has been suggested that leads to the production of terrestrial-type organic molecules in star-forming regions of the interstellar medium [25]. At higher temperatures, heavier species than atoms can diffuse more readily, but they are typically not reactive because of chemical activation energy barriers. If these stable species can be converted to radicals by photons or energetic particle bombardment, rapid diffusive reactions are possible, leading to a variety of more complex species. Laboratory experiments on surface photochemistry have been undertaken for some time, but only recently under ultra-high-vacuum conditions, which are necessary to have little contamination and ice thickness comparable to interstellar ices [79].

4.4 Application: The Molecular Oxygen Chemistry

The larger the molecules, the more chemical processes are involved in their synthesis and thus the more our knowledge of their chemistry becomes thinner. One would expect that the synthetic pathways of simple molecules such as molecular oxygen would be simple and by consequence well known nowadays. As we will show, however, this view is far from being correct and our understanding of the processes involved in interstellar chemistry is only improving little by little by combining modelling, laboratory experiments, and astrophysical observations.

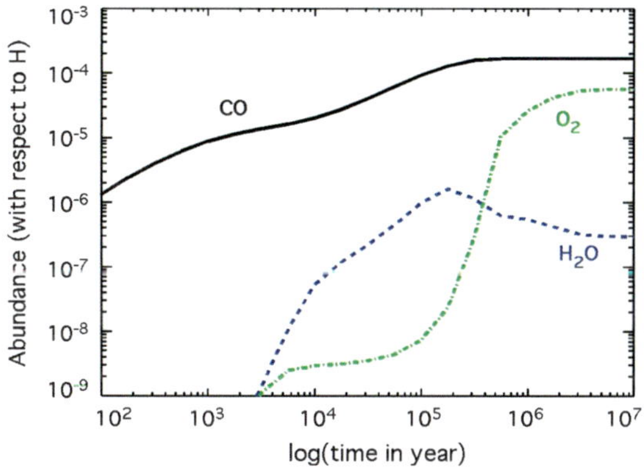

Fig. 4.7 CO, O_2 and H_2O abundance predicted by a gas-phase model under dense cloud conditions

4.4.1 Gas-Phase Synthesis

Molecular oxygen was included in chemical networks at the beginning of astrochemistry [80]. In the gas-phase, O_2 is a product of the neutral-neutral reaction $O + OH \rightarrow O_2 + H$, which is an exothermic, relatively fast reaction. The rate coefficient of this reaction has been measured at low temperature, using the CRESU apparatus (see Chap. 3) by Carty et al. [81]. They found a value of 3.5×10^{-11} cm^3 s^{-1} between 39 and 142 K. The OH radical is formed by the dissociative recombination of protonated water H_3O^+, itself formed in the following sequence of reactions:

- H_2 + cosmic-ray particle $\rightarrow H_2^+ + e^-$ + cosmic-ray particle
- $H_2^+ + H_2 \rightarrow H_3^+ + H$
- $H_3^+ + O \rightarrow OH^+ + H_2$
- $OH^+ + H_2 \rightarrow H_2O^+ + H$
- $H_2O^+ + H_2 \rightarrow H_3O^+ + H$

This scenario was first proposed by Herbst and Klemperer [80], who discussed the importance of the H_3^+ cation in the ISM (see also [82]).

Under dense cloud conditions (temperature around 10 K, atomic hydrogen density of a few times 10^4 cm^{-3} and no UV penetration), a pure gas-phase model will predict that oxygen and carbon will mainly form carbon monoxide. The rest of the oxygen will go into O_2 (see Fig. 4.7). At steady state, after 10^7 year, the predicted CO/O_2 abundance ratio is approximately 3. Such a model result is strongly dependent on the assumptions concerning the elemental abundances of oxygen and carbon. The results presented here have been obtained assuming that the C/O elemental ratio is 0.4, as typically assumed based on some observations of atomic lines of carbon and oxygen in the diffuse medium [15]. We will come back later to this problem. Following this prediction, interstellar molecular oxygen has been sought for many years.

4.4.2 Observational Constraints on O_2 Abundance

Since the 1980's, O_2 has been looked for in the interstellar medium using both ground-based and space telescopes (see [83], and references therein). First, analyses of data from the SWAS satellite gave an upper limit of about 10^{-6} in dense cold clouds [84]. After the first detection claim by Pagani et al. [83], Larsson et al. [85] announced the detection of O_2 by re-analysing data from the ODIN satellite and published a beam-diluted abundance of 5×10^{-8} relative to H_2. Using ground based observations of $O^{16}O^{18}$ and $C^{18}O$ lines, Liseau et al. [86] argued that the emitting region may be much smaller than the beam of ODIN and thus the O_2 abundance could be larger by one or two orders of magnitude. More recently, the Herschel satellite found three magnetic dipole rotational transitions of O_2 towards the H_2 Peak 1 position of vibrationally excited molecular hydrogen in Orion KL [87]. The fractional abundance of O_2 relative to H_2 was found to be (0.3–7.3) $\times$ 10^{-6}. The authors suggested the source of the O_2 to be either thermal evaporation from warm dust or the passage of a C-shock. Why the molecule is only found in such an unusual source remains a mystery.

4.4.3 Gas-Grain Interactions

Based on these new observational constraints, models have been altered in order to decrease the predicted abundance of gas-phase O_2. Problems in the rate coefficients of gas-phase reactions have been looked for. The very low temperature rate coefficient of the O + OH reaction has been investigated theoretically, but Quan et al. [88] showed that only a seriously unrealistic decrease of this rate coefficient would really impact the predicted O_2 abundance.

Other efforts have been made to investigate the interaction of gas-phase O_2 with interstellar grains (see for example [89–91]). Once O_2 is formed in the gas phase, the molecule sticks significantly on the surface of interstellar grains in approximately 10^5 year, decreasing the gas-phase abundance of O_2. Desorption induced by cosmic rays, however, is efficient enough to desorb the molecule and the gas-phase abundance remains almost unchanged if only sticking and evaporation are considered. This result is shown in Fig. 4.8, where we overlay the abundances predicted by a pure gas-phase model with a model including sticking of gas-phase species on grain surfaces and desorption of these species from the surfaces (Sect. 4.3.2). Once they are on the grain surfaces, however, molecules can react with other species present.

4.4.4 Grain Surface Reactions

At temperatures as low as 10 K, atoms can move efficiently and react with other molecules accreting from the gas-phase. Following this idea, Tielens and Hagen [90] suggested that O_2 on the surface can be successively hydrogenated by reactions

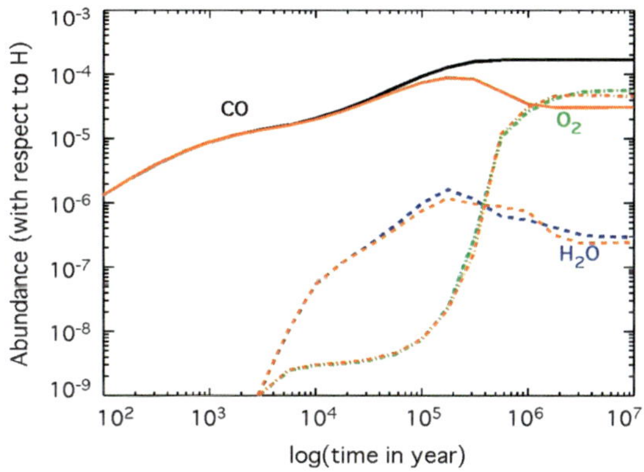

Fig. 4.8 CO, O_2 and H_2O abundances predicted by a gas-phase model under dense cloud conditions (same as Fig. 4.7) and by a gas-phase model including sticking and evaporation of species from grain surfaces (*orange lines*)

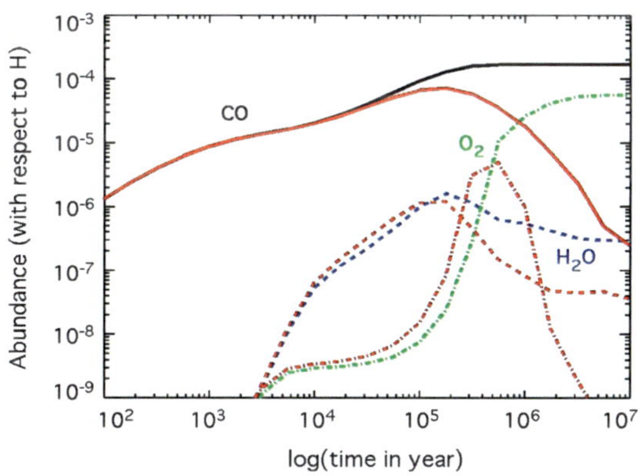

Fig. 4.9 CO, O_2 and H_2O abundance predicted by a gas-phase model under dense cloud conditions (same as Fig. 4.7) and by a full gas-grain model (*red lines*)

with atomic hydrogen to give H_2O_2 and then produce water, again by a further reaction with atomic hydrogen. Recently, this surface reaction scheme has experimentally been proven to be even more efficient than assumed in the models of Roberts and Herbst [70, 73, 74]. This formation path is, however, only one of several possibilities to transform oxygen into water on surfaces. If the direct hydrogenation of O_2 on the surface is removed, other reactions such as the hydrogenation of ozone or the reaction between OH and H_2, would take its place to remove the oxygen (whatever its form) and transform it into H_2O. Whatever the exact scenario, the predicted abundance of gas-phase O_2 is then strongly decreased at late times (see Fig 4.9). A peak of O_2 up to about 5×10^{-6} (compared to H) however remains at typical ages of dense clouds (a few $\times$ 10^5 year).

Considering the difficulty of observing O_2 in quiescent dense regions, unlike that observed by Goldsmith et al. [87], and the lack of spatial resolution of the current observations, it is difficult to reach any conclusions on the agreement between models of cold cores and the observational constraints. One may think that the scarcity of the detections despite the number of target sources (perhaps covering a large range of cloud ages) is an indication that O_2 is even less abundant than predicted by these gas-grain models. If so, work still has to be done to solve this problem but probably not with a purely chemical point of view. Chemical model predictions do indeed depend on other parameters than chemical reactions and rate coefficients. Elemental abundances are among the most important ones and remain quite uncertain (see Sect. 4.2). Observations in diffuse clouds show that the elemental C/O ratio does not vary much with the line of sight and stays around 0.4 [15]. It has been recently proposed by Jenkins [8], however, that oxygen could be slightly more depleted than carbon in denser sources, a possibility also discussed by Whittet [92]. As a consequence, the C/O elemental ratio in the gas phase of dense regions could be larger than what is typically assumed. This idea was used by Hincelin et al. [93] to explain the low O_2 abundance in dense clouds. Using a C/O ratio greater than unity with a full treatment of the gas-grain chemistry would produce an abundance of gas-phase O_2 smaller than 5×10^{-8} (compared to the total atomic hydrogen density n_H) at all times and could explain the observations directed at cold cores, without changing strongly the abundances predicted for other species.

4.4.5 Importance of Nitrogen Chemistry for O_2

Current networks for the ISM can contain more than 6,000 reactions, including both gas-phase and grain-surface processes. As discussed in Sect. 2.1.3, reduction of chemical networks can be done to allow the coupling with a dynamical model that otherwise would be very time consuming [94, 95]. Such reductions have to be performed very carefully and can only be accomplished for a specific condition. The chemistry of O_2 provides a good example. With a C/O elemental abundance of 1.2, instead of 0.5, nitrogen chemistry becomes important for the O_2 abundance although no N-bearing species are involved in its synthesis [93]. Under dense cloud conditions and with a gas-grain model (and an elemental C/O ratio of 1.2), the CN molecule is abundant and reacts with O_2 to form either O + OCN or CO + NO. The total rate coefficient of this reaction has been measured by Sims et al. [96] between 13 and 295 K, and the product branching ratios by Feng and Hershberger [97] so that the temperature dependent rate coefficients are:

- $CN + O_2 \rightarrow O + OCN$ $k_1(T) = 1.992 \times 10^{-11}\,(T/300)^{-0.63}\ cm^3\ s^{-1}$,
- $CN + O_2 \rightarrow CO + NO$ $k_2(T) = 4.98 \times 10^{-12}\,(T/300)^{-0.63}\ cm^3\ s^{-1}$.

The rates of these reactions also depend on the abundance of CN, a molecule significantly destroyed by the neutral-neutral reaction with atomic nitrogen. The

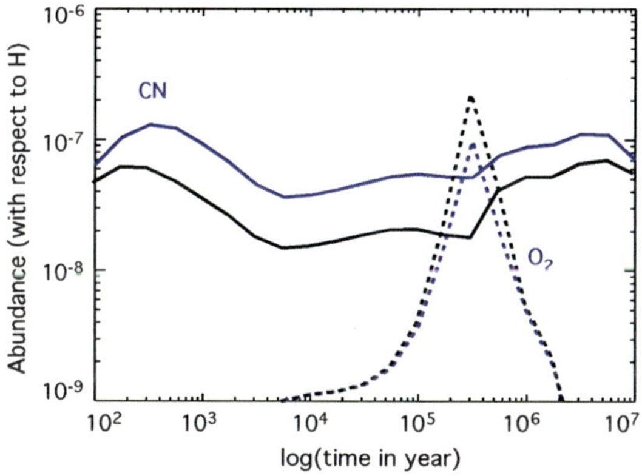

Fig 4.10 Gas-phase CN and O_2 abundances predicted by a gas-grain chemical model under dense cloud conditions (same as Fig. 4.7) and a C/O elemental ratio of 1.2, as a function of time. *Black* and *purple* curves were obtained for a C + CN rate coefficient of 3×10^{-10} cm^3 s^{-1} and 2×10^{-11} cm^3 s^{-1} at 10 K respectively

rate coefficient of the reaction N + CN has recently been revised. The value from the osu_01_2009 database [98] for N + CN $\rightarrow$ N_2 + C is 3×10^{-10} cm^3 s^{-1}, and is temperature independent. Some experimental measurements between 56 and 296 K have shown a decrease of the rate coefficient at low temperature according to the expression 8.8×10^{-11} $(T/300)^{0.42}$ cm^3 s^{-1}, which gives a rate coefficient of 2×10^{-11} cm^3 s^{-1} at 10 K by extrapolation [99]. Such a decrease of this rate coefficient has a strong effect on the abundance of CN, which is then increased, and as a consequence on O_2, which is decreased. This result is shown in Fig. 4.10, where the O_2 and CN gas-phase abundances are displayed, for the same dense cloud conditions as in the previous figures in this section, and for two N + CN rate coefficients at 10 K: the larger one from the osu database and the smaller one as extrapolated from the recent experiments (see also [93]). It can be seen in the figure that the O_2 abundance is now reduced to a maximum of 10^{-7} with the smaller rate coefficient, in better agreement with observations of cold cores. With a C/O elemental ratio of 0.5, the O_2 gas-phase peak abundance remains larger than 10^{-6} and is insensitive to the CN abundance.

4.5 Summary and Concluding Remarks

Chemical models are important for the analysis of observations of molecules in the interstellar medium and to make predictions for molecules that have not or cannot be directly observed. Modern models are able to compute the evolution of the chemical composition of a mixture of gas and dust taking into account a large number of processes including bimolecular gas phase reactions, interactions with cosmic-ray particles and UV (and X-ray) photons, interactions with interstellar grains, and grain-surface reactions. To account for all these processes, chemical networks contain thousands of reactions, each characterised by a rate coefficient associated with an uncertainty. Uncertainties in other model parameters, such as elemental abundances, temperature, density, etc., have also to be taken into account while discussing the accuracy of these models; in particular, when comparing with observations.

Depending on the astrophysical objects studied, the geometry and the physical dynamics may have to be included, making these models more and more complicated. With the start-up of a new and powerful interferometric telescope labelled ALMA, an acronym for Atacama Large Millimeter Array, with its high angular and spectroscopic resolution, and high sensitivity, the predictions of chemical models will be compared with much higher quality data. As an example, ALMA observations at high spatial resolution will resolve smaller structures in protostellar envelopes, which will rule out the use of spherical symmetry for the chemistry (see also Herbst [100]). For these reasons, these models have to be improved in the future. In addition to the coupling with better constrained physical structures, the chemistry itself has to be improved. Methods, such as sensitivity analysis, have been developed to identify the key processes in the gas phase and quantify the model accuracy. An interactive user-friendly database for gas-phase processes has even been created with the aim of improving the visibility of available data. The name of the database is KIDA, which is an acronym for KInetic Database for Astrochemistry[6]; this database is updated regularly. The most uncertain part of astrochemical models is probably all the processes related to surface reactions. Sensitivity analyses, as described in this chapter, are in principle applicable to the grain surface processes based on the rate equation method, but the number of parameters to study would be very large and drawing definitive conclusions not so easy. In addition, it is the nature of the processes themselves, rather than the parameters, that is more uncertain. Much progress has been made with experiments of surface chemistry but it remains difficult to use these experiments to improve the models because, unlike gas-phase experiments, the results of surface and ice experiments are not so easily converted from the laboratory to the very different conditions in the low-density ISM. The first step towards improvement could be the construction of a database to centralise all the information for surface reactions. The existence of this database might limit the multiplicity of gas-grain models based on different networks.

Acknowledgments V. W.'s research is supported by the French INSU/CNRS program PCMI, the Observatoire Aquitain des Sciences de l'Univers, and the Agence Nationale de Recherche (ANR-JC08−311018: EMA:INC). H.C. thanks the European Research Council (ERC-2010-StG, Grant Agreement no. 259510-KISMOL) and the Netherlands Organisation for Scientific Research (NWO) (VIDI) for financial support. E. H. acknowledges the support of the NSF (US) for his research program in astrochemistry and the support of NASA for his program in exobiology.

[6] http://kida.obs.u-bordeaux1.fr/

References

1. Bergin EA, Philipps TG, Comito C et al (2010) Herschel observations of EXtra-Ordinary sources (HEXOS): the present and future of spectral surveys with Herschel/HIFI. Astron Astrophys 521:L20. doi:10.1051/0004-6361/201015071
2. Dutrey A, Guilloteau S, Ho P (2007) Interferometric spectroimaging of molecular gas in protoplanetary disks. In: Reipurth B, Jewitt D, Keil K (eds) Protostars and planets V. University of Arizona Press, Tucson
3. Press WH, Teukolsky SA, Vetterling WT, Flannery BP (1996) Numerical recipes in Fortran 90. Cambridge University Press, New York
4. Le Bourlot J, Pineau des Forets G, Roueff E, Flower DR (1995) On the uniqueness of the solutions to the chemical rate equations in interstellar clouds: the gas-dust interface. Astron Astrophys 302:870–878
5. Wakelam V, Herbst E, Selsis F, Massacrier G (2006) Chemical sensitivity to the ratio of the cosmic-ray ionization rates of He and H_2 in dense clouds. Astron Astrophys 459:813–820. doi:10.1051/0004-6361:20065472
6. Draine BT (2003) Interstellar dust grains. Annu Rev Astron Astr 41:241–289. doi:10.1146/annurev.astro.41.011802.094840
7. Savage BD, Sembach KR (1996) Interstellar abundances from absorption-line observations with the hubble space telescope. Annu Rev Astron Astr 34:279–330. doi:10.1146/annurev.astro.34.1.279
8. Jenkins EB (2009) A unified representation of gas-phase element depletions in the interstellar medium. Astrophys J 700:1299–1348. doi:10.1088/0004-637X/700/2/1299
9. Draine BT (1990) Evolution of interstellar dust. In: The evolution of the interstellar medium; Proceedings of the conference, Berkeley, CA, June 21-23, 1989 (A91-55426 24-90). San Francisco, CA, Astronomical Society of the Pacific, 1990, p. 193–205
10. Draine BT (2009) Interstellar dust models and evolutionary implications. In: Henning T, Grün E, Steinacker J (eds) Cosmic dust – near and far ASP conference series proceedings of a conference held 8–12 September 2008 in Heidelberg, Germany
11. Hily-Blant P, Walmsley M, Pineau Des Forêts G, Flower D (2010) Nitrogen chemistry and depletion in starless cores. Astron Astrophys 513:A41. doi:10.1051/0004-6361/200913200
12. Maret S, Bergin EA, Lada CJ (2006) A low fraction of nitrogen in molecular form in a dark cloud. Nature 442:425–427. doi:10.1038/nature04919
13. van der Tak FFS, Boonman AMS, Braakman R, van Dishoeck EF (2003) Sulfur chemistry in the envelopes of massive young stars. Astron Astrophys 412:133–145. doi:10.1051/0004-6361:20031409
14. Wakelam V, Caselli P, Ceccarelli C, Herbst E, Castets A (2004) Resetting chemical clocks of hot cores based on S-bearing molecules. Astron Astrophys 422:159–169. doi:10.1051/0004-6361:20047186
15. Sofia UJ, Cardelli JA, Savage BD (1994) The abundant elements in interstellar dust. Astrophys J 430:650–666. doi:10.1086/174438
16. Wakelam V, Herbst E, Le Bourlot J, Hersant F, Selsis F, Guilloteau S (2010) Sensitivity analyses of dense cloud chemical models. Astron Astrophys 517:A21. doi:10.1051/0004-6361/200913856
17. Hassel GE, Herbst E, Bergin EA (2010) Beyond the pseudo-time-dependent approach: chemical models of dense core precursors. Astron Astrophys 515:A66. doi:10.1051/0004-6361/200913896
18. Hersant F, Wakelam V, Dutrey A, Guilloteau S, Herbst E (2009) Cold CO in circumstellar disks. On the effects of photodesorption and vertical mixing. Astron Astrophys 493:L49–L52. doi:10.1051/0004-6361:200811082
19. Crimier N, Ceccarelli C, Maret S, Bottinelli S, Caux E, Kahane C, Lis DC, Olofsson J (2010) The solar type protostar IRAS16293-2422: new constraints on the physical structure. Astron Astrophys 519:A65. doi:10.1051/0004-6361/200913112

20. Serena V, Collings MP, Dever JW, McCoustra MRS, Williams DA (2004) Evaporation of ices near massive stars: models based on laboratory temperature programmed desorption data. Mon Not Roy Astron Soc 354:1141–1145. doi:10.1111/j.1365-2966.2004.08273.x
21. Aikawa Y, Wakelam V, Garrod RT, Herbst E (2008) Molecular evolution and star formation: from prestellar cores to protostellar cores. Astrophys J 674:984–996. doi:10.1086/524096
22. Wakelam V, Herbst E, Loison JC, Smith IWM, Chandrasekaran V, Pavone B et al (2012) A kinetic database for astrochemistry (KIDA). Astrophys J Supplt Volume 199, Issue 1, article id. 21
23. Semenov D, Wiebe D, Henning Th (2006) Gas-phase CO in protoplanetary disks: a challenge for turbulent mixing. Astrophys J 647:L57–L60. doi:10.1086/507096
24. Herbst E, van Dishoeck EF (2009) Complex organic interstellar molecules. Ann Rev Astron Astr 47:427–480. doi:10.1146/annurev-astro-082708-101654
25. Garrod RT, Weaver SLW, Herbst E (2008) Complex chemistry in star-forming regions: an expanded gas-grain warm-up chemical model. Astrophys J 682:283–302. doi:10.1086/588035
26. Aikawa Y, Furuya K, Wakelam V et al (2011) Hydrodynamical-chemical models from prestellar cores to protostellar cores. In: The molecular Universe, Proceedings of the international astronomical union, IAU symposium Conference held in Toledo (Spain), June 2011
27. Prasad SS, Tarafdar SP (1983) UV radiation field inside dense clouds – Its possible existence and chemical implications. Astrophys J 267:603–609. doi:10.1086/160896
28. Prasad SS, Huntress WT (1980) A model for gas phase chemistry in interstellar clouds. II – Nonequilibrium effects and effects of temperature and activation energies. Astrophys J 239:151–165. doi:10.1086/158097
29. Gredel R, Lepp S, Dalgarno A, Herbst E (1989) Cosmic-ray-induced photodissociation and photoionization rates of interstellar molecules. Astrophys J 347:289–293. doi:10.1086/168117
30. Rimmer PB, Herbst E, Morata O, Roueff E (2012) Observing a column-dependent ζ in dense interstellar sources: the case of the horsehead nebula. Astron Astrophys 537:A7. doi:10.1051/0004-6361/201117048
31. Wagenblast R, Hartquist TW (1990) Ultraviolet pumping of molecular hydrogen in diffuse cloud shocks. Mon Not Roy Astron Soc 244:265–268
32. Loison JC, Halvick Ph, Bergeat A, Hickson KM, Wakelam V (2012) Review of OCS gas-phase reactions in dark cloud chemical models. Mon Not Roy Astron Soc Volume 421, Issue 2, pp. 1476–1484
33. Wakelam V, Herbst E, Selsis F (2006) The effect of uncertainties on chemical models of dark clouds. Astron Astrophys 451:551–562. doi:10.1051/0004-6361:20054682
34. Wakelam V, Smith IWM, Herbst E, Troe J, Geppert W, Linnartz H et al (2010) Reaction networks for interstellar chemical modelling: improvements and challenges. Space Sci Rev 156:13–72
35. Ruffle DP, Rae JGL, Pilling MJ, Hartquist TW, Herbst E (2002) A network for interstellar CO – The first application of objective reduction techniques in astrochemistry. Astron Astrophys 381:L13–L16. doi:10.1051/0004-6361:20011544
36. Rae JGL, Bell N, Hartquist TW, Pilling MJ, Ruffle DP (2002) Reduced networks governing the fractional ionisation in interstellar molecular clouds. Astron Astrophys 383:738–746. doi:10.1051/0004-6361:20011748
37. Semenov D, Wiebe D, Henning Th (2004) Reduction of chemical networks. II. Analysis of the fractional ionisation in protoplanetary discs. Astron Astrophys 417:93–106. doi:10.1051/0004-6361:20034128
38. Wiebe D, Semenov D, Henning Th (2003) Reduction of chemical networks. I. The case of molecular clouds. Astron Astrophys 399:197–210. doi:10.1051/0004-6361:20021773
39. Hasegawa TI, Herbst E, Leung CM (1992) Models of gas-grain chemistry in dense interstellar clouds with complex organic molecules. Astrophys J Suppl 82:167–195

40. Westley MS, Baragiola RA, Johnson RE, Baratta GA (1995) Ultraviolet photodesorption from water ice. Planet Space Sci 43:1311–1315
41. Öberg KI, Linnartz H, Visser R, van Dishoeck EF (2009) Photodesorption of ices. II. H_2O and D_2O. Astrophys J 693:1209–1218
42. Hasegawa TI, Herbst E (1993) New gas-grain chemical models of quiescent dense interstellar clouds – the effects of H_2 tunnelling reactions and cosmic ray induced desorption. Mon Not Roy Astron Soc 261:83–102
43. Herbst E, Cuppen HM (2006) Interstellar chemistry special feature: monte carlo studies of surface chemistry and nonthermal desorption involving interstellar grains. Proc Natl Acad Sci USA 103:12257–12262
44. Garrod RT, Wakelam V, Herbst E (2007) Non-thermal desorption from interstellar dust grains via exothermic surface reactions. Astron Astrophys 467:1103–1115
45. Charnley SB, Tielens AGGM, Rodgers SD (1997) Deuterated methanol in the orion compact ridge. Astrophys J Lett 482:L203
46. Lohmar I, Krug J (2006) The sweeping rate in diffusion-mediated reactions on dust grain surfaces. Mon Not Roy Astron Soc 370:1025–1033
47. Biham O, Furman I, Pirronello V, Vidali G (2001) Master equation for hydrogen recombination on grain surfaces. Astrophys J 553:595–603
48. Green NJB, Toniazzo T, Pilling MJ, Ruffle DP, Bell N, Hartquist TW (2001) A stochastic approach to grain surface chemical kinetics. Astron Astrophys 375:1111–1119
49. Stantcheva T, Shematovich VI, Herbst E (2002) On the master equation approach to diffusive grain-surface chemistry: the H, O, CO system. Astron Astrophys 391:1069–1080
50. Charnley SB (1998) Stochastic astrochemical kinetics. Astrophys J Lett 509:L121–L124
51. Charnley SB (2001) Stochastic theory of molecule formation on dust. Astrophys J 562: L99–L102. doi:10.1086/324753
52. Vasyunin AI, Semenov DA, Wiebe DS, Henning Th (2009) A unified monte carlo treatment of gas-grain chemistry for large reaction networks. I. Testing validity of rate equations in molecular clouds. Astrophys J 691:1459–1469
53. Lipshtat A, Biham O (2004) Efficient simulations of gas-grain chemistry in interstellar clouds. Phys Rev Lett 93(17):170601
54. Barzel B, Biham O (2007) Efficient simulations of interstellar gas-grain chemistry using moment equations. Astrophys J Lett 658:L37–L40
55. Du F, Parise B (2011) A hybrid moment equation approach to gas-grain chemical modeling. Astron Astrophys 530:A131. doi:10.1051/0004-6361/201016262
56. Caselli P, Hasegawa TI, Herbst E (1998) A Proposed modification of the rate equations for reactions on grain surfaces. Astrophys J 495:309–316. doi:10.1086/305253
57. Garrod RT (2008) A new modified-rate approach for gas-grain chemical simulations. Astron Astrophys 491:239–251. doi:10.1051/0004-6361:200810518
58. Garrod RT, Vasyunin AI, Semenov DA, Wiebe DS, Henning Th (2009) A new modified-rate approach for gas-grain chemistry: comparison with a unified large-scale monte carlo simulation. Astrophys J Lett 700:L43–L46
59. Cuppen HM, van Dishoeck EF, Herbst E, Tielens AGGM (2009) Microscopic simulation of methanol and formaldehyde ice formation in cold dense cores. Astron Astrophys 508:275–287
60. Hasegawa TI, Herbst E (1993) Three-phase chemical models of dense interstellar clouds – gas dust particle mantles and dust particle surfaces. Mon Not Roy Astron Soc 263:589–606
61. Fayolle EC, Öberg KI, Cuppen HM, Visser R, Linnartz H (2011) Laboratory H_2O:CO_2 ice desorption data: entrapment dependencies and its parameterization with an extended three-phase model. Astron Astrophys 529:A74
62. Vasyunin AI, Herbst E (2011) New chemical models for new era observations: a multiphase Monte Carlo model of gas-grain chemistry. In IAU symposium, vol 280 of IAU symposium Conference held in Toledo (Spain), June 2011

63. Öberg KI, van Broekhuizen F, Fraser HJ, Bisschop SE, van Dishoeck EF, Schlemmer S (2005) Competition between CO and N_2 desorption from interstellar ices. Astrophys J Lett 621:L33–L36
64. Acharyya K, Fuchs GW, Fraser HJ, van Dishoeck EF, Linnartz H (2007) Desorption of CO and O_2 interstellar ice analogs. Astron Astrophys 466:1005–1012
65. Bolina AS, Wolff AJ, Brown WA (2005) Reflection absorption infrared spectroscopy and temperature-programmed desorption studies of the adsorption and desorption of amorphous and crystalline water on a graphite surface. J Phys Chem B 109:16836–16845
66. Fraser HJ, Collings MP, McCoustra MRS, Williams DA (2001) Thermal desorption of water ice in the interstellar medium. Mon Not Roy Astron Soc 327:1165–1172
67. Green SD, Bolina AS, Chen R, Collings MP, Brown WA, McCoustra MRS (2009) Applying laboratory thermal desorption data in an interstellar context: sublimation of methanol thin films. Mon Not Roy Astron Soc 398:357–367
68. Collings MP, Anderson MA, Chen R, Dever JW, Viti S, Williams DA, McCoustra MRS (2004) A laboratory survey of the thermal desorption of astrophysically relevant molecules. Mon Not Roy Astron Soc 354:1133–1140. doi:10.1111/j.1365-2966.2004.08272.x
69. Öberg KI, Fayolle EC, Cuppen HM, van Dishoeck EF, Linnartz H (2009) Quantification of segregation dynamics in ice mixtures. Astron Astrophys 505:183–194
70. Matar E, Congiu E, Dulieu F, Momeni A, Lemaire JL (2008) Mobility of D atoms on porous amorphous water ice surfaces under interstellar conditions. Astron Astrophys 492:L17–L20
71. Watanabe N, Nagaoka A, Hidaka H, Shiraki T, Chigai T, Kouchi A (2006) Dependence of the effective rate constants for the hydrogenation of CO on the temperature and composition of the surface. Planet Space Sci 54:1107–1114
72. Fuchs GW, Cuppen HM, Ioppolo S, Romanzin C, Bisschop SE, Andersson S, van Dishoeck EF, Linnartz H (2009) Hydrogenation reactions in interstellar CO ice analogues. A combined experimental/theoretical approach. Astron Astrophys 505:629–639
73. Miyauchi N, Hidaka H, Chigai T, Nagaoka A, Watanabe N, Kouchi A (2008) Formation of hydrogen peroxide and water from the reaction of cold hydrogen atoms with solid oxygen at 10 K. Chem Phys Lett 456:27–30
74. Ioppolo S, Cuppen HM, Romanzin C, van Dishoeck EF, Linnartz H (2008) Laboratory evidence for efficient water formation in interstellar ices. Astrophys J 686:1474–1479
75. Oba Y, Watanabe N, Kouchi A, Hama T, Pirronello V (2010) Experimental study of CO_2 formation by surface reactions of non-energetic OH radicals with CO molecules. Astrophys J Lett 712:L174–L178
76. Ioppolo S, Cuppen HM, van Dishoeck EF, Linnartz H (2011) Surface formation of HCOOH at low temperature. Mon Not Roy Astron Soc 410:1089–1095
77. Ioppolo S, van Boheemen Y, Cuppen HM, van Dishoeck EF, Linnartz H (2011) Surface formation of CO_2 ice at low temperatures. Mon Not Roy Astron Soc 413:2281–2287
78. Bisschop SE, Fuchs GW, van Dishoeck EF, Linnartz H (2007) H-atom bombardment of CO_2, HCOOH, and CH_3CHO containing ices. Astron Astrophys 474:1061–1071
79. Öberg KI, Garrod RT, van Dishoeck EF, Linnartz H (2009) Formation rates of complex organics in UV irradiated CH3OH-rich ices. I. experiments. Astron Astrophys 504:891–913
80. Herbst E, Klemperer W (1973) The formation and depletion of molecules in dense interstellar clouds. Astrophys J 185:505–534. doi:10.1086/152436
81. Carty D, Goddard A, Kahler SPK, Sims IR, Smith IWM (2006) Kinetics of the radical-radical reaction, O(3PJ) + OH(X2P) $\rightarrow$ O_2 + H, at temperatures down to 39 K. J Phys Chem A 110:3101–3110
82. Watson WD (1973) The rate of formation of interstellar molecules by ion-molecule reactions. Astrophys J 183:L17–L20. doi:10.1086/181242
83. Pagani L, Olofsson AOH, Bergman P et al (2003) Low upper limits on the O_2 abundance from the odin satellite. Astron Astrophys 402:L77–L81. doi:10.1051/0004-6361:20030344
84. Goldsmith PF, Melnick GJ, Bergin EA et al (2000) O_2 in interstellar molecular clouds. Astrophys J 539:L123–L127. doi:10.1086/312854

85. Larsson B, Liseau R, Pagani L et al (2007) Molecular oxygen in the ρ Ophiuchi cloud. Astron Astrophys 466:999–1003. doi:10.1051/0004-6361:20065500
86. Liseau R, Larsson B, Bergman P, Pagani L, Black JH, Hjalmarson Å, Justtanont K (2010) $O^{18}O$ and $C^{18}O$ observations of ρ Ophiuchi A. Astron Astrophys 510:A98. doi:10.1051/0004-6361/200913567
87. Goldsmith PF, Liseau R, Bell TA et al (2011) Herschel measurements of molecular oxygen in orion. Astrophys J 737:96. doi:10.1088/0004-637X/737/2/96
88. Quan D, Herbst E, Millar TJ, Hassel GE, Lin SY, Guo H, Honvault P, Xie D (2008) New theoretical results concerning the interstellar abundance of molecular oxygen. Astrophys J 681:1318–1326. doi:10.1086/588007
89. Bergin EA, Melnick GJ, Stauffer JR (2000) Implications of submillimeter wave astronomy satellite observations for interstellar chemistry and star formation. Astrophys J 539: L129–L132. doi:10.1086/312843
90. Tielens AGGM, Hagen W (1982) Model calculations of the molecular composition of interstellar grain mantles. Astron Astrophys 114:245–260
91. Hollenbach D, Kaufman MJ, Bergin EA, Melnick GJ (2009) Water, O_2, and Ice in molecular clouds. Astrophys J 690:1497–1521. doi:10.1088/0004-637X/690/2/1497
92. Whittet DCB (2010) Oxygen depletion in the interstellar medium: implications for grain models and the distribution of elemental oxygen. Astrophys J 710:1009–1016. doi:10.1088/0004-637X/710/2/1009
93. Hincelin U, Wakelam V, Hersant F, Guilloteau S, Loison JC, Honvault P, Troe J (2011) Oxygen depletion in dense molecular clouds: a clue to a low O_2 abundance? Astron Astrophys 530:A61. doi:10.1051/0004-6361/201016328
94. Ilgner M, Nelson RP (2006) On the ionisation fraction in protoplanetary disks. I. Comparing different reaction networks. Astron Astrophys 445:205–222. doi:10.1051/0004-6361:20053678
95. Ceccarelli C, Hollenbach DJ, Tielens AGGM (1996) Far-infrared line emission from collapsing protostellar envelopes. Astrophys J 471:400–426. doi:10.1086/177978
96. Sims IR, Queffelec JL, Defrance A, Rebrion-Rowe C, Travers D, Bocherel P, Rowe BR, Smith IWM (1994) Ultralow temperature kinetics of neutral-neutral reactions. The technique and results for the reactions CN + O_2 down to 13 K and CN + NH_3 down to 25 K. J Chem Phys 100:4229–4241. doi:10.1063/1.467227
97. Feng W, Hershberger JF (2009) Reinvestigation of the branching ratio of the CN + O_2 reaction. J Phys Chem 113:3523–3527
98. Harada N, Herbst E (2008) Modeling carbon chain anions in L1527. Astrophys J 685:272–280. doi:10.1086/590468
99. Daranlot J, Hincelin U, Bergeat A, Costes M, Loison JC, Wakelam V, Hickson KM (2012) Elemental nitrogen partitioning in dense interstellar clouds. In: Proceedings of the National Academy of Science submitted, June 26, 2012 vol. 109 no. 26 10233-10238. doi: 10.1073/pnas.1200017109
100. Herbst E (2008) Chemistry in the ISM: the ALMA (r)evolution. The cloudy crystal ball of one astrochemist. Astrophys Space Sci 313:129–134. doi:10.1007/s10509-007-9639-9

Chapter 5
Planetary Atmospheres and Chemical Markers for Extraterrestrial Life

Lisa Kaltenegger

Abstract A decade of exoplanet research has led to surprising discoveries, from giant planets close to their star, to planets orbiting two stars, all the way to the first hot, confirmed rocky worlds with potentially permanent lava on their surfaces due to the star's proximity. Observation techniques have reached the sensitivity to explore the chemical composition of the atmospheres as well as physical structure of some detected exoplanets and to detect planets of less than 10 Earth masses (M_{Earth}) and 2 Earth radii, so called Super-Earths, among them some that may be habitable. To characterize a planet's atmosphere and its potential habitability, we explore absorption features in the emergent and transmission spectra of the planet that indicate the presence of biology. This Chapter discusses our strategy to characterize rocky exoplanets remotely, the basics underlying the concept of the Habitable Zone as well as chemical markers that indicate life through geological time.

5.1 Introduction

The current status of exoplanet characterization shows a surprisingly diverse set of giant planets. For a subset of these, some properties have been measured or inferred using radial velocity (RV) measurements, micro-lensing, transits, and astrometry. These observations have yielded measurements of planetary mass, orbital elements, planetary radii and during the last few years, physical and chemical characteristics of the upper atmosphere of some of the transiting planets. Specifically, observations of transits, that provide a radius estimate for the planet, combined with RV information, that provide a mass estimate for the planet, have provided estimates of the density of

L. Kaltenegger (✉)
Max Planck Institute for Astronomie (MPIA), Koenigstuhl 17, Heidelberg 69115, Germany

CFA, MS-20, 60 Garden street, Cambridge, MA 02138, USA
e-mail: kaltenegger@mpia.de

I.W.M. Smith et al. (eds.), *Astrochemistry and Astrobiology*, Physical Chemistry in Action, 145
DOI 10.1007/978-3-642-31730-9_5,

a subset of these planets, ranging from giant planets to rocky planets like Corot 7b [1] and Kepler 10b [2]. Due to a detection bias that provides higher sensitivity for close-in as well as massive planets, both leading exoplanet detection methods, RV and Transit, detect a large fraction of massive planets that orbit close to their host star. Due to the proximity to their host star, they receive high amounts of stellar irradiation and have subsequent high surface temperature (see e.g. [3, 4]). Direct Imaging Surveys on the other hand are currently most successful in detecting widely separated young hot planetary objects and have already detected several exoplanet candidates on wide orbits (see e.g. [5]).

Recent investigations of samples of high precision RV data have shown that between 20% and 50% of all sample stars exhibit RV variations indicating the presence of Super-Earths or ice giants [6, 7]. Among the hundreds of confirmed RV planets, already a few close-by, low mass RV planets like Gl 581 d [8], with minimum masses below 10 Earth masses, consistent with rocky planet models, orbit in the Habitable Zone (HZ) of their parent star. Exoplanets around close-by stars provide excellent targets for future atmospheric exploration with missions like the James Webb Space Telescope (JWST) and ground based telescopes like European Extremely Large Telescope (E-ELT).

RV searches and space based transit missions like ESA's Corot initially, and now NASA's Kepler mission, provide statistics of the occurrence of planets around Sun-like stars. NASA's Kepler telescope that monitors stars for planetary transits was launched in 2009 and is observing one distant stellar field monitoring about 150,000 stars continuously for 5 years with sensitivity for detecting transits down to Earth-size planets around Sun-analogue stars. Several Kepler transit planet candidates from the first data release in February 2011 [9] and about 50 planets from the February 2012 data release [10], with radii consistent with rocky planet models, orbit their host stars in the so called Habitable Zone (see discussion below), providing first statistics of the number of planets and Earth-like planets (etaEarth) in the HZ (see e.g. [11]).

The discovery of transiting planets with masses below 10 M_{Earth} and radii consistent with rocky planetary models answered the important question as to whether planets more massive than Earth could be rocky. 10 M_{Earth} and 2 Earth radii are used as estimates from planet formation theories as the upper limit for rocky planet mass and size. For comparison, Uranus has about 14.5 M_{Earth} and about 4 Earth radii. Above about 10 Earth masses a planet is thought to accumulate a substantial amount of gas that makes it akin to a gas giant with a substantial atmosphere, not a rocky planet with a thin outgassed atmosphere. Where exactly such a cut-off mass is that distinguishes rocky Super-Earths and gaseous Mini-Neptunes – if it exists at all – is an open question that mean density measurements of detected exoplanets currently explore.

Recent discoveries by ground based observations, as well as the Corot and Kepler space-missions, found planets with masses below 10 M_{Earth} and densities akin to Neptune as well as Earth, suggesting that there is not one cut-off mass above which a planet is like Neptune and below which it is rocky like Earth or Venus. Note that the term Mini-Neptune is used for small extrasolar giant planets, not mini-Uranus, even though Uranus is the less massive planet (17.1 and 14.5 Earth masses, respectively). The first planets below 10 M_{Earth} with both mass estimates and radius measurements have provided a wide range of densities (e.g., [12–14]).

Especially in the mass range below 5M_{Earth}, two planets in a multiple planet system, Kepler 11b and Kepler 11f [15], with 4.3 and 2.3 M_{Earth} have radii of 1.97 and 2.61 Earth radii and mean densities of 3.1 and 0.7 g/cm^3, respectively. These derived densities allow substantial envelopes of light gases for this mass range. For comparison, Neptune has a mean density of 1.6 g/cm^3, Earth a mean density of 5.5 g/cm^3. GJ 1214 b, the smallest transiting exoplanet found from the ground that allows for atmosphere observations, has 6.55 +/− 0.98 M_{Earth} with a radius of 2.1 Earth radii and a mean density of 1.8 g/cm^3 [16]. Atmospheric measurements (e.g. [17]) indicate hazes or high cloud cover that can block the transmitted light in such expanded planetary atmospheres (see e.g. [18]). Whether such cloud/hazes are in general common in the atmosphere of planets with low density and masses is another open question.

Observing mass and radius can distinguish between giant and rocky planets but alone cannot break the degeneracy of a rocky planet's nature due to the effect of an extended atmosphere that can block the stellar light and increase the observed planetary radius significantly from its rocky core value. Even if a unique solution would exist, planets with similar density, like Earth and Venus, present very different planetary environments in terms of habitable conditions. Therefore the question refocuses on atmospheric features to characterize a planetary environment. Designs of future space missions exist, that have the explicit purpose of detecting other Earth-like worlds, analysing their characteristics, determining the composition of their atmospheres, investigating their capability to sustain life as we know it, and searching for signs of life. They also have the capacity to investigate the physical properties and composition of a broader diversity of planets, to understand the formation of planets and interpret potential biosignatures.

In this Chapter we discuss how we can read a rocky planet's spectral fingerprint and characterize if it is potentially habitable. In Sect. 5.1 we explore the Earth as seen as an exoplanet, in Sect. 5.2 focus on low resolution biosignatures in the spectrum of an Earth-like planet, in Sect. 5.3 set the focus on the first set of measurements to characterize a potentially habitable planet. In Sect. 5.4 we discuss the concept of the Habitable Zone, Sect. 5.5 discusses the influence of the host star on the detectable features, Sect. 5.6 explores the spectral evolution of Earth through geological time, Sect. 5.7 the detectability of surface features like the vegetation red edge and Sect. 5.8 summarizes the chapter.

5.2 Characterizing a Habitable Planet (Learning from Earth)

In the coming years, ground-based as well as space missions will give us statistics on the number, size, period and orbital distance of planets, extending to terrestrial planets on the lower end of the mass range as a first step, while future space missions are designed to characterize their atmospheres. To explore characteristics

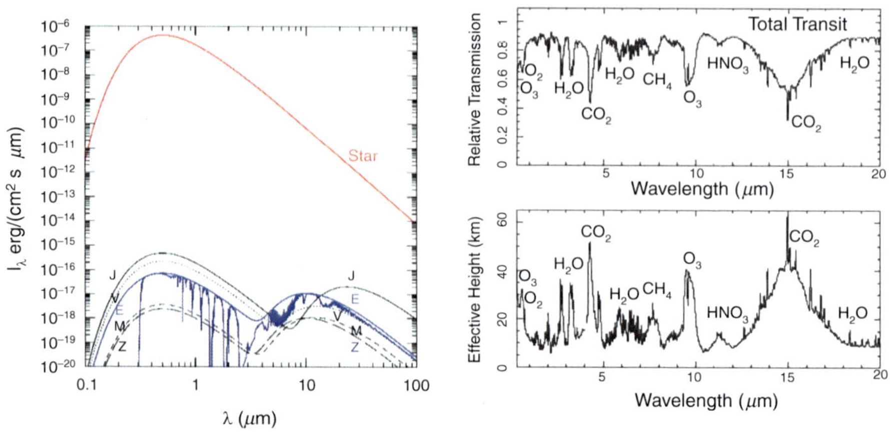

Fig. 5.1 Model of our Solar System (*left*) (assumed here to be Black Bodies) with Earth synthetic reflection and emission spectrum shown. Synthetic transmission spectra (*right*) of the Earth from VIS to IR shown. The intensity is given as a fraction of solar intensity as well as the relative height in the atmosphere. The atmospheric features are indicated [25]

of habitable exoplanets, we first look at our own planet, that is currently the only known planet to harbour life, and its remotely detectable indicators for life, that could be observed with telescopes over interstellar distances.

Different strategies exist to characterize a planet's atmosphere: direct detection resolves the planet and star individually, and transmission as well as secondary eclipse measurements subtract the stellar light from a combined star-planet detection. For directly imaged planets, in the visible part of the spectrum, we observe the starlight, reflected off the planet; in the thermal IR we observe the planet's own emitted thermal flux. An Earth-like, temperate planet is a very faint, small object close to a very bright and large object, its parent star.

For direct imaging or secondary eclipse measurements, where one observes the image of the planet, either in reflected light at visible wavelengths or its emitted flux at infrared wavelengths, the Earth-Sun intensity ratio is about 10^{-7} in the thermal infrared (at ~10 μm), and about 10^{-10} in the visible (at ~0.5 μm) (see Fig. 5.1). The contrast ratio of hot extrasolar Giant planets (EGP) to their parent stars is more favourable due to the higher temperature and larger size of the EGP while the contrast ratio of an Earth-analogue planet orbiting a cooler parent star is also much more favourable due to the cooler stellar temperature, making Earth-like planets around small stars very interesting targets for first generation planet characterisation missions.

In primary eclipse, the light of the star gets filtered through the planet's atmosphere, providing a transmission spectrum of the planet's atmosphere (see Fig. 5.1). The contrast ratio for a transmission spectrum is in a first approximation a constant fraction of the star's light over a wide wavelength range, where atmospheric absorption does not significantly block stellar transmission. For Earth, the transmission is roughly constant to first order from the visible to the infrared wavelengths. The deepest Earth atmosphere absorption feature is several 10 km, compared to the Earth's 6,375 km radius, thus indicating the measurement accuracy needed to detect atmospheric features in a transiting Earth's atmosphere. The transit of an Earth around a Sun-analogue only lasts about 12.5 h and occurs once a year. Therefore transmission spectroscopy of an Earth-like planet will have to be co-added for the first generation of space missions like JWST to provide enough signal to detect atmospheric absorption features. Earth-like planets around cooler stars will orbit closer to their host star and therefore more transits occur per Earth year but also each individual transit is shorter in accordance with Kepler's laws (see e.g. [19–21]). Hotter planets show deeper absorption features, because the atmospheric scale-height that is proportional to the depth of an absorption feature in the atmosphere of a transiting planet, scales with temperature and is inversely proportional to the mean molecular weight of the atmosphere, meaning that H/He atmospheres show deeper atmospheric features than dense atmospheres like Earth, assuming clear atmospheres.

Therefore rocky planets around cool stars, provide very interesting targets (see e.g. [22]), that can be probed for atmospheric components. Cool, low mass Main Sequence M dwarfs are also the most abundant stars in the galaxy, representing about 75% of the total stellar population. The spectrum of potentially rocky planets in the HZ of M stars, like Gl 581d (see e.g. [23] and references therein) are being used to design instruments and observation strategies that will allow us to explore the atmosphere of the first temperate rocky worlds in the near future.

Whether Earth-analogue planets around stars other than Sun-analogues exist is still an open question that will be one of the first questions we can explore with future space and ground-based missions that can characterize planetary atmospheres.

5.3 The Spectral Fingerprint of an Earth-Like Atmosphere

Spectra of the atmosphere of a planet contain information on the chemical composition of the atmosphere that allows the exploration of the planetary environment remotely. On Earth some atmospheric species exhibit noticeable spectral features in the planet's spectrum resulting directly or indirectly from biological activity: the main ones are O_2, O_3, in combination with CH_4, and N_2O (see Fig. 5.1). CO_2 and H_2O are in addition important as greenhouse gases in a planet's atmosphere and can be sources for high O_2 concentration from photosynthesis. Figure 5.1 shows the detectable atmospheric features of a habitable planet in its reflection, emission and transmission spectrum, using the Earth itself as a proxy for observations and model fits to data of spectra of the Earth.

The viewing geometry of a planet results in a different flux contribution of the overall detected signal from the bright day and dark night side. For primary eclipse, the terminator region generates the transmitted light for all wavelength ranges, for secondary eclipse the day-side generates the emitted and reflected light. For direct imaging the dayside generates the reflected light for the visible wavelength range, while both the planet's day and night-side regions contribute to the emitted flux in the thermal infrared. Therefore different wavelength ranges and observation geometries can probe different regions of a planet.

Visible combined with near-infrared wavelengths as well as infrared spectral regions contain the signature of atmospheric gases that can be observed with low resolution and can indicate habitable conditions and, possibly, the presence of a biosphere: CO_2, H_2O, O_3, CH_4, and N_2O in the thermal infrared, and H_2O, O_3, O_2, CH_4 and CO_2 in the visible to near-infrared (see e.g. [24–26], and references therein for detailed reviews).

Biosignatures is used here to mean detectable species, or set of species, whose presence at significant abundance strongly suggests a biological origin (e.g. the gas couples $CH_4 + O_2$, or $CH_4 + O_3$). It is their quantities, and detection along with other atmospheric species, in a certain context (depending, for instance, on the properties of the star and the planet) that points toward a biological origin. Sagan et al. [27] analysed a spectrum of the Earth taken by the Galileo probe as a direct image, searching for signatures of life and concluded that the large amount of O_2 and the simultaneous presence of a reducing gas like CH_4 traces are strongly suggestive of biology for a planet around a Sun-like star. O_2 or O_3 with a reducing gas like CH_4 are good biosignatures that can be detected by a low-resolution (VIS: Resolution ≥ 80, IR: Resolution ≥ 25) spectrograph. The presence or absence of these spectral features (detected collectively) will indicate similarities or differences with the atmospheres of terrestrial planets, and their astrobiological potential.

Note that the presence of biogenic gases such as $O_2/O_3 + CH_4$ may imply the presence of an active biosphere, but their absence does not imply the absence of life. Life existed on Earth before the interplay between oxygenic photosynthesis and carbon cycling produced an oxygen-rich atmosphere (see 5.6).

Our search for signs of life in exoplanets is based on the assumption that life produces the same gases as a result of metabolic processes (see Chaps. 6, 7 and 8 for a detailed discussion). Chapters 6, 7 and 8 discuss alternative chemistry for life in detail (see also [28]). Any advance in alternative life can be included at any time in models, once such organisms and their interaction with an atmosphere have been established. Life based on a different chemistry is not considered here because atmospheric signatures that indicate life for alternative life-forms are so far unknown. Therefore we assume here that extraterrestrial life would be similar to life on Earth in its use of the same input and output gases, and that it exists out of thermodynamic equilibrium as it does on Earth (see e.g. [29]).

5.4 Characterizing Planetary Environments

It is relatively straightforward to remotely ascertain Earth is a habitable planet, replete with oceans, a greenhouse atmosphere, global geochemical cycles, and life if one has data with high signal-to-noise and spatial and spectral resolution. The interpretation of observations of exoplanets with limited signal-to-noise ratio and spectral resolution as well as absolutely no spatial resolution, as envisioned for the first generation instruments, will be far more challenging and implies that we need to gather information on the planetary environment to understand what we will see.

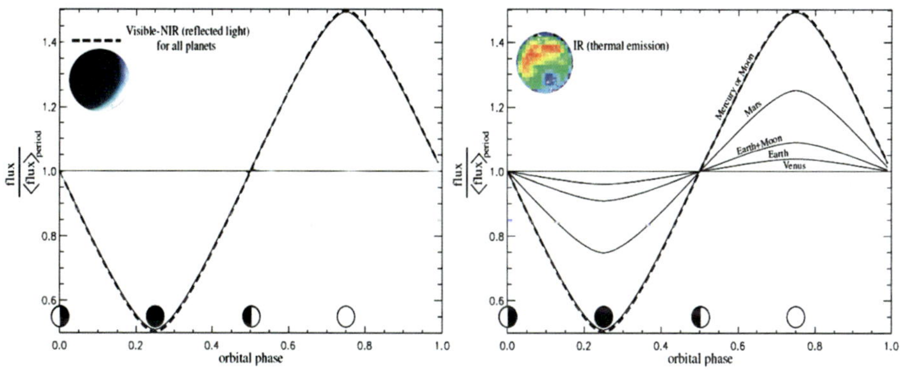

Fig. 5.2 Orbital light curve for black body planets in a circular orbit with null obliquities, with and without an atmosphere in the visible (*left*) and thermal infrared (*right*) [25]

After detection, we will focus on the main properties of the planetary system, first of all its orbital elements. The light curve of the planet in the IR can test the presence of an atmosphere by using the temperature distribution on the day and night side of the planet, which is substantial on a planet without an atmosphere but not on a planet with an atmosphere like Earth or Venus (see [30, 31] and references therein). The corresponding visible light curve variations are dominated by the change in visible reflecting surface area over the planet's orbit for both planets with and without atmospheres. The orbital flux variation in the IR can distinguish planets with and without an atmosphere in the detection phase (see Fig. 5.2) and prioritize targets for spectroscopic follow up.

Knowing the temperature and planetary radius is crucial for general understanding of the physical and chemical processes occurring on the planet (tectonics, hydrogen loss to space). A first estimate of the planetary effective temperature is obtained by calculating the stellar energy of the star that is received at the measured planet orbital distance and depends on the planetary albedo. The effective temperature is linked to the planet's surface temperature by atmospheric properties of the planet like chemical composition, cloud fraction versus height and overall pressure. The surface temperature of the planet at any distance depends on its albedo and on the greenhouse warming by atmospheric compounds. However, with a low resolution spectrum of the thermal emission, the mean effective temperature and the radius of the planet can be obtained.

The accuracy of the radius and temperature determination will depend on the quality of the spectral fit (and thus on the sensitivity and resolution of the spectrum), the precision of the Sun-star distance, the cloud coverage and also the distribution of brightness temperatures over the planetary surface. Assuming the effective temperature of our planet was due exclusively from radiation to the surface instead of an atmosphere layer (the average equilibrium temperature of Earth is about 265 K instead of the 288 K surface temperature), would only introduce a few percent error on the derived Earth radius.

The planet's radius can be estimated from thermal emission of the planet, because it is a function of the planet's temperature and surface area. A low resolution spectrum in the IR can be used to derive the effective temperature of the planet and therefore one can calculate the radiating area from the IR flux. The brightness temperature provides information on the effective temperature of the atmospheric layers responsible for the emission and can provide an idea of the surface temperature if the atmosphere is see through like on Earth and Mars.

The flux to derive the estimated radius can be measured at different points of the orbit, which should allow an estimate of the error made. When the IR light curve flux level changes significantly throughout the orbit, indicating planets without substantial atmospheres, radius determination is made difficult because most of the flux received comes from the small and hot substellar area. The ability to retrieve the radius for a planet without an atmosphere would depend on the assumptions that can be made about the orbit geometry, the rotation rate of the planet and heat capacity of the surface material. In most cases, multiple solutions will exist. Once visible flux is also measured, one can also estimate the Bond albedo of the planet, its reflectivity. In the visible ranges, the reflected flux allows us to measure the product of Bond albedo times planetary area (a small but reflecting planet appears as bright as a big but dark planet). If the planetary area is known, the Bond albedo can be derived.

Currently, radius measurements can only be performed for the fraction of planets that are geometrically aligned to our line of sight so that they transit their parent star. If the secondary eclipse of the transiting planet can also be observed (when the planet passes behind the star), then the visible reflected starlight and the thermal emission of the planet allow the values of the mean albedo and the mean brightness temperature to be retrieved, based on knowledge of the radius measurements from the primary transit. If only the mass of non-transiting planets can be measured (by radial velocity and/or astrometric observations), a first estimate of the radius can be made by assuming a bulk composition of the planet, that can then in turn be used to convert visible flux into albedo measurements.

As a next step, a crude estimate of the planetary nature can be derived, using very low-resolution information (three or four channels) (see e.g. [32]), then longer exposure times can build the signal to noise ratio (SNR) up to a low resolution spectrum, able to distinguish atmospheric absorption features. The resolution in the visible is currently suggested as 80 and in the IR as 25 to be capable of characterising the major atmospheric features in Earth's atmosphere.

The ability to associate a physical surface temperature to the spectrum relies on the existence and identification of spectral windows probing the surface. For an Earth-like planet there are some atmospheric windows that can be used in most of the cases, especially between 8 and 11 μm as seen in Fig. 5.1. Such identification is not trivial for non Earth-analogue atmospheres. Note that this window would however become opaque at high H_2O partial pressure (e.g. the inner part of the Habitable Zone (HZ) where a lot of water is vaporized) and at high CO_2 pressure (e.g. a very young Earth or the outer part of the HZ).

As a next step, a higher resolution spectrum can be used for interesting planetary targets to identify the compounds of the planetary atmosphere, constrain the temperature in the IR and radius of the observed exoplanet. In that context, we can test if we have an abiotic explanation of all compounds seen in the atmosphere of such a planet. If we do not, we can begin to consider the exciting biotic hypothesis.

5.5 The Concept of the Habitable Zone

The Habitable Zone is a concept only defined to detect life as we know it, remotely, and should be termed the Remotely detectable Habitable Zone, but the common use of Habitable Zone (HZ) is frequent. It is by no means meant as the exclusive zone where life around a star could exist. The definition of the HZ was driven by the possibility to remotely detect life as we know it. It is tied to the region around a star where water could be liquid on the surface of an Earth-analogue planet. This criterion is sometimes justified by liquid water being in our current state of knowledge, an essential part in the origins of life (see Chaps. 6, 7 and 8 for a detailed discussion). Subsurface water could provide similar conditions for life even though different energy sources would be needed than on the surface of a planet, but several options are available.

The essential argument for liquid water on the surface of a planet in the HZ is that liquid water on the surface of the planet should provide easy interaction of any gases that are produced by life with the planet's atmosphere. Such changes in the atmospheric composition and corresponding spectral features are observables that one can remotely detect and explore for signature of life over interstellar distances. To what extent subsurface life as well as life in an ocean under an ice layer could change a planet's atmosphere is currently under study. One example would be the case of a Europa-like exoplanet. We do not know currently if life could exist in Europa's oceans and have not yet detected any atmospheric gases that could indicate biota. Moving a Europa-analogue to orbit a star several light years away will not increase our ability to assess habitability on such objects remotely. Therefore the HZ is generally defined as the distance around a star, where water could be liquid on the surface of an Earth-analogue planet. Different effects like additional internal heat or other/additional effective greenhouse gases that are not part of Earth's atmosphere are not included in this model and would shift the limits of the HZ. Different aspects of what determines the boundaries of the HZ have been discussed broadly in the literature.

Habitability and the HZ are first order functions of the stellar flux at the planet's location as well as the planet's atmospheric composition. The latter determines the albedo and the greenhouse effect in the atmosphere. The inner and outer boundaries of the HZ differ for clear and cloudy conditions because the overall planetary albedo A, is a function of the chemical composition of the clear atmosphere as well as the fraction of clouds, $A = A_{clear} + A_{cloud}$.

The main differences among studies of the HZ are the imposed chemical composition and cloud fraction of the planet's atmosphere. Examples of atmospheres with different chemical compositions include the original $CO_2/H_2O/N_2$ model with a water reservoir (e.g., Earth's), or model atmospheres with high H_2/He concentrations [33] or limited water supply [34]. Two concepts are commonly used throughout the literature for cloud free [35] and cloudy atmospheres [36], assuming Earth-like planets. According to these models, the HZ is an annulus around a star where a rocky planet with a $CO_2/H_2O/N_2$ atmosphere and sufficiently large water content (such as in Earth) can host liquid water permanently on a solid surface.

This definition of the HZ makes several underlying assumptions: it assumes that the planet is rocky, water is present, the main atmospheric composition is $CO_2/H_2O/N_2$, and the abundance of H_2O and CO_2 in the atmosphere is regulated by a geophysical cycle similar to that of Earth, resulting in an H_2O- and CO_2-dominated atmosphere on the inner and the outer edges of the HZ, respectively. If the planet is not geologically active, it would not provide a feedback mechanism that can recycle atmospheric gases like CO_2 and stabilize the surface temperature against changes in stellar luminosity, like the carbon-silicate cycle on Earth. Therefore the HZ would reduce to a distance from the host star, where the stellar flux and the atmospheric composition of the planet could maintain liquid water on the planet's surface. During the evolution of the host star that brightens over time, no mechanisms are currently known that could stabilize the temperature of a geologically inactive Earth-like planet through geological times.

The limits of the HZ are calculated, assuming a H_2O- and CO_2-dominated atmosphere on the inner and the outer edges of the HZ, respectively, in accordance to the effects of the carbonate-silicate cycle. Between those limits on a geologically active planet, climate stability is provided by a feedback mechanism in which atmospheric CO_2 concentration varies inversely with planetary surface temperature. In this definition, the locations of the two edges of the HZ are determined based on the equilibrium temperature of the planet (see e.g. [35–37]). For the inner boundary of the Sun's HZ (Fig. 5.3), we consider the water-loss limit (surface temperature = 373 K) of the HZ of our Solar System for 0% and 50% cloud fraction. In this scenario, the value of limit is equal to an orbital distance of 0.95 Astronomical Units (AU) and 0.76 AU for 0% and 50% cloud fractions (f), respectively. The outer boundary denotes the distance from the star where the maximum greenhouse effect fails to keep CO_2 from condensing permanently, leading to runaway glaciation. For the outer boundary of the Sun's HZ, we consider the theoretical values of 1.67 and 1.95 AU corresponding to a cloud-free and 50% cloud-fraction CO_2 atmosphere (limits from [36]). We choose those model values because they correspond to the empirical limits based on the initial Solar flux received at the position of Venus and are slightly outside Mars' initial flux limits, with 0.75 and 1.77 AU, respectively (see [35] for details). Note that Mars' small size, did not allow this planet to maintain geological activity and a dense greenhouse atmosphere.

In this definition, the two edges of the HZ (see Fig. 5.3), depend on the Bond albedo of the planet, A, the luminosity of the star, the planet's semi major axis, D, as well as the eccentricity, e, of the orbit, and in turn the average stellar irradiation

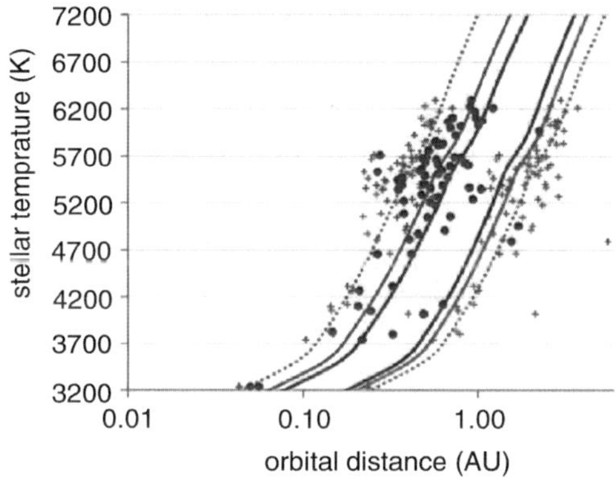

Fig. 5.3 Extent of the HZ for water loss limit for 0% and 50% cloud coverage (inner limits) and 100% cloud coverage (outer limit *dashed line*) and position of potentially habitable Kepler planetary candidates, Feb 2011 data release [9], in the HZ, individual HZ limits are indicated with *crosses* [37]

at the planet's location. A more eccentric orbit increases the annually averaged irradiation proportional to $(1 - e^2)^{1/2}$ [38]. At the limits of the HZ, the Bond albedo of a habitable planet analogue to a geologically active Earth is fully determined by its atmospheric composition and also depends on the spectral distribution of the stellar irradiation (see e.g. [36]).

However, the limits of the HZ are known qualitatively, more than quantitatively. This uncertainty is mainly due to the complex role of clouds and three-dimensional climatic effects not yet included in the modeling. Thus, planets slightly outside the computed HZ could still be habitable, while planets at habitable orbital distance may not be habitable because of their size or chemical composition. Subsurface life that could exist on planets with very different surface temperatures is not considered here, because of the lack of remotely detectable atmospheric criteria to assert habitability.

Applying this definition to the Kepler Feb 2011 data release (Fig. 5.3), assuming circular orbits and albedo corresponding to 50% cloud coverage (consistent with the empirical "Venus"-limit of the HZ), leads to 27 Kepler planetary candidates in the HZ. Among those are three planetary candidates with radii smaller than 2 Earth radii [9, 37]. The potentially rocky Kepler planet candidates in multiple systems are especially interesting objects because their mass could be determined using transit time variations to calculate a mean density and potentially confirm high density and rocky characteristics.

Due to the large mean distance to the Kepler stars and planets, the characterization of these planets atmospheres will not be explored in the near future, although it is an excellent mission to derive statistics. Several ground based searches as well as space based transit missions like TESS and PLATO, currently in design phase at NASA and ESA, respectively, are searching for rocky exoplanets in the solar neighborhood. Small planets detected around stars close to the Sun will provide the planetary targets that can be followed up with the next generation of ground and space based telescopes.

5.6 Influences on Planetary Spectra

5.6.1 *Influence of Host-Stars*

The range of characteristics of planets is likely to exceed our experience with the planets and satellites in our own Solar System by far. Models of planets more massive than our Earth – rocky Super-Earths – need to consider the changing atmosphere structure, as well as the interior structure of the planet. Also, Earth-like planets orbiting stars of different spectral type might evolve differently. Modeling these influences will help to optimize the design of the proposed instruments to search for Earth-like planets.

Using a numerical code that simulates the photochemistry of a wide range of planetary atmospheres several groups (see e.g. [39–42]) have simulated the atmospheric composition of a replica of our planet orbiting different types of star: F-type star (more massive and hotter than the Sun) and a K-type star (smaller and cooler than the Sun). The models assume similar background composition of the atmosphere as well as similar strength of biogenic sources. For an Earth-analogue planet around a spectral grid of host stars ranging in 6,250 K $< T_{eff} <$ 6,500 K at a resolution $\lambda/\Delta\lambda = 25$. Figure 5.4 shows the IR emergent modelled spectrum[1] for F, G, and K stars for clear sky emergent spectra as well as 60% global cloud coverage analogous to Earth (40% 1 km cloud layer, 40% 6 km cloud layer and 20% 12 km cloud layer). The spectra are presented relative to the planet's surface temperature black body.

The effect of clouds on the detectable spectra (see e.g. [43]) differs depending on wavelength. Clouds have high reflection in the visible and increase the albedo and therefore the overall detectable planetary flux. In the visible, the clouds themselves have different wavelength dependent albedos that further influence the overall shape of the spectrum and generally decrease the observable absorption features in the atmosphere because they block access to the lower atmosphere. In the IR, the cloud layers generally emit at lower temperatures than the surface would, decreasing the overall planetary flux, but can increase the absorption feature depth, because of the dependence of atmospheric absorption features in the IR on the temperature contrast between the absorbing/emitting layer and the continuum layer. Since on Earth clouds emit at temperatures generally colder than the surface, they can increase as well as reduce the depth of features in the IR.

Figure 5.4 shows that due to the hot stratosphere for all F stars (effective temperature, T_{eff}, greater than 6,000 K), the CO_2 absorption feature at 15 μm has a prominent central emission peak. The central peak can be thought of as an indirect feature indicating a temperature inversion in the atmosphere. The O_3 feature at 9.6 μm is increasingly difficult to resolve for hotter stellar types, despite increasing ozone abundance, due to the hotter stratosphere of the F stars. The CH_4 feature at

[1] An emergent spectrum is recorded using scattered starlight plus thermal emission from the planet's atmosphere.

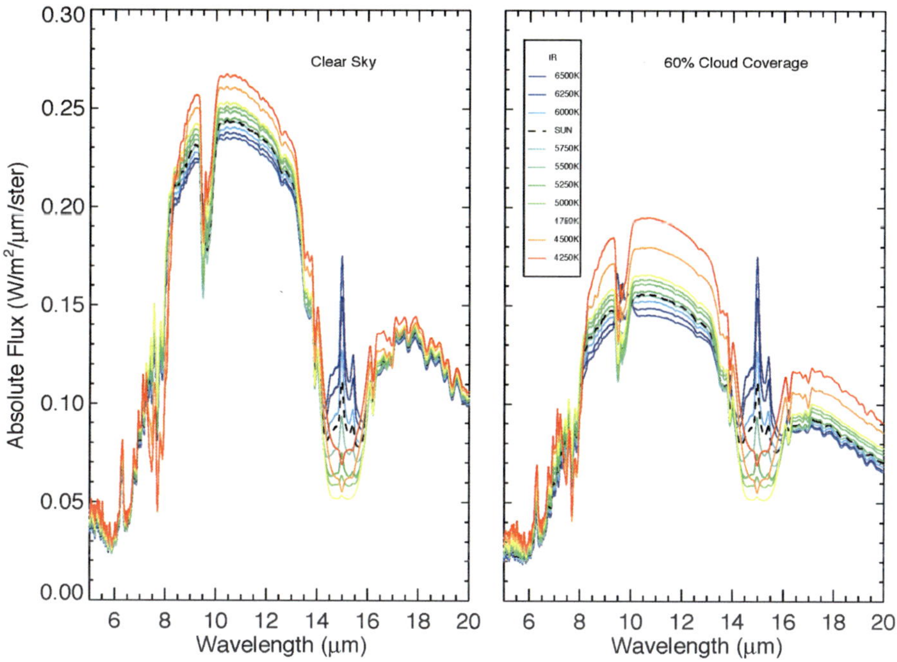

Fig. 5.4 Clear sky model (*left*) and 60 % cloud coverage (*right*) of the infrared spectral region at a resolution of 25 (corresponding to the proposed satellite designs) for an Earth-analogue planet around FGK stars

7.7 μm, while visible for F and G stars, is more prominent in the late K dwarfs (4,500 and 4,250 K) than in hotter stars due to its higher abundance from a lower UV environment. A planet orbiting a K star has a thin O_3 layer, compared to Earth's one, but still exhibits a deep O_3 absorption: because the low UV flux is absorbed at lower altitudes than on Earth which results in a less efficient warming (because of the higher heat capacity of the dense atmospheric layers). Therefore, the ozone layer is much colder than the surface and this temperature contrast produces a strong feature in the IR region. For hot F-type host stars the ozone layer is denser and warmer than the terrestrial one, decreasing the detectable IR feature. Thus, the resulting low temperature contrast produces only a weak and barely detectable feature in the infrared spectrum. This result is promising since G and K-type stars are much more numerous than F-type stars, the latter being rare and affected by a shorter stellar lifetime.

5.7 Evolution of Biosignatures over Geological Times on Earth

One crucial factor in interpreting planetary spectra is the point in the evolution of the atmosphere when its biosignatures and its habitability become detectable. The spectrum of the Earth has not been static throughout the past 4.5 Gyr. This is due to variations in the molecular abundances, the temperature structure, and the surface

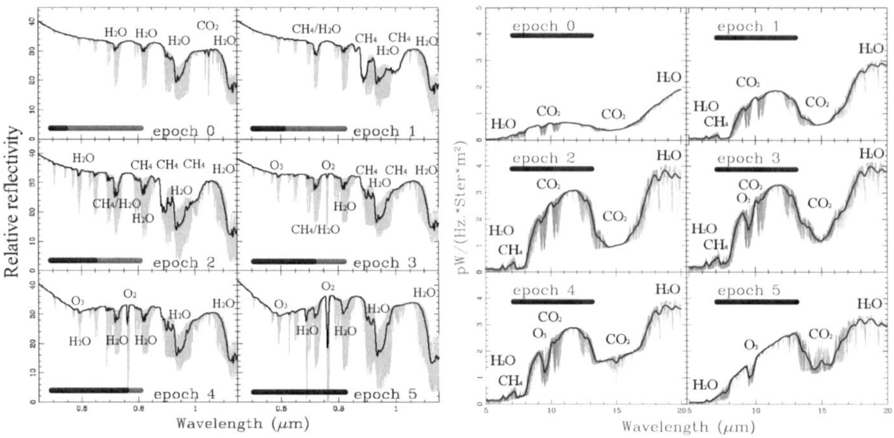

Fig. 5.5 The visible to near-IR (*left*) and mid IR (*right*) spectral features on an Earth-like planet change considerably over its evolution from a CO2 rich (epoch 0) to a CO2/CH4-rich atmosphere (epoch 3) to a present-day atmosphere (epoch 5). The *bold lines* show spectral resolution of 80 and 25 comparable to the proposed visible TPF and Darwin/TPF-I mission concept, respectively [43]

morphology over time. Earth's atmosphere has experienced dramatic evolution over 4.5 billion years (see e.g. [44, 45]), and other planets may exhibit evolution at similar or different rates.

The models shown here of the history of the Earth's atmosphere [43] are based on studies in the literature that include the concentration profiles of the spectrally most significant atmospheric molecules, as well the temperature profile, while considering the solar input and molecular oxygen concentration for specific scenarios for a 1 bar surface pressure. These studies are used to model observable spectra for Earth through geological time. The geological atmosphere model ranges from a CO_2-rich atmosphere (3.9 Gyr ago, epoch 0) to a CO_2/CH_4-rich atmosphere (epoch 3) to a present-day atmosphere (epoch 5, present-day Earth). We focus on long-lived periods in the Earth's history and we ignore relatively short-term events, such as glaciation events (the "snowball Earth") and their warm counterparts (the "hothouse Earth").

Figure 5.5 shows epochs that reflect significant changes in the chemical composition of Earth's atmosphere. The oxygen and ozone absorption features could have been used to indicate the presence of biological activity on Earth anytime during about the past 50% of the age of the solar system. At about 2.3 Gyr ago oxygen and ozone became abundant, affecting the atmospheric absorption component of the spectrum. At about 0.44 Gyr ago, an extensive land plant cover followed, generating the red chlorophyll edge in the reflection spectrum. Different signatures in the atmosphere are clearly detectable over Earth's evolution and observable with low spectral resolution (< 80).

The atmospheric composition, temperature-pressure profile as well as the composition of the surface, to the extent that it can be distinguished from clouds in the visible wavelength range, can all have a significant influence on the detectability of atmospheric and surface features. Figure 5.5 shows theoretical visible and mid-infrared spectra of the Earth at six epochs during its geological evolution.

The epochs are chosen to represent major developmental stages of the Earth, and life on Earth. If an extrasolar planet is found with a corresponding spectrum, we can use the stages of evolution of our planet to exploring it, in terms of habitability and the degree to which it shows signs of life. Furthermore we can learn about the evolution of our own planet's atmosphere and possible the emergence of life by observing exoplanets in different stages of their evolution.

To set our geological Earth atmosphere model in context with the overall Earth evolution we sketch out the conditions on Earth prior to epoch 0. The Earth formed about 4.5 billion years ago. The primitive atmosphere was formed by the release of volatiles from the interior, and/or volatiles delivered during the late bombardment period. Standard models of solar evolution predict that the Sun was 30% less luminous at 4.6 Gyr ago and has brightened monotonically since that time. Because of the faint young Sun, Earth's mean surface temperature should have been below the freezing point of seawater before about 2.0 Gyr ago if the Bond albedo was similar to today's, even if there were similar greenhouse contributions to the temperature (see, e.g. [46]). However, geological records tell us that liquid water was present by at least 3.5 Gyr ago and probably 4.0 Gyr ago. The oldest zircon crystals are as old as 4.4 Gyr, suggesting that liquid water formed even earlier than 4 Gyr ago. This argues for a higher abundance of greenhouse gases in the early atmosphere to keep the surface temperature above the freezing point of water.

Epoch 0 in our model is centred at about 3.9 Gyr ago. The atmosphere was most likely spectroscopically dominated by carbon dioxide that originated from volcanoes or the original volatile inventory, with nitrogen being the most abundant gas, and trace amounts of methane. Therefore, in our input model for this epoch we use 10% CO_2, current amounts of CH_4, and no O_2, O_3, or N_2O in the atmosphere.

Epoch 1 (about 3.5 Gyr ago) reflects the decrease of carbon dioxide and the rise of methane in the early atmosphere. Between epoch 0 and epoch 1, a significant amount of CO_2 must have been removed from the atmosphere, most likely by the transformation of carbon into carbonate rocks, although the process is still debated. The major influence of methane on the atmosphere may have begun almost as soon as life originated more than 3.5 billion years ago (see e.g. [47]). Methanogens are believed to have produced methane levels roughly 1,000 times that of today. CH_4 could have been quite abundant in an anoxic atmosphere. CH_4 has only a 10 year residence time today because it reacts with the hydroxyl radical, OH, and O^1D. In an anoxic atmosphere, OH and O^1D would have been much less abundant and CH_4 would have been destroyed in the upper atmosphere mainly by photolysis at Lyα wavelengths (121.6 nm). Under such conditions, its residence time should have been more like 10,000 year [48]. A biogenic CH_4 source comparable to the modern flux of 535 Tg CH_4 year^{-1}, which produces an atmospheric CH_4 concentration of 1.6 ppm (parts per million) today, could have generated over 1,000 ppm of CH_4 in the distant past. This is enough to have had a warming effect on climate (see e.g. [49, 50]). The atmosphere in epoch 1 consists mainly of N_2 and CO_2, with CH_4 becoming a major component. Therefore, for our model we use 1% CO_2, 0.2% CH_4, and no O_2, O_3, or N_2O in the atmosphere.

Epoch 2 (about 2.4 Gyr ago) reflects a maximum level of methane, a constant carbon dioxide concentration, and a small trace of oxygen in the early atmosphere.

The factor that limited the CH_4 abundance was likely the production of organic haze, which is predicted to form at a certain atmospheric CH_4/CO_2 ratio [49]. This haze would have created an "anti-greenhouse effect," which would have lowered surface temperatures and made life less comfortable for the predominately thermophilic methanogens, thus reducing the amount of CH_4 in the atmosphere. In our models we keep the CH_4/CO_2 ratio below unity. Primitive cyanobacteria are believed to have produced oxygen. At some point in Earth's history organisms evolved to perform photosynthesis. The oxygen produced from this reaction is responsible for most of the O_2 in Earth's present atmosphere [44]. Oxygen is toxic to methanogens. The atmosphere in epoch 2 consists mainly of N_2, about equal amounts of CO_2 and CH_4, and small amounts of oxygen. Therefore, for our model we use 1% CO_2, 0.7%CH_4, 0.02% O_2, and trace amounts of O_3 and N_2O in the atmosphere.

Epoch 3 is centred at about 2.0 Gyr ago. It reflects the rise of oxygen and decrease of methane in the early atmosphere. At some time between epochs 2 and 3, the Earth's atmosphere underwent a dramatic change (see [51] for an overview). From a variety of geological evidence, we know that significant concentrations of free O_2 began to appear in the atmosphere. This marked a sharp transition from basically anoxic to O_2-rich conditions. The introduction of O_2 into an anaerobic biosphere around 2.2 billion years ago must have represented a cataclysm in the history of life. Between epoch 2 and epoch 3, the abundance of oxygen rises in the atmosphere and Earth goes through a major glaciation event that is thought to be related to the drop in methane concentration in the atmosphere due to the rise of oxygen. Epoch 3 reflects a time after most of the reduced minerals were oxidized, and atmospheric oxygen started to accumulate in the atmosphere. The atmosphere in epoch 3 consists mainly of N_2, constant CO_2, and about equal amounts of CH_4 and O_2 (see e.g. [44] for an overview). Therefore, for our model atmospheres we use 1% CO_2, 0.4% CH_4, 0.2% O_2, and increasing trace amounts of O_3 and N_2O in the atmosphere. Global ice ages occurred at least three times in the Proterozoic era, first at 2.3 Gyr ago and again at 0.75 and 0.6 Gyr ago. The circumstances surrounding these glaciations were long unexplained, but the methane hypothesis provides compelling answers [47]. The rise in atmospheric O_2 corresponds with Earth's first well-documented glaciation, suggesting that the glaciation could have been triggered by the accompanying decrease in atmospheric CH_4.

Epoch 4 (about 0.8 Gyr ago) reflects a further rise of oxygen and further decrease of carbon dioxide in the atmosphere. Different schemes have been suggested to quantify the rise of oxygen and the evolution of life by anchoring the points in time to fossil finds. There are still many open questions. Carbon isotope data suggest that production of O_2 was occurring at rates comparable to today. Therefore, the sinks for O_2 must have been larger. If the deep oceans remained anoxic (and sulfidic) during most of the Proterozoic, then the atmospheric O_2 levels would have remained significantly lower than today, at least until 0.6–0.8 Gyr ago. We use this model for our atmosphere calculations. Epoch 4 reflects an increase in oxygen by a factor of 10 and a consequential decrease in CO_2 and CH_4. The atmosphere in epoch 4 consists mainly of N_2, 1% of CO_2, 0.04% CH_4, 2% O_2, and further increases in the trace species O_3 and N_2O.

Epoch 5 (about 0.3 Gyr ago to present-day Earth) reflects the present-day Earth's atmosphere, and also the influence of vegetation on our climate (see e.g. [52, 53]). The atmosphere consists mainly of N_2, with 0.0365% CO_2 and 21% O_2 as the second most abundant species, followed by present-day trace amounts of CH_4, O_3, and N_2O. We use this atmosphere profile to model our balloon and earthshine measurements. It shows an excellent fit to the data.

Using the geological model atmospheres, we calculate the Earth's spectra for six main geological epochs, shown in Fig. 5.5 for visible to near-infrared and thermal infrared. Major observable molecular species (H_2O, O_3, O_2, CH_4, CO_2, and N_2O) are labelled. The dark lines show a resolution of 80 in the visible and 20 in the thermal infrared, as proposed for the visible and infrared missions to characterize Earth-like planets.

Figure 5.5 shows that the changes in atmospheric signatures are detectable in both the visible and thermal infrared over Earth's evolution. In the visible to near-infrared one can see increasingly strong H_2O bands at 0.73, 0.82, 0.95, and 1.14 μm. These can be seen throughout the Earth's evolution. The strongest O_2 feature is the saturated Fraunhofer A band at 0.76 μm that can be clearly seen from epoch 3 to epoch 5. It is still relatively strong for significantly smaller mixing ratios than present Earth's, as seen in epoch 3 and epoch 4. A weaker feature at 0.69 μm cannot be seen with low resolution. O_3 has a broad feature, the Chappuis band, which appears as a broad triangular dip in the middle of the visible spectrum from about 0.45 to 0.74 μm that can be seen from epoch 3 to epoch 5. The feature is very broad and shallow and therefore requires a high signal to noise ration (SNR) for detection. Thus, in epoch 3 this feature is only marginally detectable.

Methane at present terrestrial abundance (1.65 ppm) has no significant visible absorption features, but at high abundance, as seen in epoch 1 to epoch 3, it has strong bands at 0.88 and 1.04 μm, readily detectable in early Earth's history. CO_2 has negligible visible features at present abundance, but in a high-CO_2 atmosphere of 10%, seen in the early evolution stage of epoch 0, the weak 1.06 μm band could be observed. In epoch 5 we can detect the red edge of land plants in our model. As land coverage occurred about 0.44 Gyr ago, this red edge cannot be observed before epoch 5.

In the thermal infrared, the classical signatures of biological activity are the combined detection of the 9.6 μm O_3 band in combination with the 7.66 μm CH_4 band. The 15 μm CO_2 band, and the 6.3 μm H_2O band or its rotational band that extends from 12 μm out into the microwave region indicate that a greenhouse effect warms the planet. These signatures can be detected from epoch 3 to epoch 5. O_3 is highly saturated and is thus an excellent qualitative but a poor quantitative indicator for the existence of even traces of the parent species (O_2). Ozone is a very nonlinear indicator of O_2 because the ozone column depth changes slowly as O_2 increases from 0.01 present atmosphere level (PAL) to 1 PAL. CH_4 is not readily identified in our present-day atmosphere using low-resolution spectroscopy (epoch 5), but the CH_4 feature at 7.66 μm in the thermal infrared is easily detectable for epoch 1 to epoch 4. There are three weak N_2O features in the infrared at 7.75, 8.52, and 16.89 μm. These features are strongly overlapped by CH_4, CO_2, and H_2O, so it is unlikely to become a prime target for the first generation of space-based missions searching for exoplanets that will work with low resolution, but it is a good target for follow-up missions because it is a promising biosignature.

Strong volcanism [54], as well as other geochemical cycles, could also be detected in a planet's spectrum (see e.g. [28, 55]). Such spectra will be used as part of a grid to characterize any exoplanets found and they influence the design requirements for a spectrometer to detect habitable planets.

5.8 Surface and Red Edge Features

While they efficiently absorb visible light, photosynthetic plants have developed strong infrared reflection (possibly as a defence against overheating and chlorophyll degradation) resulting in a steep change in reflectivity around 700 nm, called the red-edge. The primary molecules that absorb the energy and convert it to drive photosynthesis (H_2O and CO_2 into sugars and O_2) are chlorophyll A (0.450 μm) and B (0.680 μm). The exact wavelength and strength of the spectroscopic "vegetation red edge" (VRE) depends on the plant species and environment. On Earth around 440 million years ago, an extensive land plant cover developed, generating the red chlorophyll edge in the reflection spectrum between 700 and 750 nm. Several groups have measured the integrated Earth spectrum via the technique of Earthshine, using sunlight reflected from the non-illuminated, or "dark", side of the moon. Averaged over a spatially unresolved hemisphere of Earth, the additional reflectivity of this spectral feature is typically a few percent (see e.g. [56–58]).

Earthshine measurements have shown that detection of Earth's VRE is feasible if the resolution is high and the cloud coverage is known, but is made difficult owing to its broad, essentially featureless spectrum and cloud coverage. Space based measurements by the EPOXI mission have shown similar results [58, 59]. The high SNR produced by EPOXI, due to its proximity to Earth during the measurements, show that cloud and surface features can be distinguished with high enough SNR for Earth and in turn for Earth-like planets with second or third generation space missions that provide the collecting area needed for such high signal.

Our knowledge of the reflectivity of different surface components on Earth – like deserts, ocean and ice – helps in assigning the VRE of the Earthshine spectrum to terrestrial vegetation. Earth's hemispherical integrated vegetation red-edge signature is very weak, but planets with different rotation rates, obliquities, land-ocean fraction, and continental arrangement may have lower cloud-cover and higher vegetated fraction. Knowing that other pigments exist on Earth and that some minerals can exhibit a similar spectral shape around 750 nm [60], the detection of the red-edge of the chlorophyll on exoplanets, despite its interest, will not be unambiguous. If similar photosynthesis would evolve on a planet around other stellar types, the possible different types of a vegetation spectral signature have been modeled (e.g. [61]). Those signatures will be difficult to verify through remote observations as being of biological origin.

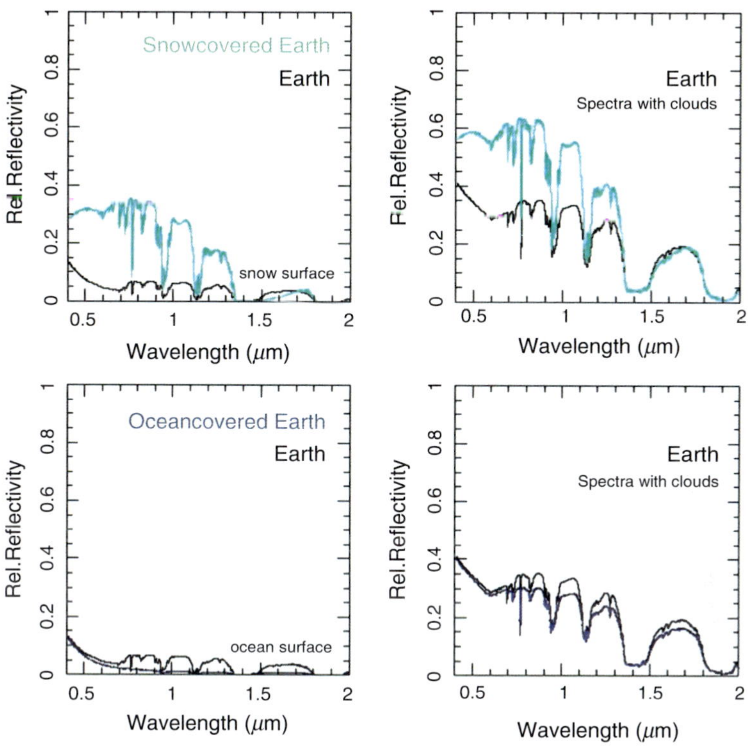

Fig. 5.6 Spectra of present-day Earth with a total ocean and snow cover without (*left*) and with (*right*) clouds for a disk averaged view. Note that the low albedo of the ocean reduces the overall flux while the high albedo of snow (as well as clouds) increases the planet's overall flux

Picking the most different reflecting surfaces (snow with a high albedo and sea with an extremely low albedo) shows the maximum effect surface coverage could have on the amount of light reflected from an exoplanet – here we assume that the whole planet surface is covered with that single material to explore if the maximum effect is detectable, the surface area is the same, and also artificially assuming similar cloud coverage and atmospheres for comparison (see Fig. 5.6).

If one could record the planet's signal with a very high time resolution (a fraction of the rotation period of the planet) and SNR, one could determine the overall contribution of clouds to the signal [62]. During each of these individual measurements, one has to collect enough photons for a high individual SNR per measurement to be able to correlate the measurements to the surface features, which precludes this method for first generation missions that will observe for a minimum of several hours to achieve a SNR of 5–10. For Earth [58, 59] such high SNR measurements show a correlation to Earth's surface features because the individual measurements are time resolved as well as having individual high SNR, making it a very interesting concept for future generations of missions.

5.9 Conclusions

Spectroscopy of the atmosphere of extrasolar planets allows us to remotely explore a planet's environment, to distinguish Mini-Neptunes from rocky Super-Earths, and to explore atmospheric compositions as well as searching for indications of habitability. Any information we collect on habitability is only important in a context that allows us to interpret what we find. To search for signs of life we need to understand how the observed atmosphere physically and chemically works. Knowledge of the temperature and planetary radius is crucial for general understanding of the physical and chemical processes occurring on the planet. These parameters as well as an indication of habitability can be determined with low resolution spectroscopy and low photon flux, as assumed for first generation space missions. Being able to measure the outgoing shortwave and longwave radiation, as well as their variations along the orbit, and to determine the albedo and identify greenhouse gases would, in combination, allow us to explore the climate system at work on other worlds.

The combination of spectral information in the visible (starlight reflected off the planet) as well as in the mid-IR (planet's thermal emission) allows a confirmation of atmospheric species, a more detailed characterization of individual planets but also to explore a wide domain of planet diversity. Ultimately future missions will allow us to probe planets similar to our own for atmospheric features indicating habitable conditions.

References

1. Léger A, Rouan D, Schneider J, Barge P, Fridlund M, Samuel B, Ollivier M, Guenther E et al (2009) Transiting exoplanets from the CoRoT space mission. VIII. CoRoT-7b: the first super-Earth with measured radius. Astron Astrophys 506:287
2. Batalha NM, Borucki WJ, Bryson ST, Buchhave LA, Caldwell DA, Christensen-Dalsgaard J, Ciardi D et al (2011) Kepler's first rocky planet: Kepler-10b. Astrophys J 729:27
3. Schaefer L, Fegley B (2009) Chemistry of silicate atmospheres of evaporating super-Earths. Astrophys J L703:L113
4. Miguel Y, Kaltenegger L, Fegley B, Schaefer L (2011) Compositions of hot super-Earth atmospheres: exploring Kepler candidates. Astrophys J L742:L19
5. Kalas P, Graham JR, Chiang E, Fitzgerald MP, Clampin M, Kite ES, Stapelfeldt K, Marois C, Krist J (2008) Optical images of an exosolar planet 25 light-years from Earth. Science 322:1345–1347
6. Bonfils X et al (2011) The HARPS search for southern extra-solar planets XXXI. The M-dwarf sample, A&A, 2011arXiv1111.5019B
7. Howard AW, Marcy GW, Bryson ST, Jenkins JM, Rowe JF, Batalha NM, Borucki WJ et al (2011) Planet occurrence within 0.25 AU of solar-type stars from Kepler, arXiv:1103.2541
8. Udry S, Bonfils X, Delfosse X, Forveille T, Mayor M, Perrier C, Bouchy F, Lovis C, Pepe F, Queloz D, Bertaux J-L et al (2007) The HARPS search for southern extrasolar planets XI. Super-Earths (5 & 8 M_Earth) in a 3-planet system. Astro Astrophy 469:43

9. Borucki WJ, Koch DG, Basri G, Batalha N, Brown TM, Bryson ST, Caldwell D et al (2011) Characteristics of planetary candidates observed by Kepler. II. Analysis of the first four months of data. Astrophys J 736:19
10. Batalha N et al (2012) Planetary candidates observed by Kepler, III: analysis of the first 16 months of data, arXiv:1202.5852
11. Traub WA (2011) Terrestrial, habitable-zone exoplanet frequency from Kepler. Astrophys J 745:20
12. Valencia D, Sasselov DD, O'Connell RJ (2007) Detailed models of super-Earths: how well can we infer bulk properties? Astrophys J 665:1413
13. Seager S, Kuchner M, Hier-Majumder CA, Militzer B (2007) Mass-radius relationships for solid exoplanets. Astrophys J 669:1279–1297
14. Grasset O, Schneider J, Sotin C (2009) A study of the accuracy of mass-radius relationships for silicate-rich and ice-rich planets up to 100 Earth masses. Astrophys J 693:722
15. Lissauer JJ, Ragozzine D, Fabrycky DC, Steffen JH, Ford EB, Jenkins JM, Shporer A, Holman MJ, Rowe JF et al (2011) Architecture and dynamics of Kepler's candidate multiple transiting planet systems. Astrophys J 197:8
16. Charbonneau D, Berta ZK, Irwin J, Burke CJ, Nutzman P, Buchhave LA, Lovis C, Bonfils X, Latham DW et al (2009) A super-Earth transiting a nearby low-mass star. Nature 462:891
17. Désert J-M, Bean J, Miller-Ricci Kempton E, Berta ZK, Charbonneau D et al (2011) Observational evidence for a metal-rich atmosphere on the super-Earth GJ1214b. Astrophys J 731:40
18. Miller-Ricci Kempton E, Zahnle K, Fortney JJ (2012) The atmospheric chemistry of GJ 1214b: photochemistry and clouds. Astrophys J 745:3
19. Kaltenegger L, Traub W (2009) Transits of Earth-like planets. Astrophys J 698:519
20. Rauer H, Gebauer S, Paris PV, Cabrera J, Godolt M, Grenfell JL, Belu A, Selsis F, Hedelt P, Schreier F (2011) Potential biosignatures in super-Earth atmospheres. I. Spectral appearance of super-Earths around M dwarfs. Astron Astrophys 529, id.A8 2011
21. Belu AR, Selsis F, Morales J-C, Ribas I, Cossou C, Rauer H (2011) Primary and secondary eclipse spectroscopy with JWST: exploring the exoplanet parameter space. Astron Astrophys 525, id.A83
22. Scalo J, Kaltenegger L, Segura AG, Fridlund M, Ribas I, Kulikov YN, Grenfell JL, Rauer H, Odert P, Leitzinger M, Selsis F, Khodachenko ML, Eiroa C, Kasting J, Lammer H (2007) M Stars as targets for terrestrial exoplanet searches and biosignature detection. Astrobiology 7:85
23. Kaltenegger L, Segura A, Mohanty S (2011) Model spectra of the first potentially habitable super-Earth – Gl581d. Astrophys J 733:35
24. Des Marais DJ, Harwit MO, Jucks KW, Kasting JF, Lin DNC, Lunine JI, Schneider J, Seager S, Traub WA, Woolf NJ (2002) Remote sensing of planetary properties and biosignatures on extrasolar terrestrial planets. Astrobiology 2:153–181
25. Kaltenegger L, Selsis F, Fridlund M, Lammer H, Beichman C, Danchi W, Eiroa C, Henning T, Herbst T, Léger A, Liseau R, Lunine J, Paresce F, Penny A, Quirrenbach A, Röttgering H, Schneider J, Stam D, Tinetti G, White GJ (2010) Deciphering Spectral Fingerprints of Habitable Exoplanets, Astrobiology 10(1):89–102
26. Meadows V, Seager S (2010) Terrestrial planet atmospheres and biosignatures. In: Seager S (ed) Exoplanets. University of Arizona Press, Tucson, pp 441–470, 526 pp. ISBN 978-0-8165-2945-2
27. Sagan C, Thompson WR, Carlson R, Gurnett D, Hord C (1993) A search for life on Earth from the Galileo spacecraft. Nature 365:715
28. Domagal-Goldman SD, Meadows VS, Claire MW, Kasting JF (2011) Using biogenic sulfur gases as remotely detectable biosignatures on anoxic planets. Astrobiology 11(5):419–441
29. Lovelock JE (1975) Thermodynamics and the recognition of alien biospheres. Proc R Soc Lond B Biol Sci 189(1095):167–180
30. Moskovitz NA, Gaidos E, Williams DM (2009) The effect of lunarlike satellites on the orbital infrared light curves of Earth-analog planets. Astrobiology 9(3):269–277

31. Selsis F, Wordsworth RD, Forget F (2011) Thermal phase curves of nontransiting terrestrial exoplanets. I. Characterizing atmospheres. Astron Astrophys 532, id.A1
32. Traub WA (2003) The colors of extrasolar planets, scientific frontiers in research on extrasolar planets. ASP Conf Ser 294:595–602
33. Pierrehumbert R, Gaidos E (2011) Hydrogen greenhouse planets beyond the habitable zone. Astrophys J 734:13L
34. Abe Y, Abe-Ouchi A, Sleep NH, Zahnle KJ (2011) Habitable zone limits for dry planets. Astrobiology 11(5):443–460
35. Kasting JF, Whitmire DP, Reynolds H (1993) Habitable zones around main sequence Stars. Icarus 101:108–119
36. Selsis F, Kasting JF, Levrard B, Paillet J, Ribas I, Delfosse X (2007) Habitable planets around the star Gliese 581? Astron Astrophys 476(3):1373
37. Kaltenegger L, Sasselov D (2011) Exploring the habitable zone for Kepler planetary candidates. Astrophys J 736:L25
38. Williams DM, Pollard D (2002) Earth-like worlds on eccentric orbits: excursions beyond the habitable zone. Int J Astrobiol 1:61
39. Selsis F (2000) Review: physics of planets I: Darwin and the atmospheres of terrestrial planets. In: Darwin and astronomy – the infrared space interferometer', Stockholm, 17–19 Nov 1999. ESA, Noordwijk, SP 451, pp 133–142
40. Segura A, Kasting JF, Meadows V, Cohen M, Scalo J, Crisp D, Butler RAH, Tinetti G (2005) Biosignatures from Earth-like planets around M Dwarfs. Astrobiology 5:706–725
41. Segura A, Krelove K, Kasting JF, Sommerlatt D, Meadows V, Crisp D, Cohen M, Mlawer E (2003) Ozone concentrations and ultraviolet fluxes on Earth-like planets around other Stars. Astrobiology 3:689–708
42. Grenfell JL, Stracke B, von Paris P, Patzer B, Titz R, Segura A, Rauer H (2007) The response of atmospheric chemistry on earthlike planets around F, G and K Stars to small variations in orbital distance. Planet Space Sci 55:661–671
43. Kaltenegger L, Traub WA, Jucks KW (2007) Spectral evolution of an Earth-like planet. Astrophys J 658:598–616
44. Kasting JF, Catling D (2003) Evolution of a habitable planet. Ann Rev Astron Astrophys 41:429
45. Zahnle K et al (2007) Geology and habitability of terrestrial planets, space sciences series of ISSI, vol 24. ISBN 978-0-387-74287-8. Springer Science+Business Media, LLC, p 35
46. Sagan C, Mullen G (1972) Earth and mars: evolution of atmospheres and surface temperatures. Science 177:52
47. Kasting JF, Siefert JL (2002) Life and the evolution of Earth's atmosphere. Science 296:1066
48. Pavlov AA, Kasting JF, Brown LL, Rages KA, Freedman R, Greenhouse R (2000) Greenhouse warming by CH_4 in the atmosphere of early Earth. J Geophys Res 105:981–992
49. Pavlov AA, Hurtgen MT, Kasting JF, Arthur MA (2003) Methane-rich proterozoic atmosphere? Geology 31(1):87
50. Mischna MA, Kasting JF, Pavlov A, Freedman R (2000) Influence of carbon dioxide clouds on early martian climate. Icarus 145:546
51. Holland R (2006) The oxygenation of the atmosphere and oceans. Philos Trans R Soc London B 361:903
52. Meadows V (2006) Modelling the diversity of extrasolar terrestrial planets. In: Aime C, Vakili (eds) Proceedings of IAU colloquium 200, direct imaging of exoplanets: science and techniques, Cambridge University Press, Cambridge, 25
53. Kaltenegger L, Henning W, Sasselov D (2010) Characterizing volcano planets. Astrophys J 140(5):1370–1380
54. Kaltenegger L, Sasselov D (2010) Detecting planetary geochemical cycles on exoplanets: atmospheric signatures and the case of SO_2. Astrophys J 708(2):1162–1167
55. Woolf NJ, Smith PS, Traub WA, Jucks KW (2002) The spectrum of earthshine: a pale blue dot observed from the ground. Astrophys J 574:430–442

56. Montanes-Rodriguez P, Palle E, Goode PR (2007) Measurements of the surface brightness of the earthshine with applications to calibrate lunar flashes. Astrophys J 134:1145–1149
57. Arnold L, Gillet S, Lardiere O, Riaud P, Schneider J (2002) A test for the search for life on extrasolar planets. Looking for the terrestrial vegetation signature in the earthshine spectrum. Astron Astrophys 392:231–237
58. Cowan NB, Agol E, Meadows VS, Robinson T, Livengood TA, Deming D, Lisse CM, A'Hearn MF et al (2009) Alien maps of an ocean-bearing world. Astrophys J 700:915
59. Livengood TA, Deming LD, A'Hearn MF, Charbonneau D, Hewagama T, Lisse CM, McFadden LA et al (2011) Properties of an Earth-like planet orbiting a sun-like Star: Earth observed by the EPOXI mission. Astrobiology 11:907
60. Seager S, Turner EL, Schafer J, Ford EB (2005) Vegetation's red edge: a possible spectroscopic biosignature of extraterrestrial plants. Astrobiology 5:372–390
61. Kiang NY, Siefert J, Govindjee, Blankenship RE (2007) Spectral signatures of photosynthesis. I. Review of Earth organisms. Astrobiology 7(1):222–251
62. Pallé E, Ford EB, Seager S, Montañés-Rodríguez P, Vazquez M (2008) Identifying the rotation rate and the presence of dynamic weather on extrasolar Earth-like planets from photometric observations. Astrophys J 676:1319–1329

Chapter 6
The Importance of Water

Philip Ball

Abstract All life on Earth needs water to survive, and special strategies are needed to cope with water scarcity, for instance because of extremes of either heat or cold. This situation has promoted the common view that water is a prerequisite for life in the universe as a whole, with important consequences for predictions about the likelihood of habitable environments. But we cannot assess that claim until we have a thorough understanding of the part that water *does* play in sustaining terrestrial life. In this chapter I will review the case for considering water to be a versatile, adaptive component of the cell that engages in a wide range of biomolecular interactions: for example, mediating protein-protein and receptor-substrate interactions, facilitating proton transport, driving hydrophobic interactions and their sensitivity to small solutes, acting as a reagent in biochemical reactions, and modulating electronic excitation energies. The chapter will aim to provide some basis for assessing water's often-alleged uniqueness as life's solvent. I conclude that, while we cannot with any confidence assert that all life must be aqueous, it is hard to identify any other solvent that could match the versatility and in particular the *responsiveness* of water in mediating the kind of molecular interactions likely to be required in any living system.

6.1 Introduction

Life on Earth is adapted to water. Although this statement is obviously true, it leaves a great deal unaddressed. Water has many properties that render it unusual among liquids; how has life adapted to these? Have those properties made adaptation easier or more complicated than it might be in other liquids? In what ways has adaptation proceeded, both at the molecular and the macroscopic

P. Ball (✉)
18 Hillcourt Road, East Dulwich, London SE22 0PE, UK
e-mail: p.ball@btinternet.com

I.W.M. Smith et al. (eds.), *Astrochemistry and Astrobiology*, Physical Chemistry in Action, 169
DOI 10.1007/978-3-642-31730-9_6,

(and indeed the geological and planetary) scales? Can life be simultaneously optimised to the many different features of water's behaviour, or is some compromise necessary? Can we say that water is well suited for life-even that it is uniquely suited for life?

For astrobiology, that last question is key. The common assumption, in searches for habitable worlds other than our own, is that water is a prerequisite. This remains an article of faith – although arguments can be made in its favour, and I will suggest a few. With disturbing frequency the claim is framed as a mere tautology: since no life on Earth can proceed (although some might be put on hold) without liquid water, it is assumed that water is a universal ingredient of life.

To ask whether water is indeed essential for life, or whether other liquids might usurp its role on other worlds, is not a modern question. The first person to consider this issue in detail was the Dutch scientist Christiaan Huygens in the late seventeenth century. Inspired, or perhaps provoked, by several speculations earlier in that century about habitation of the moon, such as Johannes Kepler's *Somnium* (1634), John Wilkins' *The Discovery of a World in the Moone* (1638), and Cyrano de Bergerac's *The States and Empires of the Moon* (1657), Huygens took a more hard-headed view of the idea in his posthumously published *Cosmotheoros* (1698), an astonishingly prescient work from today's perspective. "A Man that is of Copernicus's Opinion, that this Earth of ours is a Planet, carry'd round and enlighten'd by the Sun, like the rest of them, he wrote, cannot but sometimes have a fancy, that it's not improbable that the rest of the Planets have their Dress and Furniture, nay and their Inhabitants too as well as this Earth of ours". The discoveries since Galileo's time of moons around Jupiter and Saturn had made this position seem all the more compelling.

"As for the matter whereof the Plants and Animals there consist", Huygens went on, "we may venture to assert that their Growth and Nourishment proceeds from some liquid Principle." He suggested that the dark spots seen on the surface of Jupiter are likely to be clouds of condensed water vapour. Yet he added that "I can't say that [these planetary 'waters'] are exactly of the same nature with our Water .... For this Water of ours, in *Jupiter* or *Saturn*, would be frozen up instantly by reason of the vast distance of the Sun." In consequence, "Every Planet therefore must have its Waters of such a temper, as to be proportion'd to its heat: *Jupiter*'s and *Saturn*'s must be of such a nature as not to be liable to Frost; and *Venus*'s and *Mercury*'s of such, as not to be easily evaporated by the Sun." In other words, Huygens was in effect speculating that non-aqueous solvents served the life-sustaining roles of water on other planets.

But by the time the Harvard biochemist Lawrence Henderson considered the question of life in the universe in his book *The Fitness of the Environment* in 1913 [1], liquid water had acquired a special status as an essential precondition for life. Indeed, Henderson considered that water's apparently unique 'fitness' to act as life's matrix posed a profound question for considerations of cosmic design: why was water so well suited to this purpose? Henderson had in mind not water's highly unusual molecular-scale structure – for the hydrogen bond was not discovered until a few years later – but its unusual macroscopic properties such as high heat capacity and density anomalies. (Of course, these have their origins in water's highly unusual set of molecular characteristics.) Henderson did not answer the question he posed, and he was particularly resistant to teleological (let alone religious) answers to it. But the very act of framing the question tended to assert the validity of its assumptions: there is no other fluid but water that can provide the medium for life to arise and evolve.

With the advent of synthetic biology [2], along with chemical and biological systems for exploring 'alternative biochemistries' [3, 4], it is now conceivable that Henderson's assumption can be put to the test. We are by no means at that stage, however, and so far we must rely on speculation and hypothesis. As is not uncommon in such situations, views have become rather polarized. One camp posits a substantial list of prospective non-aqueous solvents (generally with much lower melting points than water ice) for alternative biochemistries, along with plausible-sounding arguments for why they might do at least as good a job. The other camp prefers to stay with the one fact that we are sure of – no known living organism can do without water – and make this the default position for considering life in the cosmos. Can we move beyond this dichotomy? To do so, we need at least to take a rather more sophisticated view of what it is that water does for life on Earth.

At least at the level of molecular biology, this issue has until recently been considered in a fairly simplistic fashion: water was perceived mostly as the uniform backdrop on which the molecular dramas of life are played out. But it has become increasingly clear over the past two decades or so that water is a substance that actively engages and interacts with biomolecules in complex, subtle and diverse ways. There is now good reason, for example, to regard the 'reach' of biological macromolecules such as proteins and nucleic acids as extending beyond their formal boundary by virtue of the way they shape and manipulate the shell of water that surrounds them. Moreover, the functions of such molecules depend on a delicate interplay between their own structures and dynamics and those of their aqueous environment. In such ways, the role of terrestrial life's solvent goes far beyond what has tended traditionally to be envisaged for a solvent, making it an active ingredient of the cell [5].

6.2 Water's Uniqueness?

How unique is biology's dependence on water on our planet? We do know that water can at least be *altered* without rendering life impossible. Thermophiles will cope with water hotter than 100 °C, halophiles with extreme salinity, and life in ocean trenches with water compressed to 1,000 bar. In all cases, the microscopic structure of the hydrogen-bonded liquid is somewhat different from that at the temperatures and pressures of the upper ocean. Beyond this, microbial life can adapt to deuterated water [6], and it is widely suspected that, given enough time to evolve, mammals would cope with heavy water too. Such experiences must force us to doubt that there is some 'magical' value of, say, the hydrogen-bond strength of water that makes life possible. Surely a defining feature of Darwinian evolution is its versatility in responding to the challenges and constraints of the environment. One might suspect that any system capable of such evolution will, once initiated, find a way to thrive in pretty much any environment, at least so long as it does not render conventional chemistry impossible by extremes of temperature.

The issue of water's uniqueness is often poorly posed. Some of the biological functions of water are simply the consequence of it being in the liquid state, and there is no reason to suppose that other liquids would not perform them equally well. For example, the turgor of plant cells – their inflation with liquid – contributes to the plant's mechanical stiffness. This turgor is a consequence of water uptake by osmosis, which is obviously a generic solvent property. Similarly, any liquid could in principle act as a transport and distribution medium in a vascular network.

Other properties depend on water's chemical behaviour. For example, it acts as an electron donor in photosynthesis. Not all liquids could fulfil a comparable role, but plenty could. Water's reactivity – its ability to act as a nucleophile – is exploited in many biochemical reactions. But that same characteristic makes it apt to hydrolyse biopolymers, posing one of the biggest conundrums in understanding how such polymers arose and survived for sufficient time to enable life on Earth to begin.

Water is an excellent solvent for a wide variety of species, not least for the ions that, coming from rocks and minerals on the prebiotic Earth, might plausibly have catalysed some proto-biotic reactions just as they are central to many enzymatic reactions today. This solvating power for charged species can be attributed to the screening made possible by water's large dielectric constant. On the other hand, water is in some ways not an ideal solvent for organic chemistry, as illustrated by its relative rarity in such a role in the lab. And water is by no means unique in solvating polar or charged species: several salts, including halides and cyanides, will dissolve to some extent in liquid ammonia, for example.

Water's ability to solvate ions is central to a number of biological phenomena. Not least, the genetic encoding of information in polymers depends on the solvation of charge. Because the repeating charges of the DNA backbone dominate its physical properties, it is relatively easy to change the information-bearing components of the molecule (the bases) without significantly perturbing the physical properties: DNA remains 'the same molecule' whatever message it bears [7]. But one can argue that the demands of aqueous solubility are sometimes a hindrance: the charges and polar groups needed on some metabolites to render them soluble may also make them somewhat unstable to thermal decomposition. The break-up of oxaloacetate, for example, is a problem for thermophiles [7].

Then there is water's celebrated hydrogen bonding (see also Chap. 1). As we will see, the versatility, directionality, geometry, lability and cooperativity of water's hydrogen bonding are crucial to a great many of its biological roles [5]. If a case were to be made for water's essentiality for life anywhere in the universe, hydrogen bonding would surely be the pivotal feature. While of course other liquids can engage in hydrogen bonding, including some such as formamide and ammonia that are plausible common ingredients of other worlds, none creates the delicate three-dimensional space-filling network of hydrogen bonds exhibited by water. Moreover, it is the hydrogen-bonded network that renders water ice less dense than the liquid, and which also creates the liquid density maximum at 4 °C at atmospheric pressure – characteristics which ensure that bodies of water freeze from the top down, so that an insulating layer of ice might keep the water below in a liquid state. Hydrogen bonding is also central to the formation of clathrate hydrates, which can trap volatile potential ingredients of prebiotic chemistry such as methane and other light hydrocarbons, preventing their escape and photolysis in the atmosphere.

But is hydrogen-bonding solely a 'good' attribute? It is the ability to hydrogen-bond that enables water to disrupt protein structure, for example by competing with intramolecular hydrogen bonds that help to hold the polypeptide backbone in its folded shape. Failure to protect these hydrogen bonds from water can make a globular protein prone to misfolding, denaturation and aggregation [8].

In short, it seems that most if not all of water's life-supporting attributes have a 'dark side'. Water is not some benign elixir; rather, life has had to evolve specific mechanisms for dealing with its limitations. Equally, these attributes are not all possessed by water alone – some are generic to liquids, some to polar liquids, some to hydrogen-bonded or protic liquids and so on. A common counter-argument is: maybe so, but only in water do all these properties coexist in a single substance. Such a proposition fails, however, to consider whether the properties in question must be essential in *any* chemical system one might designate as living, or whether in contrast they have been accommodated and even exploited in an opportunistic fashion by organisms evolving in an aqueous medium.

At present there seems to be no answer to that question. The purpose of the discussion here is merely to point out that the question exists. But it seems clear that, if we are to make any headway in addressing the issue of whether water is a unique solvent for life, we need to understand what functions it performs for life on Earth. Here the problem has long been not so much one of a lack of knowledge as of a presumption that there was nothing one needed to know. Most biochemistry textbooks, if they consider water in any detail at all, tend to imply that its life-supporting agency hinges on two main factors: (1) its excellent solvating power for polar and charged species; and (2) the existence of an attraction (assumed to be well-understood) between hydrophobic entities, which permits proteins to fold, enzymes to bind their substrates, and cell membranes to retain their integrity. Here I aim to show that both of these factors, while certainly crucial to the aqueous character of molecular biology, are more complex and in some respects still more mysterious than is often thought, and moreover that, in the 'life of the cell', water is involved in much else besides these roles [5]. My hope is that this will offer at least a broader perspective against which to assess the matter of water and life in the universe [9] – a question to which I return at the end of this chapter.

6.3 Water in the Cell

Most of water's 'oddness' as a liquid is well understood to be the result of its capacity to form hydrogen bonds [10]. Although there remains no complete consensus about how best to describe the molecular-scale structure of the liquid [11, 12], it is widely accepted that liquid water should be viewed as a rapidly fluctuating network of hydrogen bonds that link the molecules via approximately tetrahedral coordination (Fig. 6.1). In the conventional picture, each water molecule can in principle form hydrogen bonds to four others: two via the protons, two via the lone pairs on the oxygen atom. In ice this geometry is rigidly observed. In liquid water it is full of defects: the hydrogen bonds are being made and broken with an average lifetime of around 1 ps under physiological conditions [13], and the average number of hydrogen bonds per molecule is a little less than four because at any instant some will have bonds unsatisfied. Moreover, it is possible for the coordination number to exceed four, since hydrogen bonds can be bifurcated so as to link two molecules via a single hydrogen atom (Fig. 6.2) [14]. Although the hydrogen bond has a linear preferred geometry in which the hydrogen atom lies on the axis between the two oxygen centres, in the liquid (and in some high-pressure forms of ice) this arrangement may be distorted, which weakens the bonds. As a result of these defects, neighbouring molecules may approach more closely than they can in the ice lattice, and so there is less 'empty space' – the density of liquid water is greater than that of ice (more specifically, of the ice-I phase formed at 0 °C and ambient pressure).

The nature of water in the cell can be discussed in terms of how this three-dimensional liquid-state structure is perturbed by the presence of solutes and surfaces, such as globular proteins, salts and membranes. This perturbation is itself intrinsically dynamic, since the shapes of macromolecules and molecular assemblies fluctuate. The water that hydrates these biomolecular entities must also be reconfigured during biochemical reactions and interactions – for example, as two or more proteins join in particular unions, or as a protein binds its substrate.

While local changes to water structure have long been at least implicitly recognized to occur during the functioning of a cell, it is fair to say that they were commonly considered to be mere epiphenomena with few or no consequences for biological function, let alone as potential *driving forces* for that function. This assumption went hand in hand with the supposition that there is a clear distinction between the active biomolecules themselves and the passive 'hydration shells' around them. These hydration structures were typically considered to be made up of 'bound water' that remains in place when the protein is crystallized. Such hydration water molecules can often be precisely located by X-ray crystallography, and hydration water may constitute 30–50% of the mass of 'dry' protein powders. As we shall see, it is no longer tenable to adopt either a static or a time-averaged picture of the hydration shell, nor to consider it as a passive ingredient in biomolecular function. Biomolecules manipulate and are manipulated by their hydration environments, and that interaction has itself been shaped by (but not entirely by!) the exigencies of Darwinian adaptation. Water is in a real sense a *part of* most if not all biomolecules.

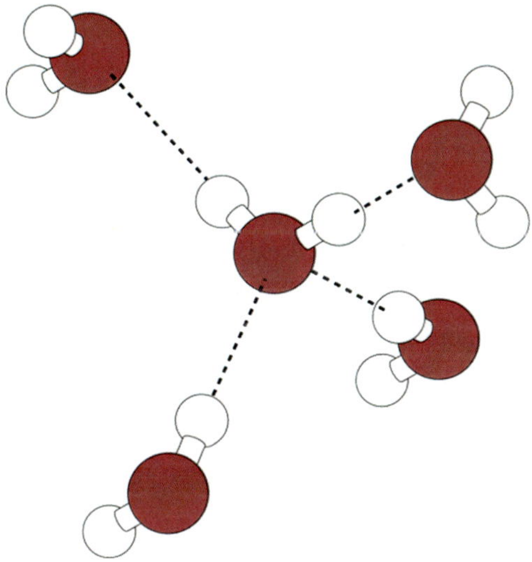

Fig. 6.1 The tetrahedral hydrogen-bonding motif in liquid water

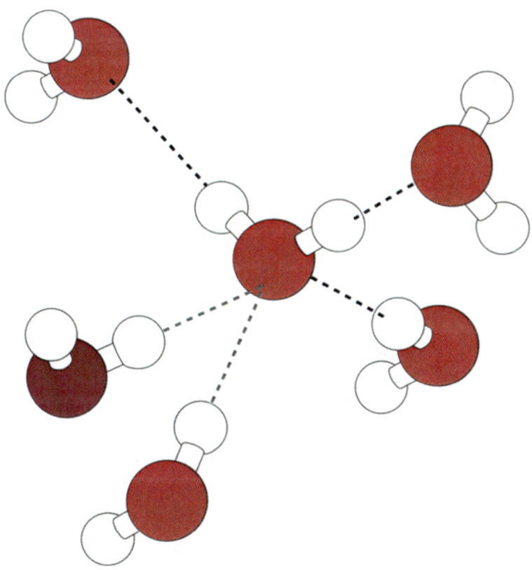

Fig. 6.2 A bifurcated hydrogen bond (*grey dashed line*) in water, which places five rather than four neighbouring molecules in the central water molecule's coordination sphere

This question of what biomolecules *do* to water in the cell is unavoidable, since there is probably rather little water that does not feel their presence. The cell is a crowded place (Fig. 6.3). Macromolecules typically occupy 5–40% of the total volume [15], so that on average they are typically separated by only 1–2 nm. This crowding may have a strong impact on the dynamics and interactions between

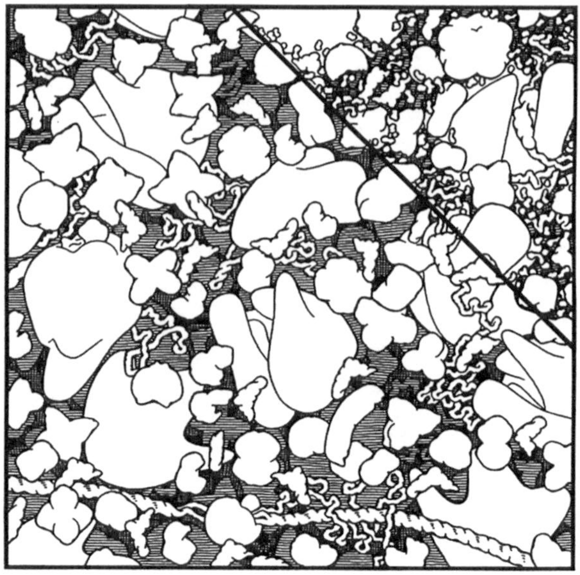

Fig. 6.3 The cell is a crowded place, as is evident in this scale drawing of the cytoplasm. Small molecules are shown only in the *upper right-hand corner* (From Goodsell D (1993) The machinery of life. Springer, New York. © Springer-Verlag)

biomolecules, for example slowing their diffusion but also increasing their binding affinities. Its effects are most probably 'allowed for' already in the adaptive evolution of protein molecules. But crowding also changes the nature of the solvent. Water in nano-confined spaces has properties different from those of the bulk liquid. The molecular mobility and the viscosity may be significantly below those of the bulk, [16, 17]. The phase behaviour is shifted depending on the geometry, separation and chemical nature of the confining walls [18] – whether they are hydrophilic or hydrophobic. Furthermore, crowding will perturb hydration relative to dilute solution, for example by potentially rendering some macromolecules imperfectly hydrated and liable to experience electrostatic interactions [19], with possibly strong effects on protein—protein binding.

The cell fluid (cytoplasm) is also inhomogeneous. Clustering of solutes has been observed even for small molecules such as methanol, and there is increasing evidence that at least some soluble proteins in such concentrated solution form relatively long-lived clusters [20, 21].

But is the fundamental, molecular nature of water itself transformed inside cells? Most biochemistry textbooks barely even consider that question – they just assume that, from the point of view of the thermodynamics and kinetics of biomolecules, one can treat the cell as a dilute solution, as if mixed up in a test tube. On the other hand, some have argued that the cytoplasm is like a gel, with the water itself acting like a sluggish fluid [22]. It has been claimed that water becomes more like the bulk liquid in diseased cells, such as cancer cells. The notion often advanced is then that the cell somehow 'tames' bulk water, making it 'more structured' in some way and thereby rendering it 'biophilic', conducive to life [23]. But there doesn't seem to be any good evidence for this. NMR and neutron-scattering studies of water dynamics in *E. coli* showed that most of the cell water has bulk-like dynamics, with perhaps 15% or so retarded by an order of magnitude [24]. This latter slow component is thought to correspond to water in biomolecular hydration shells that interacts directly with their surfaces. So most of the cell water is just like the bulk, at least in terms of its rotational dynamics.

The picture that emerges is therefore complicated. Strongly confined water is not expected to be bulk-like, and water close to interfaces can clearly have a quite different structure and dynamics to that in the bulk. But according to some measures, at least, water in the cell is not somehow transformed into a different sort of liquid. These facts are not inconsistent. Rather, they reinforce the importance of abandoning any viewpoint that posits a different *global* state for cell water – that, for example, insists on some kind of pervasive structural transformation of the hydrogen-bonded network. Instead we must consider how water responds *locally* to the presence of biomolecules, surfaces and/or small-molecule and ionic solutes. There seems currently to be no good reason to suppose that such influences extend beyond a few molecular diameters. Moreover, their effects will depend in subtle ways on the nature of solute hydrophobicity/hydrophilicity and the presence of electrical charge or of hydrogen-bonding moieties. And they will be manifested in dynamical as well as structural terms: whether water is 'altered' may depend on the timescale we are considering, and may be more apparent in fluctuations than in equilibrium averages. We must also take into account how various solutes and interfaces affect each other – whether, for example, ions are excluded from or attracted to an interface, or whether other small molecules interact directly with surfaces. Any one-size-fits-all theory of 'cell water' is almost certain to be incomplete at best and severely misleading at worst.

6.4 The Hydrophobic Effect and Its Role in Protein Folding

One of the key concepts in the interactions of biological molecules with and within water is the hydrophobic effect, which loosely characterizes the tendency of hydrophobic particles and surfaces to aggregate in aqueous solution. The phenomenon is well attested [25]. Proteins typically bury their hydrophobic residues in their interior as they fold. Hydrophobic groups on ligands are generally juxtaposed to similar groups at the surface of an enzyme's binding site. Proteins themselves associate into larger aggregates – whether, for example, as functional assemblies in the interactome, or fibrillar misfolded structures in neurodegenerative diseases – by marrying up their hydrophobic surfaces. Hydrophobic interactions drive the aggregation of lipids into membranes. Yet there is still no consensus on how these hydrophobic interactions operate [26].

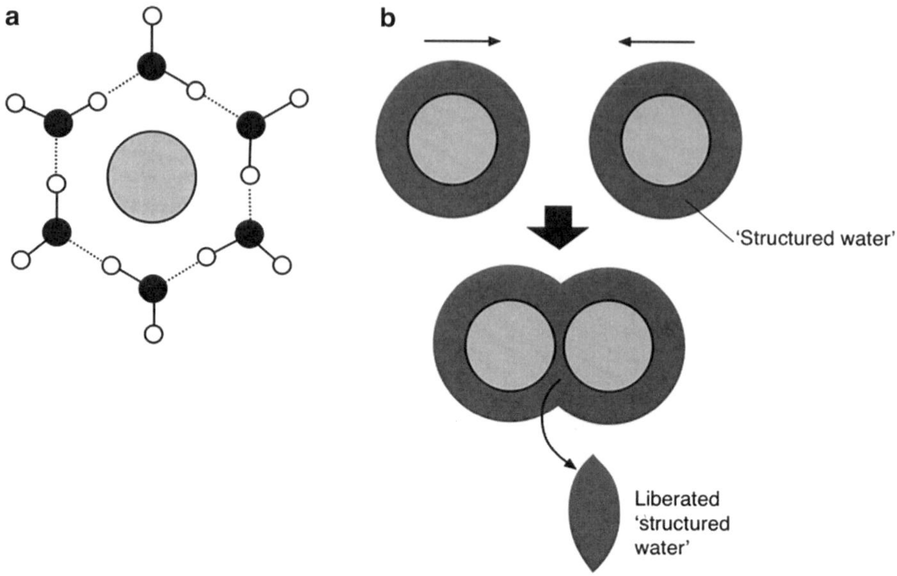

Fig. 6.4 (**a**) The 'iceberg' model of hydrophobic hydration, with a hydrophobic particle (*light grey*) surrounded by an ordered shell of water molecules. (**b**) Kauzmann's explanation of the hydrophobic interaction as an entropic effect due to the liberation of structured water in the hydrophobes' hydration shells [28]

It makes intuitive sense that a protein should tend to bury its hydrophobic side-chains. But what actually drives this apparent attraction between hydrophobic surfaces? The textbook answer draws on an idea proposed in 1945, which says that the structure of water actually becomes more orderly next to a hydrophobic surface [27]. It becomes, in other words, more like ice. This enables the water to arrange itself in a way that minimizes the loss of hydrogen bonds where the water ends and the hydrophobic surface begins. The idea, then, was that water builds a little iceberg around dissolved hydrophobic particles.

In 1959 Kauzmann suggested how this picture could lead to an attraction between hydrophobic particles in water [28]. As two particles come together, their coatings of ordered water overlap, and some of this 'structured' water is liberated into the less ordered liquid state (Fig. 6.4). So there is a gain in entropy, which makes the process favourable. According to this explanation, the hydrophobic attraction has an entropic origin.

The problem with Kauzmann's model is that there now seems to be no good evidence that it is right, and rather good reason to believe that it is wrong [26]. Specifically, it seems very unlikely that water is indeed more 'crystalline' around hydrophobic surfaces. It probably does have a different structure from that of bulk water, but it is not like a layer of ice, nor does it have some other kind of rigid, enhanced order like that of a clathrate. Rather, it appears that water molecules tend to orient their hydrogen bonds tangential to a hydrophobic surface in order to

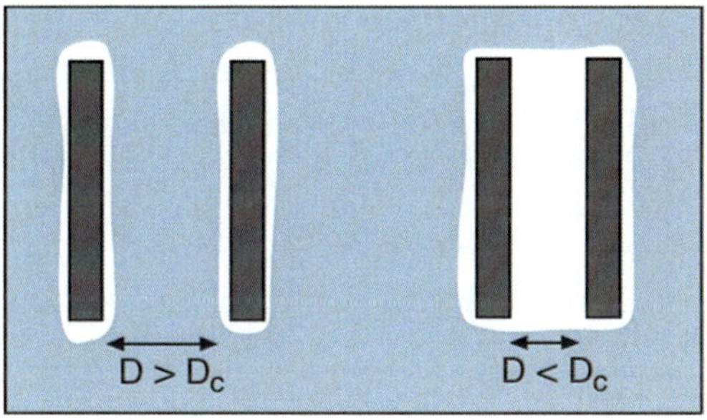

Fig. 6.5 Hydrophobic attraction in the model of Lum et al. [30]. The hydrophobic surfaces are surrounded by a thin layer of vapour (or a region of depleted water density due to fluctuations). At some critical separation D_c there is a collective drying transition in the space between the surfaces

preserve them, but remain able to exchange rapidly with neighbouring molecules more or less as in the bulk liquid. It also seems that there is a very thin layer right next to a hydrophobic surface in which the water density is significantly reduced. The width and density of this depletion layer have been much contested, but there is now a wide consensus that it is not gas-like, most probably dynamic ('flickering' between dry and wet regions) and extends perhaps just one molecular diameter or so into the liquid phase [29].

If the hydrophobic interaction is not simply an entropic effect of expulsion of 'bound' water, from where does it arise? One popular explanation argues that the hydration of small hydrophobes is qualitatively different from that of large ones [30]. In the former case, water molecules in the hydration shell might be able to adapt their orientation so as to preserve hydrogen-bonding around the cavity represented by the hydrophobic particle, while for larger, extended surfaces, breaking of hydrogen bonds is unavoidable, and the picture is then dominated by the interfacial energies. When two such surfaces come close together, at some point the water remaining between these depletion layers simply evaporates. This effect, called capillary evaporation, has a solid theoretical foundation – it means that, in effect, the boiling point of the liquid between the two surfaces is altered by confinement. If the gap between the surfaces becomes dry, the surface tension of the menisci at the edges of the space will pull the two surfaces into contact (Fig. 6.5). This is called a dewetting transition, and Lum et al. predict that it might be common between two hydrophobic surfaces when they are just 1 nm or so across [30] a length scale relevant to biological macromolecules.

Computer simulations of flat hydrophobic plates do show an abrupt dewetting transition [31]. And dewetting has also been predicted as a hydrophobic polymer folds [32], pulling the chain into a collapsed conformation. There is some experimental evidence in support of this claim. Li and Walker [33] have used an atomic-force-microscope tip to unravel a collapsed hydrophobic polymer in water one monomer at a time, and find that for polymers with large (~1 nm) monomers there is a maximum free energy of hydration at which the hydration energy changes from positive to negative – the length-scale crossover predicted by Lum et al. [30], which decreases with temperature from around 3.5 Å at 150 °C to 11.4 Å at 48 °C.

But it now looks as though dewetting is probably rather rare as a way of getting many real hydrophobic surfaces to stick together, especially those with the 'compromised' hydrophobicity typical of proteins. Computer simulations of the small protein melittin, a component of bee venom, do show a dewetting transition as the four monomers cluster to form a roughly cylindrical assembly (Fig. 6.6) [34]. On the other hand, the roughly flat faces of the two subunits of the enzyme BphC adhere to one another without sudden dewetting – the water between the two surfaces is squeezed out only gradually, molecule by molecule [35]. A survey of the protein data bank suggests that melittin is an unusual but not a unique case: dewetting is rather rare, but does happen in several other cases [36]. Just a few polar residues interspersed in the hydrophobic region of the molecules (a common situation) are generally enough to ensure that dewetting does not happen. The transition is also sensitive to the topography (roughness) of the surfaces [37] and probably to the geometry of the associated state: melittin subunits enclose a tube-like space, while that for BphC is slab-like. Many protein surfaces may simply not be hydrophobic enough, or geometrically conducive, to make dewetting happen.

But this does not mean that it is of only marginal relevance to biochemistry. Chandler and coworkers have argued that dewetting does not in fact demand the spontaneous appearance of a cavity but draws instead on the intrinsic fluctuations of water density at the water-hydrophobe interface [38]. Simulations show that these fluctuations are similar to those at a water-air interface. Patel et al. have argued that biomolecules may tune these fluctuations so that they sit close to a dewetting transition [39] – revealed not by any difference in average water density at the interface but instead more subtly, by an enhancement in rare, large fluctuations that fleetingly dry the surface. Small conformational changes can then tip the balance towards or away from the wet state, thus significantly altering the biomolecular structure and function. This tendency of biological systems to position themselves close to a phase transition and thereby to enable sensitive and pronounced responses to changes in the environment is likely to be generic (see also Lines 569-571, [lines 659–660] and Lines 815-816 below). In this case, Patel et al. [38] say that proteins such as BphC that do not aggregate by dewetting [35] lie on one side of the fluctuation-driven transition, and others such as melittin that do follow this mechanism [34] lie on the other side. In summary, there is ample reason to believe that dewetting and the consequent hydrophobic attraction are real phenomena, and that they are relevant to the behaviour of biomolecules. But they are unlikely to fully explain why hydrophobic association between these entities occurs.

6.5 Protein Stability and Denaturation

Behind the question of what makes an environment habitable for terrestrial life is the matter of how proteins retain their folded conformation. Proteins may be unravelled by extremes of heat, cold and pressure, as well as by the presence of small cosolvent molecules such as urea (denaturants). This imposes a relatively

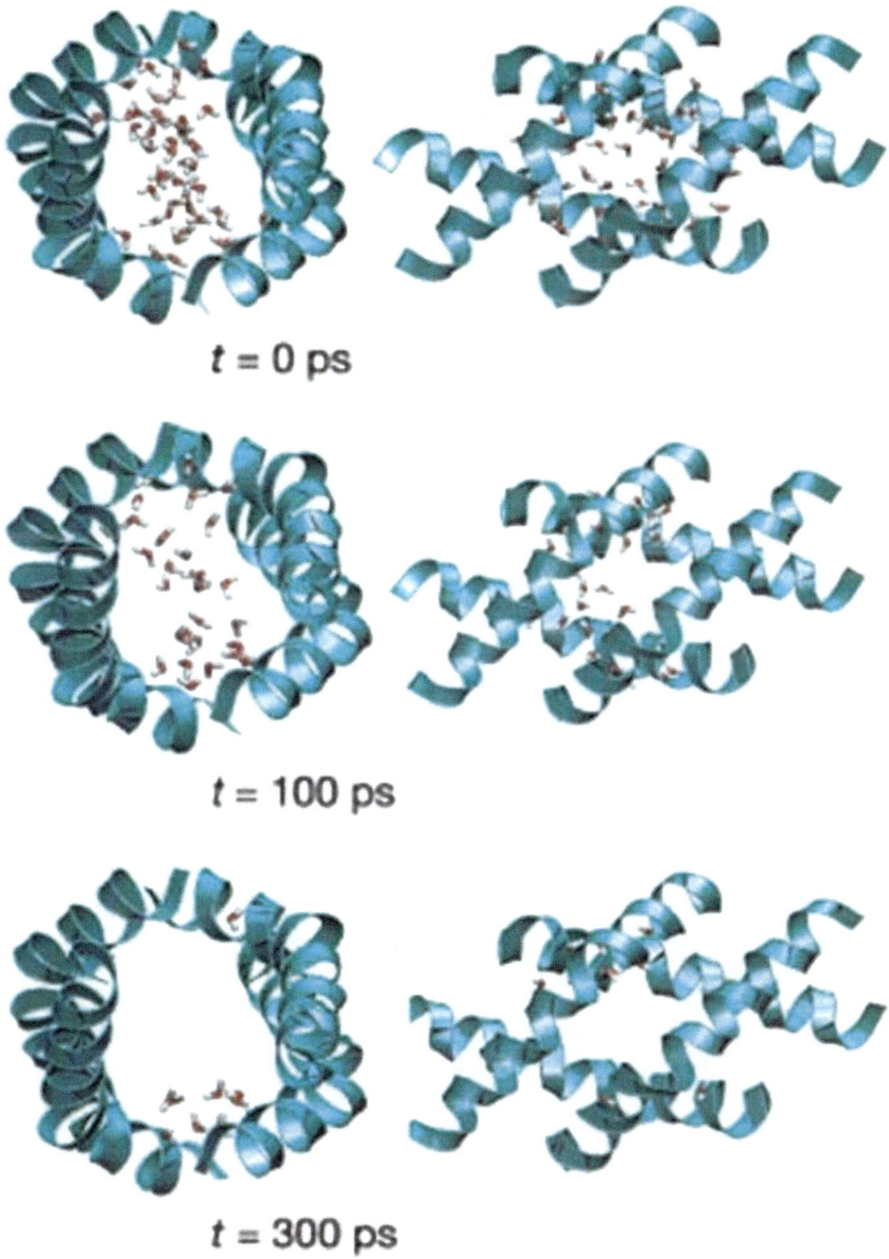

Fig. 6.6 A dewetting transition in the assembly of the four subunits of the melittin peptide. In these snapshots of the process, the roughly cylindrical central cavity abruptly empties of water from the second to the third (Reprinted with permission from Ref. [34])

small window on the 'phase diagram' of terrestrial life. The interplay between protein and solvent is the key to these effects.

The key question of protein folding is how a protein finds from the astronomical number of possible conformations precisely that one which represents the stable ground state, and does so on a non-astronomical timescale. The prevailing view is that the stable fold lies at the bottom of a unique, sharp but rather broad-mouthed energy funnel within a rugged conformational landscape of multiple shallow, metastable minima [40]. But the question of how it avoids becoming trapped in these partially (mis)folded states for any significant length of time is a subtle one [41], to answer which requires a careful consideration of water's influence. This solvent lubricates the search through conformational space, not merely as a generalized continuum fluid but at a granular scale. For example, as a globular protein folds, water molecules seem to bridge hydrophilic residues so as to form relatively long-ranged (6.5–9.5 Å) connections that guide the self-assembly process [42]. Such water bridges can be 'squeezed dry' in the later stages of the folding process: the water acts as a temporary, loose glue that holds the folded chain together until it is ready for final compaction – in effect, constraining the conformational freedom and 'smoothing' the funnel in the folding energy surface. There is a substantial improvement in the structures predicted by simulation for several proteins when these water-mediated contacts are included.

In any event, protein folding is not a miraculous 'blind search' but usually requires the assistance of molecular chaperones - for example, the large protein assemblies called chaperonins, such as the GroEL/GroES complex. This latter assembly works first by binding unfolded protein chains to a hydrophobic patch on the rim of GroEL, followed by attachment of the GroES lid which triggers an ATP-driven conformational change that removes the hydrophobic region and propels the polypeptide chain into the now-hydrophilic interior. Part of the role of these chaperonins is to enable partially misfolded proteins to unravel again and fold properly. It seems that they might do this by manipulating the nature of hydrophobic interactions within the folding chain. Patel et al. find that the interactions between hydrophobic solutes several Å to several nm across are altered close to the interface of water with self-assembled monolayers of various surface chemistries, from hydrophobic to hydrophilic [43]. In particular, the driving force for the aggregation of hydrophobic particles is smaller near a hydrophobic surface than it is in bulk. This implies that hydrophobic surfaces should act as catalysts for the unfolding of proteins. Thus, the initially hydrophobic surfaces of chaperonins should help misfolded proteins to unfold, and then conversion of the walls to a hydrophilic state releases the unfolded protein from the wall so that it might fold again inside the cavity.

Whereas a huge amount of research has gone into understanding the protein folding process, the reverse process of denaturation has been somewhat neglected. It is increasingly clear that denaturation is not merely folding in reverse – it can occur in several ways.

Heat denaturation is a fairly simple process: the thermal motions shake apart the weakly bound protein folds. Pressure-induced denaturation seems, crudely speaking, to stem from the 'squeezing' of solvent so that it enters and loosens the compact native state. And for cold denaturation, the conventional view has been that the hydrophobic groups get more soluble in water (that is, the hydrophobic interaction is weakened) as the temperature is lowered. Indeed, simple hydrophobic polymers, which collapse into compact states in water, swell at lower temperatures. That has been traditionally considered a result of the decline in the entropic driving force of the hydrophobic interaction with decreasing temperature. But it now seems the truth is more subtle. NMR measurements on four common 'model' proteins, including apomyoglobin and β-lactoglobulin, show that in neither case does cold denaturation produce a fully unfolded state [44]. Instead, the proteins remained relatively compact, as though water had penetrated within the native structure but not forced it apart. Rather than being fully exposed to the solvent, a cold-denatured protein seems to contain small clusters of water molecules interacting strongly with the peptide, which implies that it doesn't involve the classical hydrophobic effect after all.

There is now a growing view that all types of denaturation are intimately connected to changes in the way proteins are hydrated. In other words, the opening-up of the native protein is not an intrinsic property of the polypeptide chain, but comes about in collaboration with the solvent.

It was long supposed that denaturants such as urea or guanidinium chloride (GdmCl) somehow perturb water's bulk structure in a way that destabilizes the folded protein, perhaps by altering the hydrophobic interactions that keep insoluble residues buried. This idea now looks increasingly flawed. Instead, we need to look at how denaturants affect the protein's hydration shell. For lysozyme, urea displaces water from the hydration shell and penetrates into the hydrophobic core, suppressing the native fold in favour of a swollen, rather disordered 'molten globule' [45]. But GdmCl appears to behave differently: the molecules stick to the hydrophobic surfaces and give them a less water-repelling veneer [46]. Here the denaturant is acting somewhat like a surfactant, stabilizing the interface between the surface and the solvent. There is evidence that urea can act this way too: because it experiences stronger dispersion forces than water with hydrophobic surfaces, it will accumulate there and mediate interactions with water via hydrogen bonds [47]. Moreover, urea not only alters hydrophobic interactions but also disrupts the hydration of hydrophilic parts of a protein, sticking there via hydrogen bonds [48]. In effect, urea can usurp the hydrogen bonds that otherwise may help to bind the native state together. While these ideas are still being debated, the emerging picture is one in which denaturants exert their influence through direct interactions with the solute, not by restructuring the bulk solvent.

Some other small molecules, such as glycerol, have the opposite effect of stabilizing a native protein against denaturation. Some organisms use glycerol as a kind of antifreeze – but it may be that glycerol also acts here to stave off cold denaturation. The mechanism is still debated, and it is possible that glycerol in fact serves a dual purpose: inhibiting ice formation and stabilizing proteins. Bulk water—water bonding is highly disrupted by glycerol, because water binds preferentially to the polyols [49]. This could suppress ice formation. But the idea that glycerol somehow alters water's global hydrogen-bonding has been increasingly challenged [50]. The stabilization of proteins more probably results from direct interactions with these macromolecules, perhaps because the strongly polar glycerol hovering near the protein surface makes hydrophobic regions even less easily exposed. Glycerol, and also ethylene glycol and propylene glycol, stabilize proteins against mechanically induced unfolding, possibly via the formation of 'solvent bridges' that pin together hydrogen-bonding sites on the polypeptide backbone [51].

These debates supply a reminder that speculations about mixtures of water and other small molecules (ammonia, formamide) as solvents for extraterrestrial life are a much more subtle matter than simply looking for a low-freezing-point liquid. The interactions and partitioning of co-solvents should be expected to have a significant effect on the conformational stabilities of complex macromolecules, and in ways that are hard to predict.

6.6 Protein Misfolding and Aggregation

The discovery and study of protein-misfolding diseases – those triggered by prions and those caused by the accumulation of aggregated protein called amyloid – has produced a new view of protein stability. Whereas previously the energy landscape was considered to contain only one deep well, the functional ground state, it is now widely suspected that many if not most proteins may form amyloid-like misfolded states, in which exposed hydrogen-bonded groups and hydrophobic regions cause aggregation into β-sheet-like structures that comprise the basic structural units of amyloid fibrils [41]. Increasingly, this aggregation process is considered to be highly sensitive to hydration of the partially unfolded protein.

Amyloid fibrils self-assemble from interdigitated β-sheets, but amyloidogenic proteins can have strikingly varied sequences: some may be polar, for example, and others hydrophobic. There is some reason to believe, however, that water plays a central role in the assembly process in most if not all cases. In one recent simulation study [52], the association of hydrophilic sheets was mediated by one-dimensional water wires at the interface between them, which are gradually expelled. But for the hydrophobic peptides studied in these simulations, the sheets came together in something like an abrupt drying transition [30]. This happens much faster (nearly 1,000-fold) than for the hydrophilic case, since the trapped water wires for the polar peptide create a barrier to rapid assembly. Thus, although the final structures are very similar, the mechanisms and dynamics are quite different. Mediation by water, including dewetting transitions, has also been proposed in the assembly of other amyloidogenic β-sheets.

Misfolding and aggregation might be quite general consequences of a 'failure of hydration' – that is, a failure to protect intramolecular or 'backbone' hydrogen bonds (BHBs) at the surface of a protein against competition from hydration water, which threatens to intervene and break up the secondary structure [9, 53]. One way of conferring such protection is by controlling the local curvature of the surface: if the BHBs are in a sufficiently highly curved location, water cannot penetrate without compromising its own hydrogen-bonded network for purely geometric reasons.

Another way to protect BHBs is to 'wrap' them in hydrophobic groups to reduce contact with water. Many proteins have regions of proteins that are 'poorly wrapped', called dehydrons [9]. Here, processes that expel water – for example, the formation of peptide—peptide contacts – are energetically favourable. And indeed, dehydrons appear to be concentrated at sites that engage in complexation with other proteins, and may play an important role in protein—protein interactions such as the association of capsid assemblies in viruses. Dehydrons also seem to be a common feature of proteins with a propensity to form amyloid aggregates, and Fernández et al. [53] propose that destabilization of the globular fold, and consequent amyloidogenic capacity, is related to the tendency of dehydron units to promote β-sheet aggregation.

Meanwhile, the random accumulation of dehydrons in proteins due to genetic drift (in which random mutations disrupt the wrapping of BHBs) might account for the complexity of the human interactome [54]: such protein—protein associations are evolution's way of 'hiding' these defect in hydration structure. This would represent an aspect of evolution that is not fundamentally driven by natural selection, although selective forces would be expected to operate on the resulting protein associations. However, by masking the underlying problem, the increasing complexity of the interactome might simply let it accumulate to a point where it reaches a crisis. Fernández and Lynch speculate that the appearance of amyloid and prion diseases (the latter are so poorly wrapped that they may readily relinquish their functional fold) might presage such a crisis in humans and other species with small populations, which are especially prone to genetic drift [54].

Many protein functions require a significant rearrangement of the folded state, sometimes involving a marked loosening and unravelling of a compact structure (Fig. 6.7). How do proteins remain dynamic and functional without sacrificing solubility and resistance to aggregation? How, in other words, do they control their conformational changes, permitting 'a little but not too much'? Using the fruitfly acylphosphatase as a model system, De Simone et al. find that the wild-type protein has free-energy barriers that limit access to aggregation-prone conformations except under special conditions (addition of small amounts of trifluoroethanol) [55]. But this is a finely tuned affair: as De Simone et al. say, "The sensitivity of the energy surfaces of proteins to minor perturbations supports the view that there is a delicate balance between functionality, stability, and solubility, which is encapsulated by the concept of 'life on the edge'".

All this implies that the phases and dynamics of polymer folding in water are far more complex than has traditionally been supposed. If the existence of amyloid-like misfolds is a generic property of proteins, is this an *inevitable* consequence of any system that supports a unique, compact globular form – the price one pays, perhaps, for relying on biomolecular catalysts that have structure and function programmed into a linear polymeric strand? Would alternative, non-peptide polymeric entities be more or less prone to misfolds, and to what extent might that depend on the solvent, and/or on the presence of cosolvents or other osmolytes? How sensitively do the timescales and kinetics of the folding and misfolding processes depend on the solvent? A comprehensive understanding of the protein stability landscape is surely essential for any assessment of what might be required of, and what might be unavoidable in, this particular paradigm for programming molecular information and conformation in an alternative solvent system.

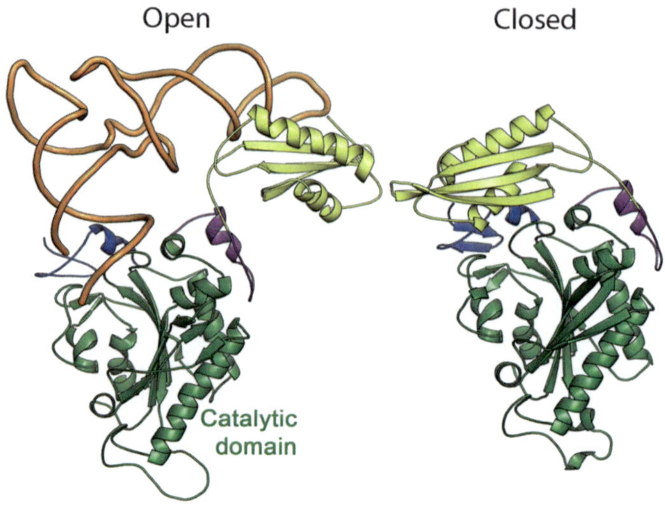

Fig. 6.7 Proteins may undergo large conformational changes during their active cycle, which might involve considerable unravelling and exposure to solvent, as illustrated here for the active 'open' and inactive 'closed' states of human mitochondrial phenylalanyl-tRNA synthetase. How is this achieved without the risk of aggregation, as may occur when proteins denature? (Reprinted with permission from Klipcan L, Moor N, Finarov I, Kessler N, Sukhanova M, Safro MG (2012). J Mol Biol 415: 527–537

6.7 Water's Role in Protein—Substrate Binding

Whether there is anything unique to polypeptides, and more specifically to those based on the 'natural alphabet' of 20 natural amino acids, as the functional molecules of life is a question that has not so far been given a definitive answer. But to the extent that one might suppose that biological function will be universally encoded in structure and enacted via conformational dynamics, we can now see that the role of the solvent is likely to be both critical and subtle.

We might posit, albeit tentatively, another potential generality for the molecular character of living systems: that information will be transmitted between molecules by selective recognition and binding. Where does the solvent fit in here?

When a protein binds its substrate, one might suppose that water is required simply to get out of the way: to vacate the active site. A more careful consideration suggests that any small-molecule solvent might provide an entropic driving force for the binding process, due to the liberation of those molecules formerly confined in the active site.

But in water, things are not so simple. On the one hand, there is now good reason to believe that water has a subtle role in the dialogue between receptor and ligand, acting as a versatile intermediary and facilitator. On the other hand, it is far from clear that entropic effects dominate this solvent-mediated discourse.

For one thing, it is hard to generalize about how the various enthalpic and entropic effects of dehydration of the cavity and the ligand balance out. Both positive and negative entropy changes have been reported previously for water entering protein cavities, and the result probably depends both on the chemistry and the geometry of the cavity. Even if there is an entropic benefit in the expulsion of bound water, the enthalpic contribution to that change is by no means obvious, and could potentially dominate over, or even counteract, any entropic gain. For example, Yu and Rick studied the entropy, enthalpy and free-energy changes on transferring a water molecule from the bulk to various types of protein cavity large enough to hold only one such molecule [56]. They found that, when there are hydrogen-bond donors and acceptors in the cavity, the thermodynamic consequences of hydrogen-bond formation are greater than those exerted, via entropic effects, by the cavity size. And using an idealized cavity—ligand combination with various permutations of surface charges on both, Baron et al. find that the signs and magnitudes of the enthalpy and entropy changes vary widely for the different cases [57]. Coincidentally, the net free energy change is similar both for binding driven electrostatically and by hydrophobic interactions. There is also a loss of entropy when the ligand is bound because this tends to suppress its own dynamics degrees of freedom.

The recognition and binding of a substrate by its receptor protein commonly involves a hydrophobic interaction, in which non-polar surfaces of the ligand and binding cavity are juxtaposed. This has been regarded as an aspect of the 'lock-and-key' complementarity between the substrate and binding cleft. The canonical signature of a hydrophobic effect in ligand binding is a negative change in the heat capacity, which is generally believed to be of entropic origin owing to the expulsion of bound water from the cleft. However, as is the case with protein folding, beyond the notion that somehow water restructuring at the interface of the ligand and binding site is involved, there is again no generally accepted explanation of exactly how this hydrophobic effect operates.

There may be a good reason for this lack of consensus. The whole discourse around the 'hydrophobic effect' in ligand binding has been for too long dominated by the notion that there is a single explanatory picture involving the expulsion of hydration water in the binding cleft, whereas in reality there may be many different types of interaction and structural change involved whose details depend on the specifics of the situation. This was evident in a detailed study of the binding between a rigid enzyme, carbonic anhydrase II, and a series of structurally related aromatic sulfonamides [58]. In some of the synthetic ligands, additional ring structures increased the hydrophobic contact area between ligand and cavity (which has a hydrophobic and a hydrophilic side). The observed changes in heat capacity revealed the 'classical' signature of a hydrophobic effect, but it was surprisingly insensitive to the hydrophobic contact area. Instead it seems to arise primarily from structural changes in the network of water molecules between the ligand and the hydrophilic side of the cavity. Thus, while at a broad level the hydrophobic effect does involve the differences in water structure close to solute surfaces, the detailed balance of entropy and enthalpy is likely to vary on a case-by-case basis that can be understood only by detailed analysis.

Water expulsion can, however, supply a thermodynamic driving force for substrate binding in some large hydrophobic cavities, from which water expulsion is relatively facile. For example, the X-ray structure of the tetrabrachion protein of the hyperthermophile *Staphylothermus marinus* reveals several hydrophobic cavities in the 70-nm-long 'stalk' segment that are all filled with water at low temperatures. Simulations of this structure [59] suggest that the two largest cavities, containing seven to nine and five water molecules at room temperature, are close to switching to a dry state at the organism's optimal growth temperature of 365 K, which may offer a docking mechanism for the binding of the nonpolar anchoring sites of two proteases present in the active form of the protein. Here again we see this idea of hydration water being tuned close to a phase transition so that small changes in the environment can elicit a pronounced response in the protein.

While hydrophobicity is (whatever its origin) certainly an important force that guides the recognition of a protein and its substrate, water seems also to be important for the interaction of hydrophilic regions of such complexes. Around 70% of interfacial residues are in fact hydrophilic. It is common to assume that such polar groups experience a direct electrostatic interaction mediated by the (continuum) solvent. But in fact the water network has a more complex role here too. For example, in the formation of the complex between the bacterial ribonuclease barnase and its inhibitor barstar, water molecules mediate and stabilize the hydrophilic interactions between receptor and substrate at the 'granular' level [60].

As a binding event proceeds, not all water is necessarily expelled from the cavity. Indeed, very often some remains, generally in locations that bridge the cavity surface and the substrate so as to make the binding very selective for a particular shape. For example, a water molecule in the binding site determines the selectivity of the ionotropic glutamate receptor, which is found in neurons in the brain and binds the neurotransmitter glutamate [61]. These receptors will also bind and be activated by an artificial mimic of glutamate called AMPA. In both cases, there is a single water molecule that bridges the receptor's surface and the bound molecule (Fig. 6.8). But the water molecule is in a different position in each case, and it is the position of this water that makes the receptor receptive to either glutamate or AMPA. Understanding this difference could be important for designing other synthetic molecules that can affect and interfere with this biochemical process in brain function.

But not all water-mediated interactions in ligand binding have a comparable degree of specificity. Some water molecules can act as a versatile, reconfigurable filler to make a protein promiscuous about which molecules it will bind to. This behaviour is displayed by the oligopeptide binding protein OppA, which will bind very small (2–5 residue) peptides with more or less any amino-acid sequence [62]. The lack of selectivity results from water in the binding site which can be expelled or admitted to 'fill up' any empty space.

Such instances present a somewhat static picture of water's roles in substrate binding. In reality, of course, this is a dynamic process that involves a continuous change in molecular conformations as binding and/or catalysis proceeds. It seems that this involves an extraordinarily subtle and 'anticipatory' use of hydration water. Grossman et al. have shown that, as a zinc metalloprotease binds its substrate to form the Michaelis complex, the water motions are retarded by coupling to the

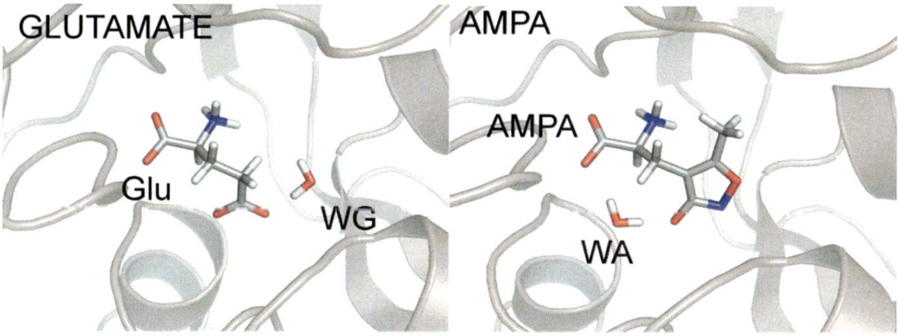

Fig. 6.8 Water molecules sit in two different positions (*WG*, *WA*) in the binding site of the ionotropic glutamate receptor protein and help two different molecules to bind: the natural neurotransmitter glutamate (*Glu*) and an artificial mimic called AMPA (From Ref. [61]. © 2011 American Chemical Society)

motions of both the enzyme and its peptide substrate [63]. Crudely put, it is as if these two components 'thicken' the water towards a more glassy form, which in turn retards the fluctuations of both the binding site and the substrate so that they can become locked securely together. This suggests changes in protein and solvent dynamics are not mere epiphenomena but play a vital role in substrate binding: they are more cause than consequence.

6.8 Transmission of Allosteric Effects

Protein—protein contacts mediated by water molecules not only can serve to assist in recognition and docking but may also play a mechanistic role in function – for example, in the allosteric regulation of oxygen binding to haemoglobin. The haemoglobin of the mollusc *Scapharca inaequivalvis* is dimeric, and the interface of the subunits contains a cluster of 17 well-ordered water molecules. Oxygenation is accompanied by loss of six of the ordered interfacial water molecules. These waters have a central role in cooperative oxygen binding, enabling allosteric interactions between the subunits by acting as a kind of transmission unit [64]. The water cluster helps to stabilize the low-affinity form of the protein, whereas a mutant form that lacks two of the hydrogen bonds from this cluster tends to adopt the high-affinity conformation instead. Thus, loss of interfacial water occasioned by oxygen binding to one of the wild-type subunits helps to promote the transition to the high-affinity conformation of the other subunit. Molecular-dynamics simulations indicate that the 11 water molecules in the oxy form exhibit slower relaxation than the 17 in the deoxy form, and that the water cluster, although rather static on ps timescales, can enhance energy transport across the interface of the subunits via vibrations [64].

These allosteric effects can also be intramolecular. The enzyme heme oxygenase contains a rather precisely defined network of water molecules that acts to transmit and coordinate the movements of the protein chain and other chemical groups during the catalytic process [65]. They become in effect part of the molecule's structure, and if the water cluster is disrupted, the enzyme loses its activity.

6.9 Chemical Participation in Enzyme Action

Water can serve enzymes in more ways than assisting the thermodynamics and dynamics of substrate binding and conformational changes. Water molecules can participate directly in chemical reactions at the binding site, for example acting as a nucleophile or a source of protons. There are countless examples; the few considered here illustrate the variety of roles that water can play.

A water molecule in the bacterial enzyme zinc lactamase, which is involved in resistance to lactam antibiotics, apparently acts as a nucleophile to initiate splitting of the lactam ring [66]. Hydrogen-bonding between this water molecule and a zinc-bound aspartate group increases its polarity and nucleophilicity, while the carboxylate group of the aspartate potentially provides a source for the proton that reacts with the cleaved ring.

In the action of DNA polymerase IV in the thermophilic archaeon *Sulfolobus solfataricus*, water molecules in the coordination sphere of the catalytic magnesium ion appear to play two important roles [67]. The enzyme adds a nucleotide to a growing DNA chain by catalysing the reaction of the terminal 3′-OH group with the α-phosphate of the new deoxyribonucleoside triphosphate, eliminating pyrophosphate. The initial, and rate-limiting, step is proton transfer from 3′-OH to phosphate, which happens via a bridging water molecule (Fig. 6.9). And the cleaving of pyrophosphate following linkage of the polynucleotide chain and the deoxyribonucleoside involves another water-mediated proton relay that protonates the γ-phosphate and partly neutralizes its negative charge.

Water plays a crucial role in the catalytic mechanism of some heme catalases, which convert hydrogen peroxide to water and oxygen. One type of catalase contains a tightly bound NADPH molecule that protects an intermediate of the ferryloxo group against deactivation to a catalytically inactive form. Here a bound water molecule supplies a hydroxyl group that binds temporarily to the porphyrin group and then assists the fast two-electron reduction of the intermediate ferryloxo species by NADPH via a series of proton shifts, to restore the catalase resting state and avoid diversion of the reaction towards the deactivated state [68] (Fig. 6.10). Proton relays and shuttles of this sort are particularly common in enzymatic catalysis.

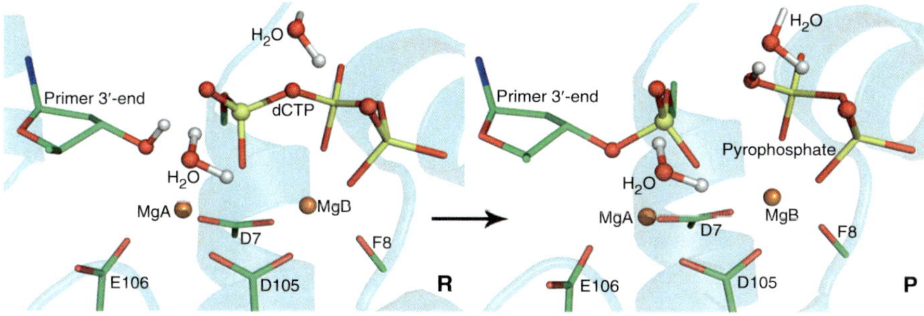

Fig. 6.9 Water mediation in catalysis (I). In the action of this DNA polymerase, the initial proton transfer to the α-phosphate of the substrate via a bridging crystal water molecule is the rate-limiting step. Subsequently, departure of the pyrophosphate is facilitated by a proton relay mechanism through mediation of water, which neutralizes the evolving negative charge. (Reprinted with permission from Ref. [67]. © 2007 American Chemical Society.)

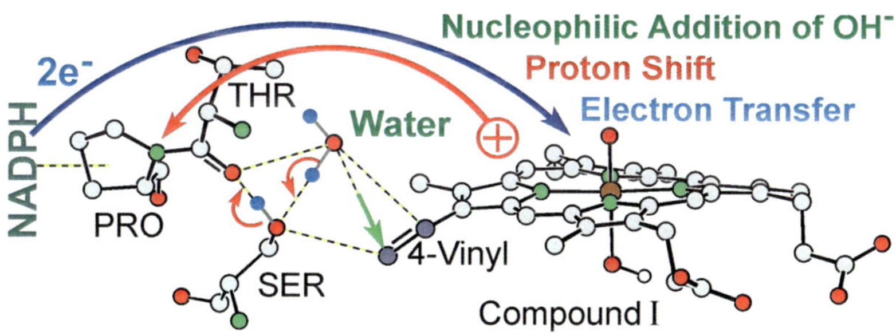

Fig. 6.10 Water mediation in catalysis (II). A bound water molecule completes a hydrogen-bonded chain that assists in proton and electron transport in heme catalase. (Reprinted with permission from Ref. [68]. © 2008 American Chemical Society.)

6.10 Water Wires and Channels

A channel wide enough to admit just one water molecule at a time can be threaded by a chain of waters hydrogen-bonded together in what is often called a water wire. These chains can transport protons very rapidly, because they can effectively be shunted along by successive water molecules flipping their arrangement of hydrogen bonds (Fig. 6.11). This is called the Grotthuss mechanism [69], and it happens in pure water too, where there are long chains like this that make up part of the usual three-dimensional hydrogen-bonded network. For this reason, hydrogen ions can move around unusually fast in water.

Hydrogen ions are often needed in biochemical reactions carried out by enzymes. Several such enzymes use water wires to shuttle the ions from the solvent through the protein and into the active site where the reaction takes place. This happens, for example, in cytochrome and peroxidase enzymes, and in bacteriorhodopsin, a light-driven proton pump used by some Archaea to transport protons across a membrane during the conversion of light energy into chemical energy.

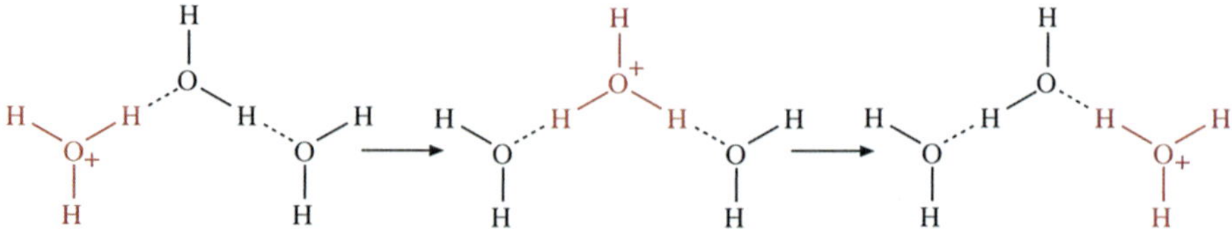

Fig. 6.11 The Grotthuss mechanism for rapid proton transport in hydrogen-bonded chains of water molecules. The hydronium ion (*red* [*grey*]) shifts essentially by the rearrangement of electronic rather than nuclear configuration

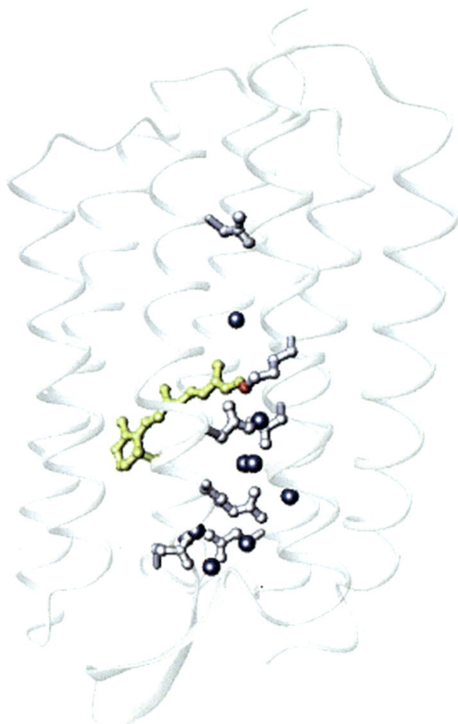

Fig. 6.12 Water molecules form a hydrogen-bonded chain (*blue spheres*) from the chromophore (*yellow*) to the extracellular surface in the central pore of the light-driven proton pump bacteriorhodopsin (Courtesy of Klaus Gerwert, Ruhr University of Bochum)

Bacteriorhodopsin is threaded by a water wire from the light-sensitive core to the exterior of the membrane; only rhodopsins that contain such strongly hydrogen-bonded water molecules are able to act as proton pumps. Following light absorption and photoisomerization of the chromophore retinal, the protein undergoes a conformational transition in which a proton is transferred from the chromophore to an aspartate residue (Asp85), accompanied by the release of a proton to the extracellular surface. Before this event, the latter proton is stored for a short time. The hydrogen-bonded network of internal water molecules – an $H_5O_2^+$ cluster – seems to be the most likely candidate for this storage site (Fig. 6.12) [70], although this

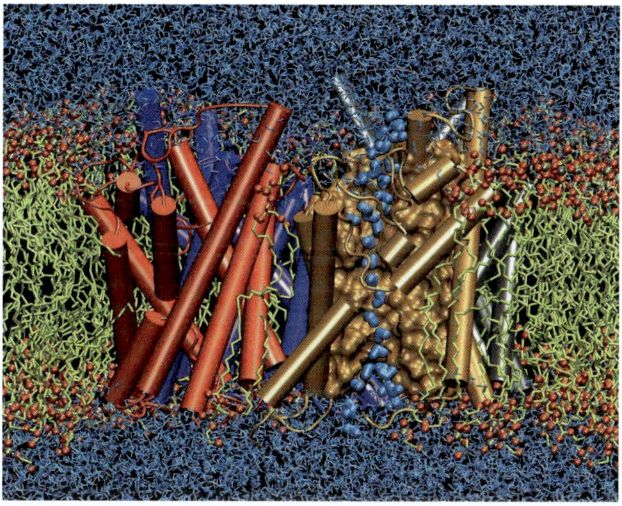

Fig. 6.13 The water-conducting channels of aquaporins are threaded by a hydrogen-bonded chain of water molecules, shown in *blue* (Courtesy of Emad Tajkhorshid, University of Illinois at Urbana-Champaign)

interpretation is still debated. Such one-dimensional water chains might serve quite generally as 'proton sponges' in proteins, not just transporting protons but also able to accept and retain them [71].

Another important class of proteins that contain water channels are the aquaporins, which regulate the flow of water in and out of cells. They will let water through but not salts or other dissolved substances, and as such, they act as molecular water filters. Water transport occurs via a chain of nine hydrogen-bonded molecules (Fig. 6.13). But if this chain were to permit rapid transmembrane proton motion, that would disturb the delicate charge balance across the membrane. So aquaporin must somehow disrupt the potential proton wire that threads through it. The mechanism has been much debated, but it now seems that the inhibition of proton transport is dominated by electrostatic repulsion by positively charged groups in a narrow constriction in the middle of the pore [72].

Protons seem to be delivered to some membrane proteins, such as the proton pump cytochrome c oxidase, via some kind of surface-enhanced, two-dimensional transport at the membrane surface. Hydrogen-bonded chains on cell-membrane surfaces appear to act as proton circuits that help guide protons from a transporter – a pump protein – to molecules that exploit the proton-motive force, such as ATP synthase. In other words, these networks act as proton-collecting antennae that improve the efficiency of proton transport. It has been proposed that this enhanced two-dimensional transport relies on ionisable (phosphate and carbonyl) groups on the lipids [73]. But fluorescence measurements of proton transfer at membranes show that proton transport can be equally fast in the absence of ionizable groups [74], and it may instead be a network of interfacial water molecules that is responsible for the rapid proton motion. As Springer et al. put it, "water structuring at the interface seems to be mandatory for providing the pathway" [74].

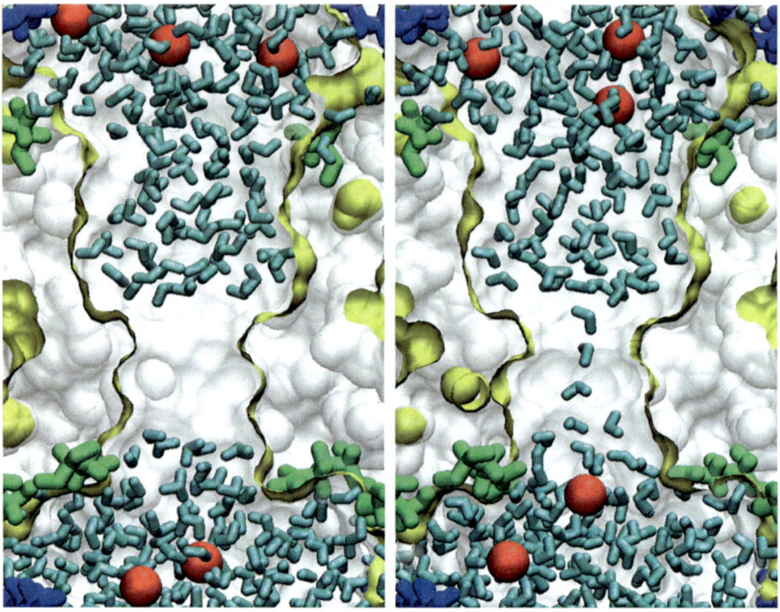

Fig. 6.14 Water dynamics in simulations of the protein channel MscS, showing snapshots in which the pore neck is 'dry' (*left*) and partly water-filled (*right*). The *red spheres* are chloride ions, and polar residues are shown in *green*. In the dry state, ions cannot pass through because of the free-energy penalty of dehydration (Reprinted with permission from Ref. [77]. © 2004 Biophysical Society)

Evidently, water inside a narrow hydrophobic channel is in a precarious state: studies with carbon nanotubes have shown that the pore can spontaneously empty if there is just a slight change in the prevailing conditions, such as a change in temperature or a mechanical deformation that narrows the channel [75]. Such capillary evaporation is well attested in theory, and is equivalent to the kind of dewetting transition discussed above [30]. Although such behaviour can be expected for any liquid confined in narrow solvophobic pores, there is pronounced cooperativity in the process for a confined hydrogen-bonded network of water molecules, which can make systems close to the dewetting transition prone to rapid fluctuations between a 'wet' and 'dry' state [76].

The emptying of narrow water channels and wires has been suggested as a mechanism for the gating of membrane protein channels that transport ions and other small solutes [75]. In the dry state, ions cannot pass even though the channel is wide enough in theory, because this would require stripping away the ion's hydration shell, which has too great a free-energy penalty. Such a 'bubble-induced' gating mechanism might also explain the anaesthetic effects of inert gases [75].

Abrupt pore-emptying has been seen in modelling studies of mechanosensitive channels, where small deformations of the membrane in which the protein channel is embedded can tip the balance between 'wet' and 'dry' states. For example, the

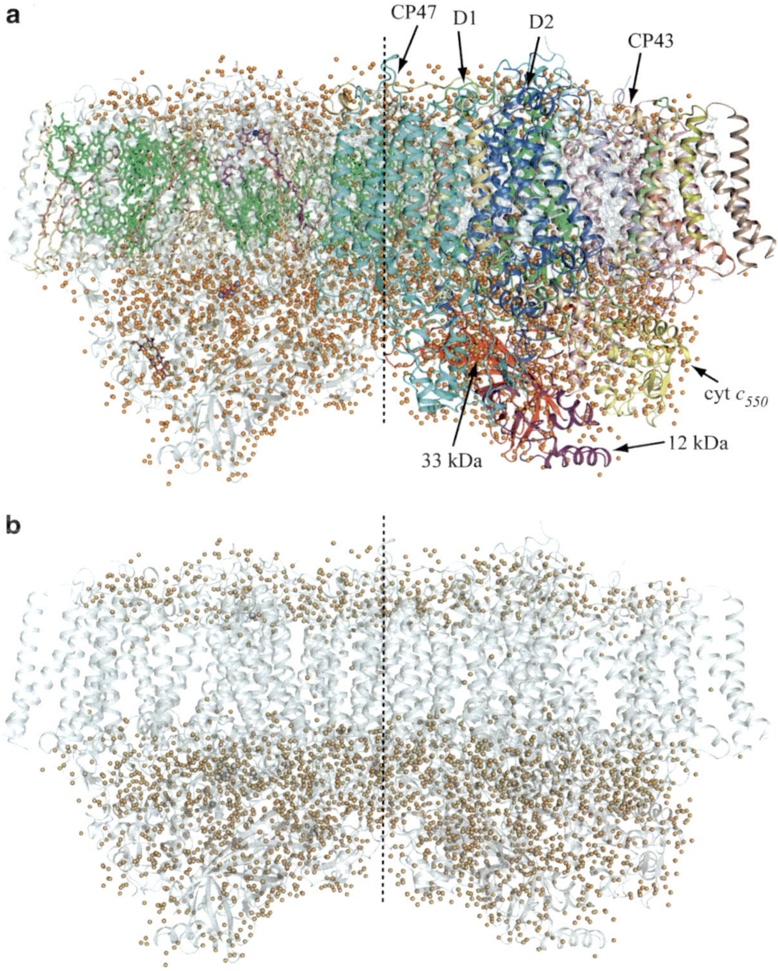

Fig. 6.15 (continued)

bacterial mechanosensitive channel MscS has a pore 7–15 Å across at its narrowest point, lined with highly hydrophobic residues, and simulations indicate that drying can be induced within the constriction (Fig. 6.14) [77]. Stretching of the cell membrane in which the protein sits distorts the protein, constricting its pore at a narrow neck, and it seems that this pushes the channel over the brink from one that can hold water to one that cannot: the slight decrease in width induces capillary evaporation, emptying the pore so that neither water nor ions can pass.

These new perspectives on biomolecular hydration are very evident in a study of photosystem II, the membrane-bound complex of proteins and pigment molecules responsible for harvesting sunlight and converting the photon energies into stored chemical energy [78] (Fig. 6.15a, b). At first glance it is tempting to regard the

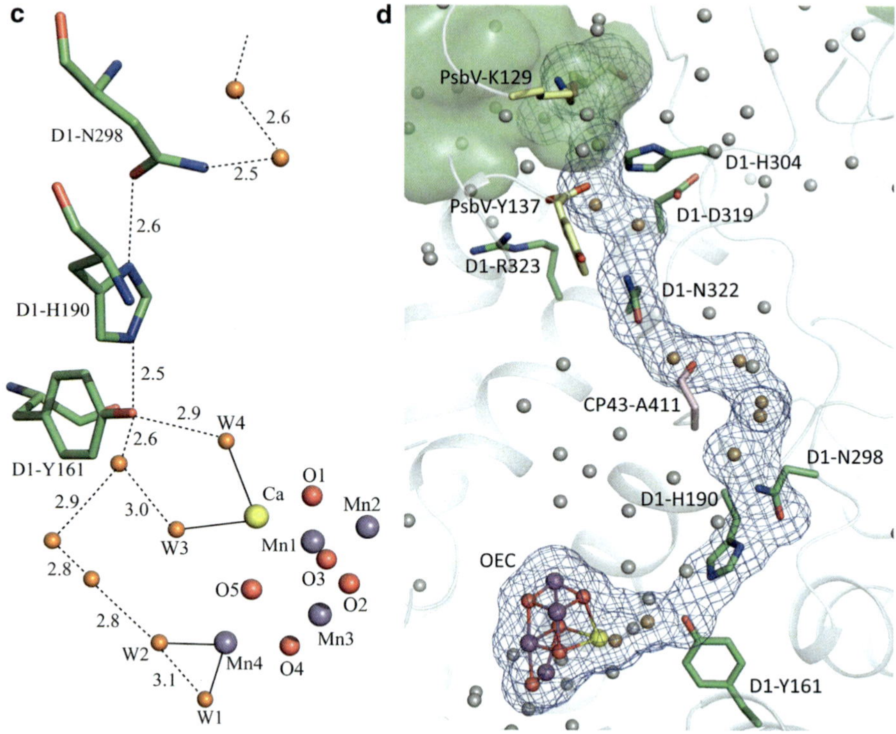

Fig. 6.15 (**a**) The hydration halo of photosystem. (**b**) The same, with protein chains shown only in *light grey* for clarity. (**c**) Water molecules (*orange*) attached to the Mn-Ca complex. (**d**) A water channel that ferries protons to the bulk aqueous phase (Reprinted with permission from Ref. [78])

1,300 or so water molecules in the hydration sphere of crystallographically defined water molecules as a generalized, disorderly halo around the functional components of the assembly. But close inspection shows this hydration sphere to be intricately structured at the local level to facilitate those functions. For example, four water molecules are attached to the manganese-calcium (Mn_4CaO_5) cluster that catalyses the oxidative splitting of water, and one or more of these probably acts as the substrate in that reaction (Fig. 6.15c). The reaction releases protons, which are thought to be transferred through the complex to the bulk phase via a channel of hydrogen-bonded water molecules and side-chain groups on the surrounding alpha helices (Fig. 6.15d). There are other such channels that are presumed to act as proton exit or water entry routes. In short, the hydration structure comprises a functional extension of the proteins and prosthetic groups, with water molecules serving biological functions.

6.11 Dynamical Aspects of Protein Hydration

As well as helping proteins hold their shape, water seems to act as a plasticizer, giving the protein molecule enough flexibility to do its job. If proteins become too dry, or of they are taken out of water and placed in other solvents, many will either fall apart or become too rigid to work. Most proteins have been reported to require about 0.4 g of water per gram of protein to achieve their normal functionality [79] (but see below).

It is generally considered that a protein needs to maintain a delicate balance between rigidity and flexibility of structure: the specificity of the folded shape is clearly central to an enzyme's substrate selectivity, but it must also remain able to adapt its shape by accessing a range of conformations without getting stuck in local energy minima [80]. One role of the solvent may be to 'inject' fluctuations into the protein to give it this conformational manoeuvrability. Alternatively, the water might compete for and loosen intramolecular hydrogen bonds. This seems to be the case for bovine pancreatic trypsin inhibitor (BPTI) and barnase, where the addition of a water molecule into the active cavities (polar and hydrophobic respectively) makes the proteins more flexible [81]. Meanwhile, water molecules in the binding pocket of scytalone dehydratase seem to play a part in the conformational flexibility that is necessary for binding of the substrate: there is evidence for cooperativity in the motions of the bound water molecules and the ligand-free protein [82].

Proteins do not take passive advantage of the molecular motions in the solvent, but rather, can be considered to adapt those motions to suit their needs. Both simulations [83] and experiments [84] show that water dynamics in the hydration layer of a peptide are anomalous with respect to the bulk. The hydration water seems to adopt a state akin to that of a glass, with a very rough potential-energy landscape and slow hopping between local potential minima. Thus, the water molecules no longer diffuse independently: their motion is dependent on that of their near neighbours. It seems plausible that this is a result of the interconnected nature of the hydrogen-bonded network, which is highly constrained close to the protein surface and so develops a degree of cooperative motion. This kind of anomalous, glass-like dynamics may be just what a protein needs to attain conformational flexibility. If proteins need thus to 'feed off' the dynamics of their solvation layer, water represents the ideal solvent because its hydrogen-bonded network makes it ideally suited to being 'moulded' by the protein into a glassy state.

There is some evidence to support the idea that the dynamics of a protein can in fact be 'slaved' to those of the solvent. Below about 200–220 K, proteins seem to 'freeze' into a kinetically arrested state that has genuine analogies with a glass [85]: the protein atoms undergo harmonic vibrations in local energy minima, but no diffusive motion. This glasslike dynamical transition coincides with dynamical changes in the solvent [86]. The origin of the transition is still much debated [87, 88]. Nonetheless, the solvent and protein motions appear to be intimately coupled [89], so that as a protein is warmed through its dynamical transition temperature the dynamics of the hydration shell 'awaken' motions in the macromolecule. This behaviour may depend on the formation, at a critical 'water coverage' on the protein surface, of a fully connected hydrogen-bonded network of water molecules [90]. In other words, the collective dynamics become 'activated' in a two-dimensional percolation transition. For lysozyme molecule this threshold corresponds to 50% of the protein surface (about 66% coverage of the purely hydrophilic regions) being covered with water. That hydrophilic coverage is essentially identical to the percolation threshold for clusters formed on two-dimensional square and honeycomb lattices.

It may therefore be best to regard a protein and its hydration water as a single dynamic entity with a rough energy landscape [83]. As Gallat et al. [91] say, "hydration water, rather than being a mere epiphenomenon, is an integral part of the biologically active protein. It is the delicately tuned give-and-take between a biological macromolecule's structural dynamics and its hydration water ... that enables macromolecular function in a cellular context."

Not all proteins modify water dynamics to the same degree, however. This interaction is graded according to the degree to which the protein needs to adopt a particular conformation. Intrinsically disordered proteins such as the human tau protein involved in Alzheimer's disease shows tight coupling between the dynamics of the hydration water and that of the peptide chain, characterized by a slowing of the water dynamics relative to the bulk [91]. For globular folded proteins, on the other hand, this coupling is weaker. And for membrane proteins it is almost absent. Thus it seems that the water dynamics are adjusted to suit different classes of proteins.

A part of the uncertainties and controversies over dynamical aspects of protein hydration can be attributed to the range of techniques used to investigate the issue – or to put it another way, to the lack of any single technique uniquely suited to investigating the wide range of length and time scales involved. Computer simulations struggle to accommodate the longest relaxation times for some molecular motions, and while NMR and inelastic neutron scattering are in many respects excellent probes with a wide dynamical range, they may not be so sensitive to the rather long-ranged correlations that seem to exist between some motions of the water molecules and peptide chains. Terahertz spectroscopy is now emerging as a highly promising tool for studying the latter, as illustrated for example in Ref. [63]. The experimental lacunae are exemplified, however, by the suggestion that the protein dynamical transition might in fact be an experimental artifact caused by finite instrumental energy resolution [92].

Despite the clear evidence for strong coupling between protein and solvent dynamics and its significance for protein function, it should be noted that the idea that proteins 'only work in water' has acquired something of a dogmatic status that is not fully warranted. It is well known that some enzymes can retain functionality both in non-aqueous solvents and in a vacuum. Although in these instances the proteins commonly retain some tightly bound water molecules on their surfaces, Lopez et al. [93] have reported significant catalytic activity of pig liver esterase in near-anhydrous conditions – not merely in a non-aqueous medium, but in a 'dry' powder with just 3 ± 2 water molecules per enzyme molecule. This leads those authors to propose the challenging idea that "one of the biological requirements of water may lie with its role as a diffusion medium rather than any of its more specific properties." It remains an open question whether this result can be generalized to all enzymes – the one studied here might just happen to be particularly rigid. Nevertheless, these findings should warn against too ready an assumption that water is indispensible to the functioning of any complex protein-like macromolecule. My own view is that current evidence points to a strong and sensitive general dependence of protein function on both the structure and dynamics of the hydration environment, which in turn are intimately connected to the hydrogen-bonded network of the water molecules and their interactions with donors and acceptors at the protein surface. But a tendency to assume that this is always a precondition of protein function has militated against a thorough investigation of what proteins can and cannot do in truly anhydrous conditions. In short, we risk here the aquacentric prejudice that has coloured a great deal of the discussion about potential extraterrestrial biochemistries.

6.12 Water and Nucleic Acids

Compared with the attention given to hydration in determining protein structure and function, the role that water plays in the properties of nucleic acids has been rather neglected. It is often overlooked that the famous and rather beautiful double-helical structure of DNA is not intrinsic to that molecule but relies on a subtle balance of energy contributions present in aqueous solution. Without water to screen the electrostatic repulsions between phosphate groups, the classic, orderly helix is no longer viable. Thus DNA undergoes conformational transitions, and even loses its double helix, in some apolar solvents [94]. Even though the double helix is not lost entirely in the gas phase, under those conditions it has none of the familiar elegance and order [95].

In water, this double helix results from a delicately poised balance of forces. Single-stranded DNA seems to be shorter in water than in non-aqueous solvents, because of water bridges between bases [96]. These hydrogen bonds are relatively weak, and if they were much stronger they might in fact inhibit the formation of the double helix altogether. If that is so, water seems here to function in a 'Goldilocks' mode: some hydration is essential for a stable double helix, but not too much.

DNA in the crystalline state has a highly ordered hydration shell. A-T sequences have a 'spine of hydration' in which one layer of water molecules bridges the nitrogen and oxygen atoms of bases in the minor groove, while a second layer bridges water molecules in the first layer [97]. Moreover, this 'spine' persists in aqueous solution, with water residence times in the minor groove of more than 1 ns, comparable to those of 'buried' water molecules in globular proteins. More generally, the residence times of hydration water molecules around DNA show an analogously broad distribution to those around proteins, and fluctuations in the hydrogen-bond network happen on fast (fs to ps) timescales [98].

The sensitivity of B-DNA (the normal biologically relevant conformation) hydration to sequence suggests that the arrangement of water molecules might effectively transmit sequence information to locations remote from the bases themselves. There now seems to be good evidence that hydration structures are indeed used by DNA-binding proteins, as well as some small synthetic DNA-binding molecules, as part of the recognition process. In particular, the energetics of water release from sequence-specific hydration structures might be expected to influence the binding strengths. For example, the interaction of the *lac* repressor protein with the lac operon site on DNA in the presence of glutamate (which is known to influence protein—DNA interactions) differs between specific and non-specific binding primarily in that the former incurs release of bound water from the DNA [99]. Robinson and Sligar suggest that the loss of sequence specificity of the restriction enzyme *Eco*RI in the presence of certain solutes could be explained by the fact that water mediates the protein—DNA interaction, and that this influence is suppressed under conditions of decreased water activity [100]. They conclude that "water mediation may constitute a general motif for sequence-specific DNA recognition by restriction enzymes and other DNA-binding proteins."

How such effects depend in general on the particular sequence-specific hydration structures of DNA is still not clear. Fuxreiter et al. [101] found that these structures influence the water release on binding of the restriction enzyme *Bam*HI to its cognate sequence GCATCC and to similar but non-cognate sequences. The entropic and enthalpic changes due to water release from the protein—DNA interface are one of the key driving forces of the interaction, and simulations show that this release is highly dependent on sequence, so that a given DNA sequence has a 'hydration fingerprint' that determines the binding energetics.

The hydration structure of DNA can also play a functional role by determining its conformation. The conformational state of double-stranded DNA in solution is very sensitive to hydration: at low hydration, the most biologically relevant B form undergoes conformational transitions to other forms. The stabilization of the B form occurs very close to the hydration level at which water clusters in the primary hydration shell link up to form a fully connected (percolating) cluster in the major groove [102]. There is an almost identical percolation threshold for A-DNA, but in that case it corresponds to the appearance of a spanning water network in the *minor* groove [103]. It isn't yet clear whether this near-coincidence of thresholds arises from chance or from some deeper physical cause. In any event, these hydration structures may hold the key to transitions between the A and B conformations, particularly in so far as these are governed by the presence of ions, which may alter the hydration structures and thus the relative stabilities.

That hydration of DNA affects its conformation is also evident from the fact that dehydration stabilizes the A phase. Rather remarkably, quantum effects might be largely responsible for this. The zero-point motions of protons are entirely responsible for the binding of water to A-DNA, and in more hydrated conditions a change in the zero-point kinetic energy is sufficient in itself to motivate the transition to B-DNA [104]. Whether the protons concerned are those of water molecules in the hydration shell or those in the DNA's H-bonds (or perhaps a bit of both) is not yet clear.

Some of the water molecules in the minor groove of DNA can be substituted by cations, which induce electrostatic effects that can influence molecular curvature, alter the width of the groove, and affect the duplex melting temperature. Shui et al. [105] suggest that these ‘extrinsic’ influences on DNA deformation far outweigh any ‘intrinsic’ contributions owing to sequence-specific base—base interactions. They conclude that the hydration structure, and the presence of monovalent cations within it under physiological conditions, are essential for stabilizing the native B configuration of DNA – which is broadly consistent with the suggestions of Brovchenko et al. [102] above.

RNA appears to be more highly hydrated than DNA. As in DNA, G-C pairs are more hydrated than A-U(T) pairs, and the hydration structures around the former are better defined. The folding of RNAs into their functional forms resembles in many ways that of proteins: both macromolecules have hydrophilic and hydrophobic segments in their chain-like structures, and both may engage in intramolecular hydrogen-bonding in the folded state. But the distribution of the two types of component is more regular in RNA – all the bases are, aside from their hydrophilic substituents, hydrophobic, while the sugar—phosphate backbone is uniformly polar. This regular structure leads to correlated collapse of RNA strands into a compact form [106], which is more likely to trap water molecules between hydrophobic bases than is the less cooperative collapse of proteins, where hydrophobic residues are more sparsely distributed. Simulations suggest that this trapped water is expelled late in the folding process, so that there remains considerable potential for water-mediated interactions as compaction proceeds [106]. In this respect, it seems that specific water molecules buried within the folding macromolecule can play an important role in mediating compaction, as has been proposed for proteins [42].

All these results should lead us to infer that the celebrated ability of nucleic acids to store and transmit chemical information cannot be considered an intrinsic property of these molecules, but becomes possible – or at least, biologically viable – only via the influence of water. This does not, of course, imply that a genetic system of replication, inheritance and evolution is inconceivable in other solvents, and indeed some of the primitive self-replicating synthetic molecules reported to date work in non-aqueous solution [107]. But it does suggest that a chemical replicating system that approaches the sophistication of nucleic acids, controlled by selective regulatory molecules and subject to influences such as local curvature and conformation, might well have to make use of subtle interactions and structural arrangements that involve the solvent too.

6.13 The Influence of Salt

It is challenging enough to include the specific effects of water molecules and water structure in studies of the behaviour of biological macromolecules. But the cytoplasm’s fluid component is not of course pure water: it is an electrolyte, and that changes everything. The effect of dissolved ions on biomolecular shape, interaction and function seems to be enormously complex, and at present remains rather poorly understood. The starting point for much of the discussion is Franz Hofmeister's experiments in the late nineteenth century on the influence of different salts on protein aggregation [108]. He found that some ions promote protein solubilization (salting-in), while some promote aggregation and precipitation (salting-out). Anions and cations can be arranged in so-called Hofmeister series according to the strengths of the salting-out effect:

$$H_2PO_4^- > SO_4^{2-} > F^- > Cl^- > Br^- > NO_3^- > I^- > ClO_4^- > SCN^-$$

$$Mg^{2+} > Li^+ > Na^+ \sim K^+ > NH_4^+$$

The traditional explanation for the Hofmeister series invokes the concept of 'structure-making' and 'structure-breaking' ions. The basic idea is that large, low-charge ions such as I^- and NH_4^+ disrupt 'water structure' – they are structure-breakers – while small or highly charged ions such as F^- and Mg^{2+} are structure-makers, imposing order on the hydrogen-bonded network. Then salting-out and salting-in of proteins are explained on the basis of entropic changes induced in their hydration shells by the addition of ions, or alternatively, of a reduction in the strength of hydrogen bonding of water molecules complexed to dissolved ions. The classical hypothesis is that salting-out arises from a competition for solvation between the salt and the protein, in which an ion's ability to sequester waters of solvation is somehow connected to its effect on water structure. Thus the structure-making effect of small or highly charged ions depletes proteins of hydration water and causes precipitation.

Yet there is little consensus – and sometimes plain contradiction – about what structure-making and -breaking actually entails. Does structure-making render water denser, or does it, in making water more ice-like, actually reduce the density? More to the point, this concept lacks any real evidence from experimental studies of the structure of electrolyte solutions that significant changes to the bulk hydrogen-bonded network of water really do occur in the presence of salts.

Instead of trying to understand the Hofmeister series on the basis of 'global' changes in solvent structure induced by ionic solutes, it seems far more logical, and also more consistent with current experimental and theoretical studies, to consider the effects that these ions have on the local hydration of protein residues or other hydrophobes. It now seems likely that Hofmeister effects must be understood in terms of these specific and often rather subtle interactions between ions and proteins or other biomolecules [109].

Ions do not, in general, simply disperse homogeneously throughout the solution so as to create a kind of 'mean-field' solvent for other large solutes such as macromolecules. Rather, many ions tend to segregate preferentially at either hydrophilic or hydrophobic surfaces. Traditionally, ions have been considered to be excluded from the air-water interface because electrolytes increase surface tension. But recent studies show that the picture is not so simple [110, 111]: while it may hold for hard (non-polarizable) ions such as sodium and fluoride, large soft ions such as iodide (and to a lesser extent, bromide and chloride) can accumulate

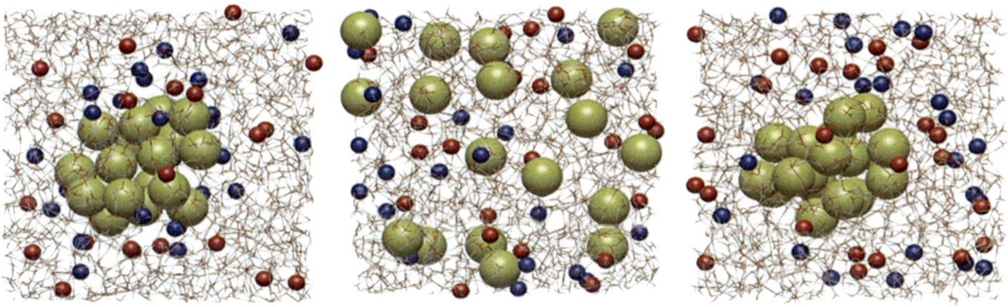

Fig. 6.16 The distribution of ions around hydrophobic (Lennard-Jones) particles in water. The hydrophobes are *yellow*, positive ions are *red* and negative ions are *blue*. Low-q ions (*left*) are adsorbed preferentially at the particle surfaces, leading to micelle-like clusters of hydrophobic particles surrounded by ions, which prevents further aggregation and precipitation. High-q ions (*right*) tended to be depleted at the particle surfaces, which again leads to the formation of clusters. In the intermediate-q case (*centre*) there is neither adsorption nor depletion, and the hydrophobes remain individually dispersed (Reprinted with permission from Ref. [112]. © 2006 American Chemical Society)

preferentially at the surface. Since one might expect the interface of water with a hydrophobic surface to mimic in many respects that with air, this inhomogeneity of solutes at a surface could have significant implications for the hydrophobic interaction and the solvation of proteins.

As an example of the kind of complexity that might stem from the inhomogeneity of ion distributions, Zangi and Berne have considered how ions interact with small hydrophobic particles 0.5 nm across [112] (Fig. 6.16). They found that ions with high charge density (q) induce salting out by promoting stronger hydrophobic interactions that cause particle aggregation. But low-q ions could have either a salting-out or a salting-in effect, depending on their concentration. These effects are related to preferential absorption or exclusion of the ions at the particle surfaces, but not in any simple, monotonic fashion. High-q ions tended to be depleted at the surface of the hydrophobic particle clusters, but are tightly bound to water elsewhere, thereby decreasing the number of water molecules available for solvating the particles. Low-q ions are absorbed preferentially at the particle surfaces, and at high ionic concentrations this can lead to salting-in because the hydrophobic particles form clusters surrounded by ions in a micelle-like arrangement. At lower concentrations, the ions are unable to solubilize aggregates in this way, and salting-out occurs.

Analogous partitioning of ions at the surfaces of nanoscale hydrophobic plates should alter the hydrophobic interaction between them [113]. But it seems that Hofmeister effects may have a different origin, and thus a different character, for small and large hydrophobic particles: whereas in the former case there is an increase in hydrophobic aggregation for both high-q and low-q but not medium-q ions (except at high concentrations), for hydrophobic plates the relationship is monotonic, with an increasing tendency towards salting-in as the ion charge density decreases. In both cases, however, the mechanism is somewhat subtle and dependent on the direct ion—hydrophobe interaction, and need not invoke the vague notion of 'water structure'.

Preferential segregation of ions may apply equally to hydronium and/or hydroxide, rendering the air-water interface either acidic or basic. But there is no consensus about which of these situations obtains in general, with both theoretical and experimental evidence to support both contentions. One suggestion is that the positive charge on the oxygen atom of H_3O^+ renders it a poor hydrogen-bond acceptor, and indeed makes it rather hydrophobic, so that this ion acts as an amphiphile and aggregates at the interface [114]. It is estimated that this could lower the pH of the water surface to around 4.8 or less [115], a finding that might be expected to have significant implications both for hydration of hydrophobic patches on biomolecules and for rates of hydrolysis at the interface. Surfactant behaviour of hydronium ions could even be expected to stabilize water—hydrophobe interfaces. On the other hand, zeta potential measurements seem to indicate that both the air—water and oil—water interfaces are basic, due to specific adsorption of hydroxide [116, 117]. At present, these discrepancies remain to be resolved.

6.14 Conclusions

It will now hopefully be clear that water is itself a biomolecule, in so far as it can be found playing a wide variety of roles in biochemical processes. It maintains macromolecular structure and mediates molecular recognition, it activates and modulates protein dynamics, it provides a switchable communication channel across membranes and between the inside and outside of proteins. Faced with this diverse array of functions for water in biology, the temptation in an astrobiological context is to start searching for non-aqueous solvents that might be capable of fulfilling the same roles – as proton wires, say, or as hydrogen-bonded bridging groups in macromolecular interactions.

But a search for molecules that can do what water does is not really the right way to address the questions posed at the outset. All too often, the question of whether non-aqueous solvents could host living systems has been interpreted as a matter of finding liquids (or other fluids) that can stand in for water, without first considering whether making the respective functions of water a *sine qua non* of life is not a little anthropocentric. I recall some years ago at a conference on this very question [118] hearing the claim that life could not exist outside water because only water offered fast proton transport via Grotthuss-like water wires. The speaker felt no obligation to explain why facile proton transport should be regarded as an essential requirement for life. Especially given our present, incomplete state of knowledge about pivotal biochemical concepts such as the hydrophobic interaction, it is not obvious that any one of the functions of water in biology has to stand as an irreducible aspect of a 'living system'.

We must also be careful to distinguish the roles of water that depend on its 'special' status as a three-dimensional structured liquid from ones that are more generic. Many of these roles do seem to depend, to a greater or lesser degree, on the uniqueness of the H_2O molecule, in particular its ability to engage in directional, weak bonding in a way that allows for reorientation and reconfiguration of discrete and identifiable three-dimensional structures. On the other hand, some of water's functions in biology are those of a generic polar solvent. Certainly, solvophobicity and solvophilicity are rather general properties, although it is far from clear that they will have precisely the same characteristics in a solvent that lacks the three-dimensional, directional associations of water molecules. Some peptide-like molecules, for example, can collapse into fairly well-defined, compact shapes in non-aqueous solvents, dictated by both hydrogen bonding and solvophobic effects.

Moreover, life in water has some notable drawbacks [7] – perhaps most notably the solvent's reactivity, raising the problem of hydrolysis of polymeric structures and of basic building blocks such as sugars. How the first pseudo-biological macromolecules on the early Earth avoided this problem is still something of a puzzle. It is also unclear whether a solvent capable of engaging in hydrogen-bonding might (before evolution has brought about much fine-tuning) help or hinder the use of this valuable, reversible supramolecular interaction for defining complex structures in macromolecules and their aggregates.

Behind all such discussions is the ambiguity about the very concept of life. Attempts to enunciate the irreducible *molecular-scale* requirements for (as opposed to the emergent characteristics of) something we might recognize as life have been rather sporadic [7, 118], and are often hampered by the difficulty of looking at the question through anything other than aqua-tinted spectacles. From the point of view of thinking about non-aqueous astrobiological solvents, a review of water's roles in terrestrial biochemistry surely raises one key consideration straight away: it is not sufficient, in this context, to imagine a clear separation between the 'molecular machinery' and the solvent. There is a two-way exchange of behaviours between them, and this literally erases any dividing line between the biological components and their environment.

The key questions here are, then, necessarily vague. But the more we understand about the biochemical aspects of water, the less likely it seems that another solvent could mimic its versatility, sensitivity and responsiveness, for example to distinguish any old collapsed polypeptide chain from a fully functioning protein. It is perhaps this notion of *responsiveness* that emerges as the chief characteristic from a survey of water's biological roles. It can be manipulated in three dimensions to augment the influence of biomolecules. It can receive and transmit their dynamical behaviours, and at the same time it can impose its own influence on solute dynamics so that some biomolecular behaviours become a kind of intimate conspiracy between solute and solvent. This adaptive sensitivity seems to facilitate the kind of compromise between structural integrity and reconfigurability that lies at the heart of many biomolecular processes, including molecular recognition, catalytic activity, conformational flexibility, long-range informational transfer and the ability to adapt to new environments. It is easy to imagine – but very hard to prove! – that such properties are likely to be needed in any molecular system with sufficient complexity to grow, replicate, metabolize and evolve – in other words, to qualify as living.

In these respects it does seem challenging to postulate any solvent that can hold a candle to water – not so much in terms of what it *does*, but in terms of *the opportunities it offers for molecular evolution*. This is by no means to endorse the dictum of NASA that astrobiologists need to 'follow the water'. But hopefully it might sharpen the question of where else we might look.

References

1. Henderson L (1913) The fitness of the environment. Macmillan, New York
2. Ferber D (2004) Microbes made to order. Science 303:158–161
3. Chin JW, Cropp TA, Anderson JC, Zhang Z, Schultz PG (2003) An expanded eukaryotic genetic code. Science 301:964–967
4. Kool ET (2002) Replacing the nucleobases in DNA with designer molecules. Acc Chem Res 35:936–943
5. Ball P (2007) Water as an active constituent in cell biology. Chem Rev 108:74–108
6. Katz JJ, Crespi HL (1966) Deuterated organisms: cultivation and uses. Science 151:1187–1194
7. Benner SA, Ricardo A, Carrigan MA (2004) Is there a common chemical model for life in the universe? Curr Opin Chem Biol 8:672–689
8. Fernández A, Scott R (2003) Dehydron: a structurally encoded signal for protein interaction. Biophys J 85:1914–1928
9. Lynden-Bell RM, Morris SC, Barrow JD, Finney JL, Harper CL (eds) (2010) Water and life. CRC Press, Boca Raton
10. Franks F (2000) Water: a matrix of life. Royal Society of Chemistry, Cambridge
11. Wernet Ph, Nordlund D, Bergmann U, Cavalleri M, Odelius M, Ogasawara H, Näslund LÅ, Hirsch TK, Ojamäe L, Glatzel P, Pettersson LGM, Nilsson A (2004) The structure of the first coordination shell in water. Science 304:995–999
12. Head-Gordon T, Johnson ME (2006) Tetrahedral structure or chains for liquid water. Proc Natl Acad Sci USA 103:7973–7977
13. Luzar A, Chandler D (1996) Hydrogen-bond kinetics in liquid water. Nature 379:55–57
14. Sciortino F, Geiger A, Stanley HE (1991) Effects of defects on molecular mobility in liquid water. Nature 354:218–221
15. Ellis RJ, Minton AP (2003) Join the crowd. Nature 425:27–28
16. Major RC, Houston JE, McGrath MJ, Siepmann JI, Zhu X-Y (2006) Viscous water meniscus under confinement. Phys Rev Lett 96:177803
17. Li T-D, Gao J, Szoszkiewicz R, Landman U, Riedo E (2007) Structured and viscous water in subnanometer gaps. Phys Rev B 75:115415
18. Henderson D (ed) (1992) Fundamentals of inhomogeneous fluids. CRC Press, Boca Raton
19. Hassan S, Steinbach P (2011) Water-exclusion and liquid-structure forces in implicit solvation. J Phys Chem B 115:14668–14682
20. Stradner A, Sedgwick H, Cardinaux F, Poon WCK, Egelhaaf SU, Schurtenberger P (2004) Equilibrium cluster formation in concentrated protein solutions and colloids. Nature 432:492–495
21. Gliko O, Pan W, Katsonis P, Neumaier N, Galkin O, Weinkauf S, Velikov PG (2007) Metastable liquid clusters in super- and undersaturated protein solutions. J Phys Chem B 111:3106–3114
22. Pollack GH (2001) Cells, gels, and the engines of life. Ebner & Sons, Seattle

23. Chaplin M (2006) Do we underestimate the importance of water in cell biology? Nat Rev Mol Cell Biol 7:861–866
24. Halle B, Persson E (2008) Cell water dynamics on multiple time scales. Proc Natl Acad Sci USA 105:6266–6271
25. Tanford C (1980) The hydrophobic effect, 2nd edn. Wiley, New York
26. Blokzijl W, Engberts JBFN (1993) Hydrophobic effects. Opinions and facts. Angew Chem Int Ed 32:1545–1579
27. Frank HS, Evans MW (1945) Free volume and entropy in condensed systems. III. Entropy in binary liquid mixtures; partial molal entropy in dilute solutions; structure and thermodynamics in aqueous solutions. J Chem Phys 13:507–532
28. Kauzmann W (1969) Some factors in the interpretation of protein denaturation. Adv Protein Chem 14:1–63
29. Ball P (2003) How to keep dry in water. Nature 423:25–26
30. Lum K, Chandler D, Weeks JD (1999) Hydrophobicity at small and large length scales. J Phys Chem B 103:4570–4577
31. Wallqvist A, Berne BJ (1995) Computer simulation of hydrophobic hydration forces on stacked plates at short range. J Phys Chem 99:2893–2899
32. ten Wolde PR, Chandler D (2002) Drying-induced hydrophobic polymer collapse. Proc Natl Acad Sci USA 99:6539–6543
33. Li ITS, Walker GC (2011) Signature of hydrophobic hydration in a single polymer. Proc Natl Acad Sci USA 108:16527–16532
34. Liu P, Huang X, Zhou R, Berne BJ (2005) Observation of a dewetting transition in the collapse of the melittin tetramer. Nature 437:159–162
35. Zhou R, Huang X, Margulis CJ, Berne BJ (2004) Hydrophobic collapse in multidomain protein folding. Science 305:1605–1609
36. Hua L, Huang X, Liu P, Zhou R, Berne BJ (2007) Nanoscale dewetting transition in protein complex folding. J Phys Chem B 111:9069–9077
37. Giovambattista N, Lopez CF, Rossky PJ, Debenedetti PG (2008) Hydrophobicity of protein surfaces: separating geometry from chemistry. Proc Natl Acad Sci USA 105:2274–2279
38. Patel AJ, Varrily P, Chandler D (2010) Fluctuations of water near extended hydrophobic and hydrophilic surfaces. J Phys Chem B 114:1632–1637
39. Patel AJ, Varilly P, Jamadagni SN, Hagan MF, Chander D, Garde S (2012) Sitting at the edge: how biomolecules use hydrophobicity to tune their interactions and function. J Phys Chem B. doi:10.1021/jp2107523
40. Onuchic JN, Luthey-Schulten Z, Wolynes PG (1997) Theory of protein folding: the energy landscape perspective. Annu Rev Phys Chem 48:545–600
41. Dobson CM (2003) Protein folding and misfolding. Nature 426:884–890
42. Zong C, Papaoian GA, Ulander J, Wolynes PG (2006) Role of topology, nonadditivity, and water-mediated interactions in predicting the structures of α/β proteins. J Am Chem Soc 128:5168–5176
43. Patel AJ, Varilly P, Jamadagni SN, Acharya H, Garde S, Chandler D (2011) Extended surfaces modulate hydrophobic interactions of neighboring solutes. Proc Natl Acad Sci USA 108:17678–17683
44. Davidovic M, Mattea C, Qvist J, Halle B (2009) Protein cold denaturation as seen from the solvent. J Am Chem Soc 131:1025–1036
45. Hua L, Zhou R, Thirumalai D, Berne BJ (2008) Urea denaturation by stronger dispersion interactions with proteins that water implies a 2-stage unfolding. Proc Natl Acad Sci USA 105:16928–16933
46. England JL, Pande VS, Haran G (2008) Chemical denaturants inhibit the onset of dewetting. J Am Chem Soc 130:11854–11855
47. Zangi R, Zhou R, Berne BJ (2009) Urea's action on hydrophobic interactions. J Am Chem Soc 131:1535–1541
48. Bennion BJ, Daggett V (2003) The molecular basis for the chemical denaturation of proteins by urea. Proc Natl Acad Sci USA 100:5142–5147

49. Towey JJ, Soper AK, Dougan L (2011) Preference for isolated water molecules in a concentrated glycerol-water mixture. J Phys Chem B 115:7799–7807
50. Pioliti R, Sapir L, Harries D (2009) The impact of polyols on water structure in solution: a computational study. J Phys Chem A 113:7548–7555
51. Dougan L, Genchev GZ, Lu H, Fernández JM (2011) Probing osmolyte participation in the unfolding transition state of a protein. Proc Natl Acad Sci USA 108:9759–9764
52. Reddy G, Straub E, Thirumalai D (2010) Dry amyloid fibril assembly in a yeast prion peptide is mediated by long-lived structures containing water wires. Proc Natl Acad Sci USA 107:21459–21464
53. Fernández A, Kardos J, Scott LR, Goto Y, Berry RS (2003) Structural defects and the diagnosis of amyloidogenic propensity. Proc Natl Acad Sci USA 100:6446–6451
54. Fernández A, Lynch M (2011) Non-adaptive origins of interactome complexity. Nature 474:502–505
55. De Simone A, Dhulesia A, Soldi G, Vendruscolo M, Hsu S-TD, Chiti F, Dobson CM (2011) Experimental free energy surfaces reveal the mechanisms of maintenance of protein stability. Proc Natl Acad Sci USA 108:21057–21062
56. Yu H, Rick S (2010) Free energy, entropy, and enthalpy of a water molecule in various protein environments. J Phys Chem B 114:11552–11560
57. Baron R, Setny P, McCammon JA (2010) Water in cavity-ligand recognition. J Am Chem Soc 132:12091–12097
58. Snyder PW, Mecinovic J, Moustakas DT, Thomas SW III, Harder M, Mack ET, Lockett MR, Héroux A, Sherman W, Whitesides GM (2011) Mechanism of the hydrophobic effect in the biomolecular recognition of arylsulfonamides by carbonic anhydrase. Proc Natl Acad Sci USA 108:17889–17894
59. Yin H, Hummer G, Rasaiah JC (2007) Metastable water clusters in the nonpolar cavities of the thermostable protein tetrabrachion. J Am Chem Soc 129:7369–7377
60. Ahmad M, Gu W, Geyer T, Helms V (2011) Adhesive water networks facilitate binding of protein interfaces. Nat Commun 2:261
61. Sahai MA, Biggin PC (2011) Quantifying water-mediated protein-ligand interactions in a glutamate receptor: a DFT study. J Phys Chem B 115:7085–7096
62. Tame JRH, Sleigh SH, Wilkinson AJ, Ladbury JE (1996) The role of water in sequence-independent ligand binding by an oligopeptide transporter protein. Nat Struct Biol 3:998–1001
63. Grossman M, Born B, Heyden M, Tworowski D, Fields GB, Sagi I, Havenith M (2011) Correlated structural kinetics and retarded solvent dynamics at the metalloprotease active site. Nat Struct Mol Biol 18:1102–1108
64. Gnanasekaran R, Xu Y, Leitner DM (2010) Dynamics of water clusters confined in proteins: a molecular dynamics simulation study of interfacial waters in a dimeric hemoglobin. J Phys Chem B 114:16989–16996
65. Rodriquez JC, Zeng Y, Wilks A, Rivera M (2007) The hydrogen-bonding network in heme oxygenase also functions as a modulator of enzyme dynamics: chaotic motions upon disrupting the H-bond network in heme oxygenase from *Psuedomonas aeruginosa*. J Am Chem Soc 129:11730–11742
66. Krauss M, Gilson HSR, Gresh N (2001) Structure of the first-shell active site in metallolactamase: effect of water ligands. J Phys Chem B 105:8040–8049
67. Wang L, Yu X, Hu P, Broyde S, Zhang Y (2007) A water-mediated and substrate-assisted catalytic mechanism for *Sulfolobus solfataricus* DNA polymerase IV. J Am Chem Soc 129:4731–4737
68. Sicking W, Korth H-G, de Groot H, Sustmann R (2008) On the functional role of a water molecule in clade 3 catalases: a proposal for the mechanism by which NADPH prevents the formation of compound II. J Am Chem Soc 130:7345–7356
69. Agmon N (1995) The Grotthuss mechanism. Chem Phys Lett 244:456–462
70. Garczarek F, Gerwert K (2006) Functional waters in intraprotein proton transfer monitored by FTIR Spectroscopy. Nature 439:109–112

71. Mathias G, Marx D (2007) Structures and spectral signatures of protonated water networks in bacteriorhodopsin. Proc Natl Acad Sci USA 104:6980–6985
72. Chen H, Ilan B, Wu Y, Zhu F, Schulten K, Voth GA (2007) Charge delocalization in proton channels, I: the aquaporin channels and proton blockage. Biophys J 92:46–60
73. Brändén M, Sandén T, Brzezinski P, Widengren J (2006) Localized proton microcircuits at the biological membrane-water interface. Proc Natl Acad Sci USA 103:19766–19770
74. Springer A, Hagen V, Cherepanov DA, Antonenko YN, Pohl P (2011) Protons migrate along interfacial water without significant contributions from jumps between ionizable groups on the membrane surface. Proc Natl Acad Sci USA 108:14461–14466
75. Beckstein O, Biggin PC, Sansom MSPA (2001) A hydrophobic gating mechanism for nanopores. J Phys Chem B 105:12902–12905
76. Rasaiah JC, Garde S, Hummer G (2008) Water in nonpolar confinement: from nanotubes to proteins and beyond. Annu Rev Phys Chem 59:713–740
77. Anishkin A, Sukharev S (2004) Water dynamics and dewetting transitions in the small mechanosensitive channel MscS. Biophys J 86:2883–2895
78. Umena Y, Kawakami K, Shen J-R, Kamiya N (2011) Crystal structure of oxygen-evolving photosystem II at a resolution of 1.9 Å. Nature 473:55–60
79. Rupley JA, Careri G (1991) Protein hydration and function. Adv Protein Chem 41:37–172
80. Purkiss A, Skoulakis S, Goodfellow JM (2001) The protein-solvent interface: a big splash. Phil Trans R Soc Lond A Math Phys Eng Sci 359:1515–1527
81. Olano LR, Rick SW (2004) Hydration free energies and entropies for water in protein interiors. J Am Chem Soc 126:7991–8000
82. Okimoto N, Nakamura T, Suenaga A, Futatsugi N, Hirano Y, Yamaguchi I, Ebisuzaki T (2004) Cooperative motions of protein and hydration water molecules: molecular dynamics study of scytalone dehydratase. J Am Chem Soc 126:13132–13139
83. Bizzarri AR, Cannistraro S (2002) Molecular dynamics of water at the protein-solvent interface. J Phys Chem B 106:6617–6633
84. Russo D, Murarka RK, Copley JRD, Head-Gordon T (2005) Molecular view of water dynamics near model peptides. J Phys Chem B 109:12966–12975
85. Rasmussen BF, Stock AM, Ringe D, Petsko GA (1992) Crystalline ribonuclease A loses function below the dynamical transition at 220 K. Nature 357:423–424
86. Reat V, Dunn R, Ferrand M, Finney JL, Daniel RM, Smith JC (2000) Solvent dependence of dynamic transitions in protein solutions. Proc Natl Acad Sci USA 97:9961–9966
87. Tarek M, Tobias DJ (2002) Role of protein-water hydrogen bond dynamics in the protein dynamical transition. Phys Rev Lett 88:138101
88. Chen S-H, Liu L, Fratini E, Baglioni P, Faraone A, Mamontov E (2006) Observation of fragile-to-strong dynamic crossover in protein hydration water. Proc Natl Acad Sci USA 103:9012–9016
89. Tournier AL, Xu J, Smith JC (2003) Translational hydration water dynamics drives the protein glass transition. Biophys J 85:1871–1875
90. Smolin N, Oleinikova A, Brovchenko I, Geiger A, Winter R (2005) Properties of spanning water networks at protein surfaces. J Phys Chem B 109:10995–11005
91. Gallat F-X, Laganowsky A, Wood K, Gabel F, van Eijk L, Wuttke J, Moulin M, Härtlein M, Eisenberg D, Colletier J-P, Zaccai G, Weik M. (2012) Dynamical coupling of intrinsically disordered proteins and their hydration water: comparison with folded soluble and membrane proteins. Biophys J 103:129–136
92. Magazù S, Migliardo F, Benedetto A (2011) Puzzle of protein *dynamical transition*. J Phys Chem B 115:7736–7743
93. Lopez M, Kurkal-Siebert V, Dunn RV, Tehei M, Finey JL, Smith JC, Daniel RM (2010) Activity and dynamics of an enzyme, pig liver esterase, in near-anhydrous conditions. Biophys J 99:L62–L64
94. Turner DH (2000) Conformational changes. In: Bloomfield VA, Crothers DM, Tinoco I (eds) Nucleic acids. Structure, properties and functions. University Science, Sausalito, pp 259–334
95. Rueda M, Kalko SG, Luque FJ, Orozco M (2003) The structure and dynamics of DNA in the gas phase. J Am Chem Soc 125:8007–8014

96. Cui S, Albrecht C, Kühner F, Gaub HE (2006) Weakly bound water molecules shorten single-stranded DNA. J Am Chem Soc 128:6636–6639
97. Kopka ML, Fratini AV, Drew HR, Dickerson RE (1983) Ordered water structure around a *B*-DNA dodecamer: a quantitative study. J Mol Biol 163:129–146
98. Szyc L, Yang M, Elsaesser T (2010) Ultrafast energy exchange via water-phosphate interactions in hydrated DNA. J Phys Chem B 114:7951–7957
99. Ha JH, Capp MW, Hohenwalter MD, Baskerville M, Record MT Jr (1992) Thermodynamic stoichiometries of participation of water, cations and anions in specific and non-specific binding of *lac* repressor to DNA: possible thermodynamic origins of the 'glutamate effect' on protein-DNA interactions. J Mol Biol 228:252–264
100. Robinson CR, Sligar SG (1993) Molecular recognition mediated by bound water: a mechanism for star activity of the restriction enzyme endonuclease *Eco*RI. J Mol Biol 234:302–306
101. Fuxreiter M, Mezei M, Simon I, Osman R (2005) Interfacial water as a 'hydration fingerprint' in the noncognate complex of *Bam*HI. Biophys J 89:903–911
102. Brovchenko I, Krukau A, Oleinikova A, Mazur AK (2006) Water percolation governs polymorphic transitions and conductivity of DNA. Phys Rev Lett 97:137801
103. Brovchenko I, Krukau A, Oleinikova A, Mazur AK (2007) Water clustering and percolation in low hydration DNA shells. J Phys Chem B 111:3258–3266
104. Reiter GF, Senesi R, Mayers J (2010) Changes in the zero-point energy of protons as the source of the binding energy of water to *A*-phase DNA. Phys Rev Lett 105:148101
105. Shui X, Sines CC, McFail-Isom L, VanDerveer D, Williams LD (1998) Structure of the potassium form of CGCGAATTCGCG: DNA deformation by electrostatic collapse around inorganic cations. Biochemistry 37:16877–16887
106. Sorin EJ, Rhee YM, Pande VS (2005) Does water play a structural role in the folding of small nucleic acids? Biophys J 88:2516–2524
107. Tjivikua T, Ballester P, Rebek J Jr (1990) Self-replicating system. J Am Chem Soc 112:1249–1250
108. Kunz W, Henle J, Ninham BW (2004) 'Zur Lehre von der Wikrung der Salze' (about the science of the effect of salts): Franz Hofmeister's historical papers. Curr Opin Colloid Interface Sci 9:19–37
109. Tobias D, Hemminger J (2008) Getting specific about specific ion effects. Science 319:1197–1198
110. Jungwirth P, Tobias DJ (2006) Specific ion effects at the air/water interface. Chem Rev 106:1259–1281
111. Petersen PB, Saykally RJ (2006) On the nature of ions at the liquid water surface. Annu Rev Phys Chem 57:333–364
112. Zangi R, Berne BJ (2006) Aggregation and dispersion of small hydrophobic particles in aqueous electrolyte solutions. J Phys Chem B 110:22736–22741
113. Zangi R, Hagen M, Berne BJ (2007) Effects of ions on the hydrophobic interaction between two plates. J Am Chem Soc 129:4678–4686
114. Petersen MK, Iyengar SS, Day TJF, Voth GA (2004) The hydrated proton at the water liquid/vapor interface. J Phys Chem B 108:14804–14806
115. Buch V, Milet A, Vácha R, Jungwirth P, Devlin JP (2007) Proc Natl Acad Sci USA 104:7342
116. Beattie JK, Djerdjev AM, Warr GG (2009) The surface of neat water is basic. Faraday Discuss 141:31–39
117. Creux P, Lachaise J, Graciaa A, Beattie JK, Djerdjev AM (2009) Strong specific hydroxide ion binding at the pristine oil/water and air/water interfaces. J Phys Chem B 113:14146–14150
118. Daniel RM, Finney JL, Stoneham M (2004) The molecular basis of life: is life possible without water? Phil Trans R Soc Lond B 359:1143–1328

Chapter 7
The Boundaries of Life

Charles S. Cockell and Sophie Nixon

Abstract The boundaries of life are set by the physical and chemical limits beyond which functions associated with life, including growth and reproduction, cannot occur. Although these limits might appear to be specific to terrestrial life, thermodynamics and the basic biophysical properties of carbon-based molecules mean that the boundary of life using carbon as a molecular backbone and water as a solvent (the 'biospace') is likely to be universal, although exhibiting small variations depending on the particular molecular architecture adopted by life. Entirely novel biospaces using different chemistries (e.g. ammonia as a solvent) might be possible, although there is currently no empirical evidence for these alternative life chemistries.

7.1 Introduction

The complexity of biological systems and the extraordinary diversity of organisms generated by the process of biological evolution often leads to an assumption that biological systems cannot be universal, that every 'experiment' in biological evolution (if other experiments exist) will lead to different outcomes. In this chapter, we shall briefly review what we know about life in extreme environments and we will discuss the possibility that although the organisms generated by biological evolution may well be very different in different places in the universe, the 'Biospace' (the space bounded by extreme physical and chemical conditions), of carbon-based life that uses water as a solvent, is universal. However, different chemistries of life could lead to different 'biospaces', although no empirical evidence for these other biospaces currently exists.

C.S. Cockell (✉) • S. Nixon
School of Physics and Astronomy, University of Edinburgh, King's Buildings, Edinburgh EH9 3JZ, UK
e-mail: c.s.cockell@ed.ac.uk

I.W.M. Smith et al. (eds.), *Astrochemistry and Astrobiology*, Physical Chemistry in Action,
DOI 10.1007/978-3-642-31730-9_7,

Although the purpose of this chapter is not to debate the definition of life, we assume that life exhibits several functions, including metabolism and replication and that the limits of life are established by the conditions that prevent these functions from occurring. Thus, the discussion in this chapter focuses on how physical extremes influence components of cells that allow them to grow, reproduce and evolve.

7.2 Adaptations to Extremes

Life has adapted to a remarkable variety of extremes, all of which illustrate various principles of physical chemistry in action in extreme environments. Needless to say, the literature on these extremes is enormous and covers extreme parameters such as: pH, temperature, radiation, pressure, salt tolerance (water activity), heavy metal resistance and so on. A review of each of these extremes would take us into the realms of a book and indeed there are books reviewing the various adaptations of extremophiles [1]. It is also pertinent to point out that there is now a vast amount of literature on the species that can survive in these different extremes.

Here, our purpose is to illustrate the physical chemistry of how organisms can tolerate extremes and how this may lead to definable boundaries for life on Earth and potentially elsewhere. Rather than being comprehensive for all extremes, we focus on low and high temperatures, pressure, water activity and pH and describe the evidence for the limits to life, the adaptations that organisms use to survive these extremes and the physico-chemical basis of some of them. This chapter is also not intended to be a comprehensive reference of the vast number of organisms that have been studied in extremes, although species names are mentioned where it is useful.

7.2.1 Low Temperatures

A large number of Earth's environments experience low temperatures, for example sea ice, the deep oceans and polar regions. The lower temperature limit for metabolically active life is controversial and it is worth exploring the challenge in a little more detail here as it offers a very good example of the difficulties in defining the limits of physical chemistry that allows for actively-metabolising life. Although the data described here are specific to low temperature studies, the general problem of finding the exact boundary between life and 'non-life' is illustrated well.

Micro-organisms can survive prolonged exposure at temperatures below $-20°C$, but convincing evidence for microbial reproduction at such temperatures is lacking. Direct measurements of increasing cell numbers over time using microscopy or spectrophotometry have not demonstrated microbial replication below temperatures of $-12°C$ to $-13°C$ for cultivated micro-organisms in glycerol-amended or supercooled media (*Psychromonas ingrahamii* [2]; *Colwellia psychrerythraea* strain 34H [3]) or below $-12°C$ for an uncharacterized microbial population from winter Arctic sea-ice in an organic-rich brine [4]. A photograph of a dividing microorganism in a winter sea-ice brine inclusion at $-15°C$ (Junge et al. [5], their Fig. 3), if accepted, represents the lowest temperature at which microbial reproduction has been directly demonstrated.

As temperatures get lower, direct approaches become more challenging, partly because of the long time-scales required to get interpretable results (e.g. for the two isolates above, measured doubling times were 5–10 days [6] and Bakermans et al. [7] estimated doubling times of 20–39 days for permafrost bacteria at $-10°C$). The relevant variables at low temperatures are difficult to model in the laboratory or control in the field with these approaches. Particularly given the small fraction ($<1\%$) of micro-organisms yet cultured in the laboratory, many investigators have adopted less direct approaches to measuring microbial activity at low temperatures, including the use of radioactive tracers to quantify incorporation of precursors into DNA, protein or lipids, application of fluorogenic dyes to determine electron transport chain activity, or measurements of gas fluxes or excess gas to infer respiration.

It is important to realize that these approaches measure metabolism that may not be associated with microbial replication and that attempts to infer or rule out replication are seldom made or justified by the measurements. Enzymatic activities have been observed at lower temperatures than microbial replication. It has even been suggested that biochemical reactions may persist in aqueous solution down to temperatures approaching the glass transition point of $-140°C$ [8], a hypothesis for which there is some support (reviewed in[9]; see also [10]). Bearing in mind these caveats, experiments using radiolabelled precursors of DNA, protein or lipids show incorporation of these into biomolecules to temperatures as low as $-20°C$ [6, 10]. Due to the technical challenges of making accurate measurements at such temperatures, some of these results have been called into question. For example, Warren and Hudson [11] suggested that Carpenter et al. [12] measured activity at warmer temperatures than they realized due to an experimental artefact. It is likewise difficult to understand why rates at $-20°C$ were comparable to those of an intended negative control at $-80°C$, contrary to thermodynamic expectations, in the work of Junge et al. [10].

Other studies support microbial activity in the vicinity of $-20°C$. Using soil and permafrost organisms cultured on solid media, Panikov and Sizova [13] measured production and/or uptake of CO_2 by bacteria to $-17°C$ and, transiently (for a few weeks), by a eukaryotic consortium to $-35°C$, with prolonged metabolism only at temperatures $\geq -18°C$. Respiratory activity, determined using a redox-sensitive fluorogenic dye, was reported in sea-ice bacteria at $-20°C$ [14]. Methane production attributable to micro-organisms in permafrost has been measured at $-16.5°C$ [15]. Release of CO_2 from tundra soils down to $-18°C$ [16] and $-39°C$ [17] has also been ascribed to biological activity based on the observed kinetics and, in the latter case, the elimination of production following sterilization. Finally, anomalous concentrations of gases in ice cores have been used to infer metabolic activity in glacial ice to temperatures as low as $-40°C$ [18].

One important unresolved question is how much metabolism over what time scales is necessary for a micro-organism to replicate. In particular, is the metabolism at temperatures below $-15^{\circ}C$ sufficient to support reproduction? While some studies have used models to convert measured metabolic rates into presumptive growth rates, such inferences have not been confirmed with direct measurements, and so leave open whether the observed level of metabolism is actually indicative of an ability for a whole cell to replicate.

The above discussion illustrates a general problem – that chemical processes associated with life can occur in environments when there is little possibility for the activity of whole organisms. Enzymes may show activity at low temperatures at which whole cells are not active. In other words, some chemical reactions can proceed at extremes where it is not possible for all chemical reactions that would support life to proceed. This makes it difficult to quantify the exact boundaries of many physical and chemical extremes for life.

One currently prevalent approach to this question at low temperatures, first proposed by Morita [19], is to classify metabolisms into three categories, (1) sufficient for growth, (2) sufficient for maintenance, and (3) sufficient for survival, with the latter understood as providing only enough energy to counter macromolecular damage. This classification gained a measure of support when Price and Sowers [20] surveyed available information on microbial metabolism and, from the generated Arrhenius plots, determined that the data fell into three roughly linear groups, which, they proposed, accorded with Morita's definitions. In the case of what they called "survival metabolism," they found that the measured rates were comparable to extrapolated rates of amino acid racemization and DNA depurination as a function of temperature, consistent with Morita's definition. The meaning of the other two categories, and whether or not they accord with Morita's definitions, is less clear. For example, no empirical evidence was presented demonstrating that energy levels characterized as sufficient only for maintenance metabolism actually precluded growth. Nor were energy levels characterized as sufficient for growth necessarily shown to be sufficient for replication.

Generating data that could validate these categories is far from trivial. While demonstration of replication is experimentally tractable, and lower limits for survival metabolism can be set by chemical decay processes (e.g., rates of amino acid racemization and DNA depurination), how does one empirically establish that "maintenance metabolism" is insufficient for growth or replication? In particular, what would the appropriate time scale be for such a demonstration? The high degree of scatter among the data that Price and Sowers [20] associated specifically with maintenance metabolism might alternatively be interpreted to suggest an energetic continuum without clearly delineated borders between the three categories or at least different boundaries determined by the particular reactions and their relative importance in different organisms. Thus, the limits to life at cold temperatures may not be defined by a highly defined physico-chemical boundary, but rather the boundary may be blurred and depend upon the definition adopted for life at cold temperatures.

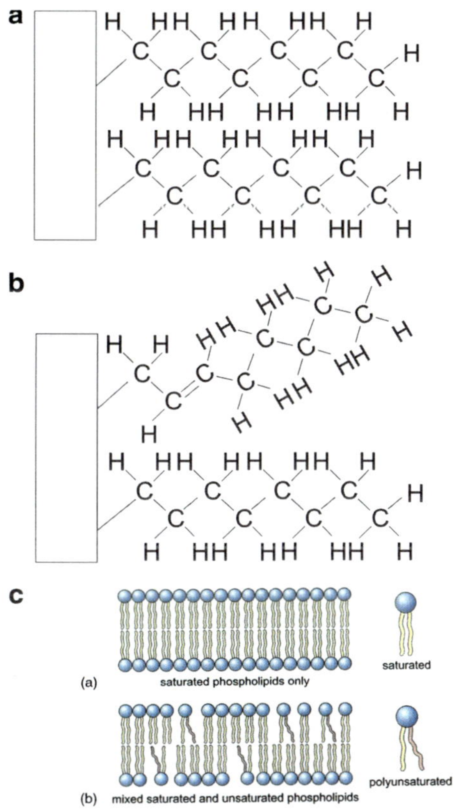

Fig. 7.1 Unsaturated fatty acids introduce kinks into phospholipids that are one mechanism by which cell membranes can adapt to low temperatures and high pressures by increasing membrane fluidity. (**a**) Depiction of the structure of a phospholipid containing saturated lipids. (**b**) Introduction of a single double bond creates a kink in the lipids. (**c**) Overall effect on the membrane structure of (a) and (b). Unsaturated phospholipids make the membrane less well-stacked (more fluid)

Life has adapted a number of ways to cope with low temperatures which are nicely reviewed in D'Amico et al. [21]. One example provides an excellent example of how adapting to extreme challenges often involves simple adaptations in physical chemistry. A challenge in living at low temperatures is maintaining membrane fluidity. As illustrated in Fig. 7.1, a membrane that contains saturated fatty acids will be relatively stiff. However, a membrane that contains more unsaturated fatty acids will be relatively more flexible. The introduction of double bonds within the fatty acid tail has the effect of introducing 'kinks' into the structure which make them stack together in a less compact configuration. This configuration introduces flexibility into the membrane and enhances its fluidity at low temperatures. An analogous situation can be found in fatty acids familiar in household situations. A typical olive oil has about 65% of its bonds as unsaturated fatty acids, which explains why it remains liquid at room temperature and even when stored in a refrigerator. A typical butter, by contrast, has about 15% unsaturated fatty acids. The fatty acids, because they stack together more compactly, render it a solid when refrigerated.

7.2.2 High Temperatures

Micoorganisms that grow at very high temperatures have mainly been isolated from hydrothermal vents in the deep oceans where the high pressures prevent water from boiling at 100°C (normal boiling point at 1 atm). Microorganisms have been isolated from deep oceans vents whose optimum growth temperature is 103°C (falling within the category of 'hyperthermophiles', organisms whose optimum growth temperature is above 80°C). An example of such a deep vent organism is *Pyrolobus fumarii*, whose doubling time at 112°C was ~8 h, but showed no growth at 115°C. Kashefi and Lovley [22] report growth of an organism at 121°C, which grows optimally at 105–107°C and can survive short periods of time at 130°C.

The physico-chemical limits to life at high temperatures are still not understood. High temperature adaptations evolved in life include the capacity to produce thermostable enzymes [23]. There are a number of modifications of macromolecular chemistry that allow proteins and other molecules to function at high temperatures. In proteins, the increase in abundance of disulphide bonds between peptide chains is one mechanism by which proteins are stabilised. The binding of monovalent and divalent salts is another method by which the folding of proteins can be stabilised against high temperatures. However, the large diversity of thermostable proteins that have been reported has led some to question whether the stability of proteins is really the defining limit for the upper temperature limit for life [24]. Smaller molecules of biological importance, such as thermolabile amino acids (e.g. cysteine and glutamic acid), ATP (adenosine triphosphate) and other low molecular weight metabolites have very short half-times at high temperatures (amino acids on the order of seconds or minutes at temperatures above 250°C) and the stability of these molecules may be a defining factor in the high temperature limit.

The best characterised thermostable enzyme has been DNA polymerase from *Thermus aquaticus*, a thermophile isolated from Lower Geyser Basin in Yellowstone National Park, USA. The organism has an optimum growth temperature of 70°C, but a range of 50–80°C. The DNA polymerase of this organism (referred to as Taq polymerase) has a temperature optimum for operation of 80°C. During the artificial replication of DNA (polymerase chain reaction or PCR) it is necessary to separate the two strands of DNA so that each can become a template for DNA polymerase to be used to synthesise the matching strand. Originally DNA polymerases were used from *Escherichia coli* to perform the strand synthesis, but the high temperatures required to separate the strands rendered the DNA polymerase inactive, requiring new enzyme at each step. The thermostable enzyme obviates the requirement for this and has found widespread use in PCR.

Heat shock proteins seem to play an important role in stabilising cell components at high temperatures; at 108°C about 80% of the soluble protein in extracts of the hyperthermophilic organism *Pyrodictium occultum* were a heat-inducible chaperone protein [25]. With this protein fully induced the organism was able to survive 1 h of being autoclaved (2 bar and 121°C).

Adaptations in the nucleic acids and membrane lipids of high temperature organisms have also been reported. Hyperthermophiles have gyrase enzymes (a unique type I DNA topoisomerase) that positively supercoil DNA strands and increase its stability against high temperature denaturation. In lipids, the formation of ether links across lipids within the membrane is one mechanism by which membrane stability can be improved. Membrane lipids of the hyperthermophile, *Thermus maritima* contain a glycerol ether lipid, 15, 16-dimethyl-30-glyceryloxy-triacontanedioic acid [26], which is thought to increase the stability of the membrane against hydrolysis. All archaea, many of whose representatives are hyperthermophilic, contain ether lipids, which provide resistance to high temperatures and acidic pH. These adaptations illustrate again how adaptations to environmental extremes may be defined by the ability to adapt the chemistry of molecules for stability against the given extremes, in this case changes to the chemistry of membrane structures.

7.2.3 High Pressure Environments

Microorganisms are found in many high pressure environments. Perhaps the best characterised are the deep oceans. Microbes are found at a depth of 11 km. Here pressures are approximately 1,100 atm (110 MPa). The locations are habitats for barophilic microorganisms that require high pressures to grow and barotolerant organisms that are able to adapt to high pressures, but whose representatives are also found at atmospheric pressures. A challenge to living at high pressures is maintaining membrane fluidity to allow for the transfer of solutes into the cell and waste products to be exuded through transporters and channels. This challenge is met by increasing membrane fluidity by the incorporation of unsaturated fatty acids into the membrane structure [27]. Indeed, this adaptation to high pressures is similar to the adaptation adopted in cold environments, where membrane fluidity is a challenge. The adaptation of cells in this way illustrates in a very powerful way that similar chemical adaptations in life may serve adaptations to different environmental extremes. Provided the adaptation results in the required biochemical tolerance in the given extreme then convergent evolutionary adaptations to different extremes can occur.

Examples of barophiles include *Moritella* and *Shewanella* species that laboratory experiments suggest can grow at 70 MPa, but not at pressures of 50 MPa [28], suggesting they are true barophiles. Similar bacteria have been obtained from the Mariana Trench. More remarkable evidence has been presented for growth using a diamond anvil apparatus. *Shewanella oneidensis and Escherichia coli* were reported to remain physiologically active at pressures up to 1,680 MPa for up to 30 h [29]. As for low temperatures, the pressure limits for life remain unknown.

7.2.4 Low Water Activity Environments

Water is the essential solvent for all life on the Earth. Several measures of water availability exist and are used by various scientific disciplines, but water activity (a_w) is perhaps the most universal. The a_w of pure, liquid water is 1.0; factors that decrease a_w below 1.0 include solutes, desiccation, and subzero (°C) temperatures. The a_w of a substance can be quantified by measuring the relative humidity (e.g., by thermocouple psychrometry) of an atmosphere in equilibrium with the substance and applying the following relationship: $a_w = rh/100$, where rh = percent relative humidity.

Microbes can survive under conditions that are much drier and/or saltier than those that enable growth. In fact, it is difficult to establish an absolute lower a_w limit below which all microbes perish, although the surface soil in the driest parts of the Atacama Desert may approach that limit [30].

Studies from food science provide the lowest well documented a_w permitting growth of microorganisms. Some yeasts and moulds have been shown to reproduce, albeit slowly, in sucrose solutions as concentrated as 83% (2.4 M), for which $a_w = 0.62$ [31]. The food spoilage fungus *Xeromyces bisporus* is the world's record holder for growth in a sugar solution at the lowest water activity ($a_w = 0.61$; [32]).

Evaporite deposits are sources of solutes that are the most important to consider with reference to microbial proliferation. Saturated brines of NaCl (~5.2 M), where the a_w = approximately 0.75, are relatively common on Earth; and diverse microorganisms, including members of the Bacteria, Archaea, and Eukarya, are known to maintain stable populations in these hypersaline environments by regularly undergoing cell division [32].

Physiological studies of these microorganisms have shown that although they require high NaCl concentrations in their surrounding environments, e.g. for maintaining membrane stability, Na^+ ions inhibit intracellular enzymatic function. Cells that are adapted to growth in NaCl brines exclude Na^+ ions from their cytoplasms and accumulate intracellular compatible solutes, e.g., KCl, amino acids, or sugars to balance their internal a_w to that of the extracellular solution [33]. This adaptation requires expenditure of cellular energy for transport and/or synthesis of solutes. Despite the worldwide distribution of these saturated NaCl brines and their microbial inhabitants, they likely represent the most extreme solute-induced a_w that permits microbial reproduction.

Other evaporites, which occur with even lower a_w than that of saturated NaCl, include KCl, $CaCl_2$, and $MgCl_2$. Extreme environments containing nearly saturated brines of these salts exist on Earth and have been suggested to contain indigenous microorganisms. The Dead Sea is one such environment that has received extensive microbiological study. Dead Sea brine has 340 g/L total dissolved solids that are dominated by $CaCl_2$ and $MgCl_2$, with lesser amounts of NaCl and KCl; the a_w is approximately 0.67 [34], a value that is seemingly only slightly more extreme than that for NaCl brines. Nonetheless, actual microbial reproduction in Dead Sea brine has not been demonstrated. While there is evidence of microbes being present in the

Dead Sea and some microbes have been isolated, actual proliferation of these salt-adapted microbes likely occurs exclusively during periods following dilution by rainwater. Lack of growth in the Dead Sea may be caused by the chaotropic nature of $MgCl_2$, i.e., its tendency to destabilize biological macromolecules [35]. Don Juan Pond, a $CaCl_2$-rich lake in an Antarctic Dry valley has solute concentrations approaching 50% (wt./vol.) and $a_w = \sim 0.45$. As with the Dead Sea, microbes have been reported here and associated with that a certain degree of controversy with regard to the reproduction of microbes under such conditions [32].

More extreme even than the Dead Sea are deep anoxic basins in the Mediterranean Sea with $MgCl_2$ concentrations that can approach saturation (5.0 M, $a_w = \sim 0.3$), as in Discovery Basin. There, microorganisms have been isolated and detected by direct microscopy. Genetic analysis suggests a brine microbial community extremely distinct from both the overlying interfacial and normal seawater communities, and metabolic activities including uptake of glutamic acid, methane production and sulfate reduction have been measured in the brines, where the latter two rates were higher than in overlying (including interfacial) waters [36]. The question nonetheless arises whether these microbes are actually reproducing in the low a_w conditions of the deepest $MgCl_2$ brines or whether reproduction is restricted to the overlying sea water with lower a_w.

Hallsworth et al. [35] tested the biological effects of $MgCl_2$, with emphasis on the potential for life in a vertical chemocline extending from sea water in the Mediterranean to the underlying, nearly saturated $MgCl_2$ brine of the Discovery Basin. They reported that, in addition to decreasing a_w, $MgCl_2$ is extremely chaotropic, being even more destabilizing to biological macromolecules than ethanol, urea, and guanidinium-HCl and comparable to phenol, all of which are commonly used in biological laboratories to disrupt molecules or to prevent biological activities. Although van der Wielan et al. reported measurable phosphatase and aminopeptidase activity in 5 M $MgCl_2$, Hallsworth et al. [35] found that the enzyme glucose-6-phosphate dehydrogenase was completely inactivated at $MgCl_2$ concentrations greater than 1 M. Growth of microbial isolates from samples collected from the Discovery basin chemocline was completely inhibited above 1.26 M $MgCl_2$ ($a_w = 0.916$). In contrast to the measurable methane production and sulfate reduction in 5 M $MgCl_2$ reported by van der Wielen et al. [36], Hallsworth et al. [35] were unable to detect associated genetic activity in the chemocline at $MgCl_2$ concentrations greater than 1.88 M ($a_w = 0.845$) for sulfate reduction and 2.23 M ($a_w = 0.801$) for methanogenesis. The authors conclude that microbial activities are unlikely at $MgCl_2$ concentrations greater than 2.3 M ($a_w = 0.79$) or above 2.5 M $MgCl_2$ ($a_w = 0.76$) if a kosmotropic (stabilizing) solute such as NaCl is present. The chaotropic nature of $MgCl_2$, which appears to be more inhibitory than the effect of cellular water loss, may explain the patterns of microbial abundance, activity, and growth in the Dead Sea as well as in the deep Mediterranean basins. However, since the conclusions of Hallsworth et al. [35] are based primarily on culture-based studies, they do not completely rule out the existence of growth in concentrated $MgCl_2$ brines by uncultivated extremophiles. While the indications of life in these environments are somewhat controversial, the current, best working hypothesis is that actual microbial proliferation occurs only in less saturated solutions that are separated in space or time from the most concentrated brines. This hypothesis, however, still awaits confirmation.

Desiccation, i.e. loss of liquid water from a solid surface or from a porous medium (soil, food, martian regolith, etc.) imposes a related but different water stress on microorganisms. There is not a specific-ion effect as with solutes containing Na^+ or Mg^{2+}; however, there is an indirect effect from the decreasing thickness of water films. As the water films at solid–liquid interfaces become thin and discontinuous, the advection and diffusion of solutes as well as the mobility of microorganisms diminish significantly. The indirect effect of this water loss is to deprive microorganisms of necessary dissolved nutrients. Few if any solids can provide all of the energy and nutrients required for cellular reproduction, and so cells that are immobilized within thin water films are essentially starved. This indirect effect of desiccation is seen at relatively high a_w values in soils, where metabolic activities are below detection at $a_w < 0.86$ and cell division of even desert-adapted soil microbes has not been observed below $a_w = 0.88$ [37].

Thin water films on mineral surfaces have been proposed as possible habitats for martian microbes [38]. The extent to which 'adsorbed' water (thin layers of unfrozen water bound to surfaces in monolayers only a molecules thick) is accessible to microorganisms is unknown. However, any microbes that metabolize in such thin films would presumably have to expend energy for acquisition and retention of liquid water against a gradient. While neither growth nor even appreciable metabolic activities have been detected among terrestrial microbes in contact with these thin water films, either in soil or food, it is an intriguing idea that bears further investigation.

The a_w of liquid water in equilibrium with ice at sub-zero (°C) temperatures is dependent only on the temperature, as the solute concentration will adjust such that the solution will have the same water activity as ice, if the ice-water equilibrium is maintained [39]. The a_w values of liquid water in equilibrium with ice at −20°C, and −40°C are 0.82 and 0.67, respectively. Microbes existing in these supercooled waters suffer the combined effects of the low temperature and loss of cellular water. Response to the latter may require expenditure of cellular energy for accumulation of intracellular compatible solutes as one example of a microbial response to such conditions.

7.2.5 Extremes of pH

Few biochemical reactions can occur at extreme pH values and for that reason most terrestrial biology has evolved to maintain a near-neutral cell pH, whatever the external pH environment. Therefore, despite the nomenclature of 'acidophiles' and 'alkinophiles', strictly speaking none of these organisms are adapted to extreme pH's, but are rather tolerant of extreme environmental pH values. An exception is *Acetobacter aceti*, which maintains an internal acid pH.

A wide range of bacteria, archaea and eukaryotes have been found that grow at pH values less than 1 [e.g. 40]. Acidophiles maintain a near-neutral pH inside the cell by pumping protons out of the cell, which requires energy expenditure. This is sometimes achieved by the production of large numbers of proton pumps and other membrane transporters that increase the efficiency of proton pumping across the cell membrane. Highly acidic conditions within the cytoplasm result in protein denaturation, providing a selection pressure for acidophiles to evolve low proton permeabilities across cell membranes to prevent internal cell damage. Despite mechanisms to pump out protons, many acidophiles have adaptations to tolerate low pHs, including the evolution of acid-stable proteins. In most acid stable proteins (such as pepsin), there is an abundance of acidic residues which minimizes destabilization at low pH caused by a build-up of positive charge. Other mechanisms include minimization of solvent accessibility of acidic residues or binding of metal cofactors such as iron. In a specialized case of acid stability, the NAPase protein from *Nocardiopsis alba* was shown to have relocated acid-sensitive salt bridges away from regions that play an important role in the unfolding process [41].

Organisms that tolerate high pH values are less well understood, but suffer from the problem of low proton concentrations to maintain an efficient proton gradient across the cell membrane, necessitating the evolution of efficient proton transport mechanisms across the cell membrane. The hydroxyl ion, generated at high pH, is damaging to molecules on account of its property as a powerful nucleophile [42]. Natural environments where these organisms are found are in soda lakes, such as Mono Lake in the USA. A diversity of bacteria and cyanobacteria have been isolated from environments with a pH of 10.5.

7.3 Synergies Between Extremes

Very few organisms exist in a single physical or chemical extreme and in the natural environment most organisms are subjected to multiple extremes in any given environment. Some of these factors may interact in unintuitive and synergistic ways that depend upon the chemical basis of their effect. For example, although only a few studies have considered how growth temperature affects the water activity range of growth for a micro-organism, they show that, at least for some microorganisms, the range is extended when the organism is grown at lower temperature (e.g. for *Glaciecola punicea* ACAM 611^{T} at 8 versus 15°C [43], *Clostridium algoriphilum* Strain 14D1 at −5 versus 5°C [44]; tested growth temperatures did not affect the water activity range of *Gelidibacter* sp. Strain IC158 [43]). In this context, it is important to bear in mind that microorganisms are usually characterized under a very limited set of conditions; the interplay of conditions, and the corresponding hyperdimensional space of physical and biotic variables relevant to microbial growth and survival, has been little explored.

Other examples of these potential complications have been illustrated by the increase in *Salmonella* virulence following growth on the International Space

Station [45] and by the enhanced resistance of *Salmonella* to osmotic, acid and thermal stresses following simulated space flight [45]. As a consequence, it is not always correct to assume that challenges act additively, when they can precondition or select organisms to endure further challenges.

7.4 Water as a Solvent

We have explored some examples of the physical limits to life, how those limits may be established and some of the chemical adaptations that life has acquired to deal with those extremes (see also Chap. 6). All of these extremes assume that living organisms use water as a solvent and are made of carbon. In the following sections we explore why water and carbon are considered to be the best solvent and chemical backbone to life, respectively, and compare them briefly to some other suggestions for alternative solvents and compounds. This enquiry provides a basis to consider whether the limits to life familiar to scientists on the Earth can really be said to be universal.

Life requires a solvent for chemical reactions to occur. This solvent must have a number of characteristics which include: (1) a liquid phase that is similar to the environmental conditions in which the biota is to exist, (2) a viscosity and density that allows molecules essential to biological function to be maintained at sufficient concentrations in the cells, (3) an environment that allows chemical reactions to occur, but in which the conditions also allow complex molecules to be synthesised that are not broken down by chemical reactions.

Of the characteristics that make water a particularly suitable solvent for life, its dipole moment is an important one [46].

The dipole moment of water (1.85 D) is such that the molecule readily dissolves both salts and organic molecules. Salts are important to life as a source of cations and anions such as K^+, Na^+, Fe^{3+}, Cl^-, all of which play a role in enzyme active sites, in stabilising membranes and as sources of energy. The dissolution of salts in water as charged ions contributes to the ability of water to act as a medium for chemical reactions that require charged species. The high dipole moment of water also allows for the dissolution of small organic compounds such as amino acids, used in protein synthesis. This property allows water to act as a mediator of organic complexation reactions.

Although water is a good solvent for polar compounds, it does not readily dissolve non-polar compounds, particularly large organic molecules, allowing for their stability, which is crucial for maintaining the integrity of molecules such as membrane phospholipids and enzymes, which are able to retain their folding in water.

The polarity of water allows water molecules to bond together through hydrogen-bonding. This property accounts for the wide temperature range of water, temperatures which overlap with environmental conditions on the surface and in the sub-surface of the Earth. Without these polar characteristics, the small

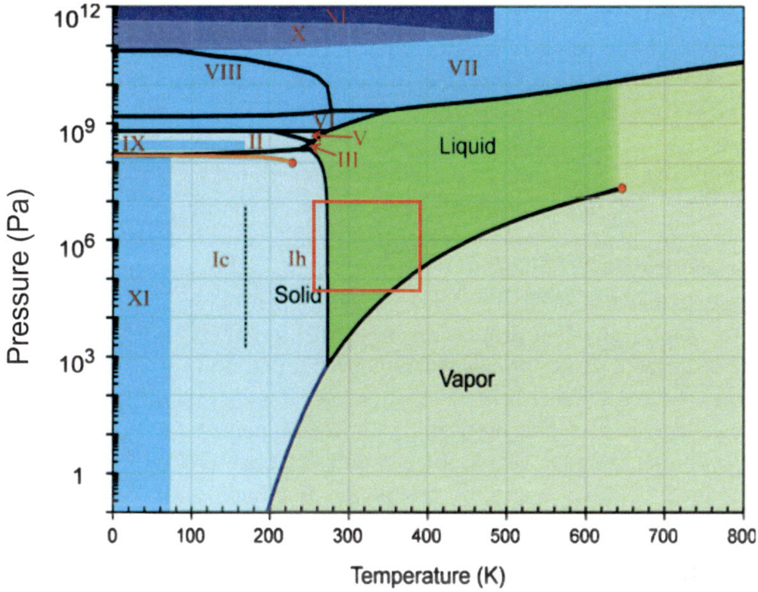

Fig. 7.2 Phase diagram of water showing space within which terrestrial biology operates (*red box*)

molecular weight of water would allow for a much smaller range of temperature conditions in the liquid state. Figure 7.2 shows the phase diagram of water with a region depicting the conditions experienced on the surface and in the sub-surface of the Earth.

The physico-chemical properties of water account for many of its beneficial uses as a biological solvent. Water has a high heat of vaporisation, which promotes a stable liquid phase inside organisms and stabilises temperatures, enhancing the ability of organisms to cope with fluctuating temperature regimens. By contrast, a high heat of vaporisation also implies high energy lost during evaporation, which is used by organisms to achieve evaporative cooling against high temperatures in the environment.

Perhaps one of the most feted properties of water that has been implicated in its biological usefulness is the lower density of ice than water (at least Ice I, the form of ice encountered at the Earth's surface). The property that ice has of floating on water when frozen provides protection for organisms in water bodies that can remain in liquid water beneath the ice layer. However, many microorganisms can resist freezing, and the wood frog (*Rana sylvatica*), similarly to other northern frogs that hibernate close to the surface in soil or leaf litter, can tolerate the freezing of its blood and other tissues [47]. Urea is accumulated in tissues in preparation for overwintering, and liver glycogen is converted in large quantities to glucose in response to internal ice formation. These molecules act as cryoprotectants to limit the formation of ice and to reduce osmotic shrinkage of cells. Frogs can survive freeze/thaw events during winter if not more than about 65% of the total body water freezes. Although the wood frog is an unusual example, it is clear that evolutionary strategies do exist to tolerate freezing and that although the physical attribute that ice has to float on water may appear to favour life, it is not required for life to exist on the Earth.

The wood frog illustrates a more general observation that care should be taken in assigning water 'essential' properties that make it fine-tuned for biological use. Although organisms have evolved specific characteristics that allow them to adapt to the physico-chemical properties of water, these adaptations should not be taken as evidence that water is finely tuned for biological compatibility. Hypothetical changes to waters' physical properties might still remain within the bounds of the adaptive capabilities of the process of biological evolution.

Water does have some properties that are not conducive to life. It is a reactive solvent, causing the dissociation of many macromolecules, for example it disrupts hydrogen bonding because of its polar characteristics, making it unconducive to protein folding. It is reactive with phosphatidyl molecules and results in a short half-life of molecules such as ATP, used by cells as a source of energy [48]. Its reactivity with DNA is conducive to DNA damage, contributing to the evolution of DNA repair capabilities which are also used to repair damage caused by radiation.

7.5 Alternative Solvents to Water

The possibilities for alternative solvents to liquid water have for a long time been discussed by the astrobiology community [e.g. 46]. Both polar and non-polar solvents have been mooted as possible water substitutes. Although water is the most abundant solvent in the universe, other solvents might be plausible in planetary environments with different physical and chemical regimens. Unfortunately few of these alternatives is open to empirical investigation since, as yet, there are no planetary bodies in our Solar System that show compelling evidence to be plausible environments where alternative solvents might be used and that could be investigated. Additionally, our knowledge of how the origin of life occurred is not sufficiently complete to be able to test alternative solvents in the laboratory as possible alternative solvents for the origin of life. Nevertheless, some physical properties of alternative solvents might make them candidates.

It has been pointed out previously [49] that many terrestrial enzymes can still be active in non-polar solvents such as ethers and benzene and that about 20% of the human genome encodes trans-membrane proteins that require the non-polar environments inside membranes to operate. Although these observations do not provide any direct evidence for the possibility of non-aqueous solvents being potentially successful media for biochemistry, they show that even terrestrial biochemistry can operate in non-aqueous solvents and in the case of trans-membrane proteins, actually require these environments.

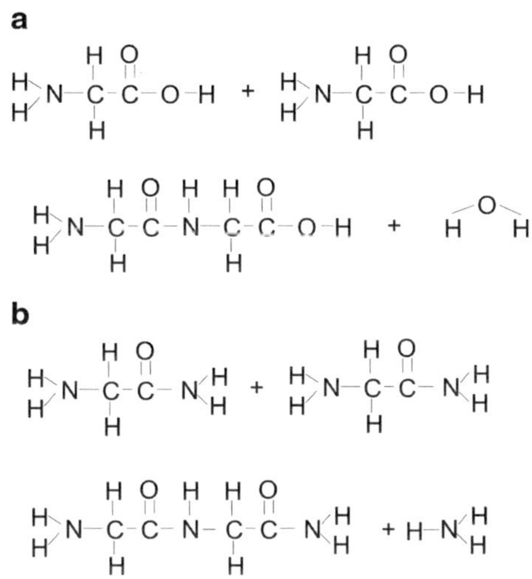

Fig. 7.3 Reactions showing speculative chemical reactions for formation of equivalent protein 'peptide' bonds in ammonia (**b**), rather than liquid water (**a**) biochemistry

7.5.1 Ammonia as a Solvent

Ammonia has been one of the most discussed alternatives to water. Early discussions on the topic focused on the possibility of peptide bonds mediated by a $-CONH_2$ group as opposed to the $-COOH$ group in amino acids. In Fig. 7.3 the terrestrial peptide bond formation reaction is shown and the corresponding reaction that might occur in liquid ammonia. Indeed Firsoff [50], who discussed this possibility in early papers, speculated that the presence of nitrogen in the peptide bond might be a remnant of early ammonia chemistry. Other analogue chemical groups have been envisaged for ammonia solutions. Ammonophosphates have been suggested as replacements to phosphate in speculative ammonia chemistries, forming amide bonds with carbon molecules.

Comparing the physical properties of ammonia to those of water yields some insights into its possible comparative advantages and disadvantages [51]. Ammonia is about four times less viscous than water and so molecules diffuse through it more efficiently with implications for chemical reactions. However, ammonia has a lower heat of vaporisation than water and so is less able to accommodate temperature fluctuations. Perhaps the greatest difference with water is that ammonia has a smaller range of temperature in which it remains liquid.

Ammonia offers the potential for ammonia-philic and -phobic solutions, analogously to water and so could allow for partitioning in analogue cell membrane structures. Alternative energy transduction pathways have been suggested [49], whereby electrons, rather than protons, are used to generate energy in ammonia solutions. In terrestrial biochemistry, proton gradients established across cells (the chemiosmotic theory) by the metabolic pumping of protons out of cells during electron transfer reactions within the cell membrane are used to drive protons out of the cell. The movement of protons back into the cell drives the trans-membrane ATPase, which results in the formation of ATP, subsequently used in energy yielding reactions. Conceivably these reactions could occur in liquid ammonia, however, free electrons are stable in liquid ammonia (unlike in liquid water where they rapidly form hydroxyl ions). Alkali metals dissolve in liquid ammonia to form metal^{+}/electron solutions that can be stable for months, and allowing the solution to conduct. A biochemistry could conceivably use free electrons to drive energy transducting reactions. These electron-rich solutions also underline a more general point, which is that novel chemical reactions and biologically-useful properties may be available in some solvents that are not available in liquid water.

Ammonia presents other challenges for life, most notably the extremely high pH at which ammonia solutions form. Ammonia solution of 1% or greater have pHs greater than 11.0 and biochemistries would require adaptation to these conditions.

7.5.2 *Other Solvents*

Few alternative solvents offer the flexibility of either water or ammonia, but some have been suggested [46].

Sulfuric acid has been suggested as an alternative solvent on Venus, where high sulphuric acid concentrations in the clouds (>95%) make it a candidate for biochemistry. Factors in favour of sulphuric acid include its high dipole moment, but its high viscosity (over 20 times higher than H_2O) would reduce its effectiveness as a medium for molecular diffusion. Its high reactivity with carbon-containing molecules and hygroscopic characteristics also make it an implausible solvent for any chemistry resembling terrestrial chemistry.

Hydrofluoric acid has been discussed as another alternative solvent. Its similar heat of vaporisation to water and wide temperature range at which is remains liquid make it a possible candidate, with fluorine replacing oxygen in many molecular structures. However, its low cosmic abundance and high reactivity with organic carbon molecules make its use limited.

Houtkooper and Schulze-Makuch [52] suggest hydrogen peroxide as a potential solvent to improve the characteristics of water as a solvent in some environments. They point out its strong oxidising properties, which make it incompatible with many carbon-containing molecules and the formation of free radicals contribute to its ability to damage macromolecules. However, mixed with liquid water, supercooled liquids are formed, which might mitigate freeze-damage to organisms. The hygroscopic nature of mixtures of hydrogen peroxide with water makes them potential water scavengers.

7.6 Carbon as a Building Block for Life

The previous sections considered why, from a physico-chemical standpoint, water is an ideal solvent for life. Now we consider the building-block element of life. The basic building block element of all terrestrial life is carbon [42]. Why is this so? Onc important reason is the sheer versatility of molecules that carbon can form. Several characteristics of carbon contribute to its molecular versatility including: (1) its electronic structure which allows for the formation of single, double or triple bonds, (2) the high activation energies of bonds resulting in stable molecular structures, (3) high carbon-carbon bond strengths compared to other bonds that increase the stability of carbon-containing molecules.

Molecules containing carbon range in two dimensional structure from chains to rings. The simplest carbon molecules are alkanes ($C_2H_{(2n + 2)}$) (n = 1 is CH_4, methane, n = 2 is C_2H_6, ethane). The substitution of hydrogen atoms results in functional groups of wide use in different biochemical functions, for example: esters: –COO–R– (where R is an alkyl group) used in membranes; amino ($–NH_2$) and carboxyl (–COOH) groups found in amino acids and which react to form the peptide bond responsible for protein structure; $–PO_4$, organic phosphates found in DNA and other molecules; alcohols (−OH) used by microorganisms in energy yielding reactions etc.

Carbon atoms are also arranged in rings as complete carbon rings or carbon rings joined together by other atoms such as oxygen. The benzene ring, a six carbon ring, forms the basic structure of many biologically important molecules from nucleotides in DNA to sterols such as cholesterol, which is a precursor to vitamins and a range of hormones.

The ability to form complex and stable chains of molecules is undoubtedly the single most important characteristic of carbon in natural systems. All three of the major components of life – lipids for cell membranes, proteins for cell components and the information storage system, DNA, are based on the chain-forming capacities of carbon. In the case of lipids, the fatty acids that make up their 'tails' are composed of long chains of carbon that are either saturated (single) bonds or unsaturated (which means that they contain double bonds that impart a 'kink' to the chain). Earlier in the chapter it was explained how unsaturated chains allow for organisms to adapt to cold temperature or high pressures. These adaptations to environmental extremes are therefore made possible by a simple change of a single bond to a double bond within the lipid chain; showing how quite complex adaptations to environmental extremes are made possible by very basic physico-chemical changes.

Proteins are chains of repeating units made up of two carbon atoms and one nitrogen atom, the C–N bond constituting the peptide bond between the different amino acids that make up the protein chain. Each repeating carbon unit has a different functional group depending on what amino acid it is. The protein structure is possible because of the similar bond energies between the C–N bond and the C–C bond.

Nucleic acids are formed by repeating units of three carbon atoms, one oxygen atom, a phosphorus atom and another oxygen atom. This backbone allows for the addition of pyrimidines and purines that form base pairs across two of these repeating units, resulting in the DNA double helix.

The ability to form chains is one of the most important characteristics of a carbon-based biochemistry. The complexity of life depends upon repeating units that can be formed into structures capable of carrying out complex functions from information storage to cell membrane formation. Carbon fulfils this need.

Apart from biomolecule construction, carbon also performs many useful functions in energy acquisition in cells. The ability of carbon to change between valence states from −4 (e.g. methane) to +4 (carbon dioxide) makes it a versatile electron donor and acceptor in many energetic reactions. This is particularly the case amongst microbes. Many microorganisms use organic carbon as a source of electrons in energy transduction, for example using carbon compounds in aerobic respiration and to drive reactions such as iron reduction under anaerobic reactions. Microorganisms are able to use a wide variety of carbon compounds from methane (in methanotrophy) to aromatic compounds (for example used by *Geobacter* species in iron reduction in soils contaminated with industrial waste).

The likelihood of carbon as a common building element for life (if it exists elsewhere) is supported by its distribution in the Universe. Carbon is a common element and the discovery of a variety of carbon-containing compounds in meteorites, including amino acids, nucleotides including adenine, guanine and a large variety of other compounds shows that the building blocks associated with terrestrial life are universally distributed, as observation further supported by observations of the formation of alcohols and other carbon compounds in the interstellar medium. These data show that the universe contains a large number of sources of the building blocks associated with terrestrial life that could conceivably drive biochemistry elsewhere.

7.7 Silicon as an Alternative Building Block

Of alternative elemental building blocks to life, silicon has been most widely discussed by the chemistry community [46, 49]. As a p-block element of group IV, below carbon in the periodic table it shares many common chemical characteristics (see Chap. 1). Amongst the similarities between the elements that are of biological relevance are: both form sp^3 hybrid orbitals with tetrahedral structures leading to similar structures in many of their compounds, both elements have high melting and boiling points, both are solids at standard temperatures and pressures and both elements are in the mid-range of electronegativities.

Silicon is generally more reactive than carbon, which is attributed to three characteristics. Firstly, silicon has accessible *d*-orbitals, which allows for higher coordination numbers than is possible with carbon, and which allows reactions such as hydrolysis to occur at lower energies. In carbon, similar reactions must be mediated by free radicals. For example, silane combusts spontaneously in air even at 0°C, whereas its carbon analogue, methane, remains completely stable, even in pure oxygen. Secondly, many silicon bonds with other elements are weaker than in carbon, requiring less energy to break them. Finally, silicon is more electropositive than carbon, leading to strongly polarised bonds with other non-metals which are much more susceptible to both nucleophilic and electrophilic attack. Although the more reactive nature of silicon may at first appear to be a disadvantage in any potential biochemistry, this high reactivity might make it more conducive to biochemical reactions at low temperatures (for example in liquid nitrogen), which would probably be required to provide a non-aqueous solvent in which silicon biochemistry could evolve.

Although there are similarities, both elements have some significant differences which affect compound formation. The Si-Si bond strength is lower than the C–C bond strength resulting in a lower energy of vaporisation. The larger radius of silicon accounts for its weaker bond strengths (see Chap. 1), which means it less readily forms complex compounds. In particular, bond angles of silicon compounds are generally larger because of its larger size, meaning that silicon cannot form molecules analogous to aromatic compounds in carbon biochemistry. Aromatic compounds are found throughout carbon biochemistry and give huge versatility to the complexity of compounds that can be assembled and their bonding interactions, for example base pairs within DNA. Few silicon compounds contain double and triple bonds, which are common in carbon compounds and account for many chemically important properties, for example the difference between saturated and unsaturated fatty acids. In the case of silicon, the formation of single bonds with other atoms often leads to unreactive states. For example, fully oxidised silicon forms silica, a highly unreactive compound which makes up quartz, whereas carbon forms a double bond with oxygen to produce carbon dioxide, a gas which has a diversity of uses in biochemistry, not least as an easily accessible form of carbon for life, but also as an electron acceptor in energy yielding reactions such as methanogenesis. Indeed, in most settings under standard temperatures and pressures silicon forms unreactive silicates, which on Earth are found in a wide diversity of different rock types.

A more plausible chemistry involving silicon is a hybrid system with carbon. Silanes have the ability to form branched chains of molecules. Although silicon cannot easily form a six-ring structure with delocalised electrons like benzene, it can form a six-ring structure (siloxene) in which oxygen atoms hold together the silicon atoms. By replacing the hydrogen atom in silanes with organic groups, greater stability is achieved since the silicon-hydrogen bond is weak. Chained silanes are analogous to long chain hydrocarbons. These organosilicon compounds are thermally stable and are chemically inert. Although laboratory experiments can be performed to generate these compounds, rarely are they found in nature. One problem with silicon it its reactivity in oxygen environments which leads to the formation of inert silicates, as witnessed in the Earth's rocks. However, if oxygen is replaced by S or N, a variety of chained silicon compounds can be formed.

Silicon also forms stable tetra-, penta- and hexa-cordinated compounds with Si-N, -C and -O bonds and can form stable covalent bonds with nitrogen, phosphorus, sulphur, the halogens and many other elements that are associated with the generation of molecular diversity in carbon biochemistry. Other analogies with carbon chemistry exist. Silicon can form ring structures analogous to sugars (in silicone chemistry). Cage systems such as silsesquixones can be functionalised with a wide diversity of sidegroups to allow for a remarkable diversity of molecules. Siloxenes are polymeric structures with the general formula $Si_6H_6O_3$. They are made of cyclosilane rings with attached-OH groups and are made up of monomeric units with analogies to carbohydrates and whose ring structures have analogies to chlorophylls.

Schulze-Makuch and Irwin [46] consider the constraints on different types of silicon-based life. For silanes, they point out that chemical reactions would have to occur in an oxygen and water-free environment to prevent reactions of silanes leading to inert silicates. However, it should also be pointed out that chemical disequilibria can generate conditions where silicon-containing compounds can exist even in environments under highly oxidising conditions. Silicon carbide, for example, is found as a mineral in the Earth's crust and is found in meteorites. An additional problem is the solvent to be used in the absence of water. Methane is one possibility. However, if planetary conditions are such that oxygen and water have been driven off from the planetary surface to allow silane biochemistry, then methane would also be unlikely to be present, in which case higher molecular weight compounds could act as alternative solvents.

Other speculations have considered silicate-based life forms at high temperatures when the melting of rocks increases their reactivity. Feinberg and Shapiro [48] discuss the possibility of lavobes and magmobes, organisms that inhabit molten rocks and make use of the formation of layered silicates where cation exchange reactions could occur within the silicates. Information would be encoded within structural defects within the minerals. Although these ideas are intriguing, no evidence exists for them in the terrestrial rock record despite the fact that the earth has had abundant habitats containing molten rocks throughout its history.

7.8 Thermodynamic Limits to Life

So far we have considered the limits to life as set by physical and chemical extremes and explored some of the physical chemistry behind those extremes and what makes water as a solvent and carbon as a building block element so favourable. However, ultimately, for an organism to be able to grow and reproduce at an extreme, it must be able to gather sufficient energy, not only to grow and reproduce, but to repair molecules and synthesise new ones required for biochemistry to function at a given extreme. Therefore, one way to examine the limits to life is to consider the limits to energy acquisition. In this section we will consider some of the physical chemistry behind the thermodynamic limits to life (see also Chap. 9).

The universal demand for energy is a commonly used descriptor of life; all life forms require energy to fulfil the specific functions that power metabolic processes. Whether harnessed from the physical energy of the sun, the chemical energy of matter, or the biological energy taken in from other organisms, life draws energy from its surroundings and converts it to a biologically-useful form; ATP. Any process in which energy is converted from one form to another must be thermodynamically feasible in order to proceed. Defining this feasibility across a range of reactions and environmental conditions provides a powerful tool with which to define the energetic limits to life.

Most relevant to this discussion are the limits to chemosynthetic life; microorganisms that harness chemical energy from the environment by catalysing redox (reduction-oxidation) reactions out of thermodynamic equilibria to fuel cellular metabolism. These microorganisms are common to many extreme environments on Earth. Electrons are acquired from the electron donor (substance being reduced) and passed to the terminal electron acceptor (substance being oxidised) via a membrane-associated electron transport chain which ultimately establishes a proton gradient across the membrane. Across this gradient electron transport phosphorylation occurs, generating ATP from ADP. Chemosynthetic micro-organisms are capable of exploiting an extraordinary range of redox couples, from iron oxidation to uranium reduction. Each combination of electron donors and acceptors is characterised by different inherent thermodynamic constraints, in addition to those set by the environment in which they are found. Pressure, temperature, pH and the concentration of each redox constituent all exert influence on the thermodynamic feasibility. As such, the ability to define the extent of this feasibility for any given environment using thermodynamic tools is an essential element of understanding the overall limits of life in extreme environments.

7.8.1 Gibbs Energy

How can the energetic limits to life be quantified? The most powerful diagnostic tool for assessing the feasibility of microbially-mediated redox reactions is Gibbs energy (G), a measure of the ability of a system to do work. In Chap. 1, thermodynamics was discussed from a chemical perspective. Here we discuss its application to energy-gathering in microorganisms. Gibbs energy is based on the combined effects of two fundamental thermodynamic functions, enthalpy (H) and entropy (S). The former represents the heat content of a system, and the latter represents the degree of disorder. The change in Gibbs energy (ΔG) for a given reaction can be used to accurately determine whether the reaction will proceed or not. In this respect, ΔG can be thought of as a measure of the 'tendency to react', and is indicated by the sign of the value: if ΔG is less than zero, the reaction will proceed spontaneously, but not necessarily rapidly. If it is greater than zero, the reaction will not proceed unless energy is supplied from outside the system; and if ΔG is zero, the reactants and products will exist side by side in a state of equilibrium. Furthermore, the magnitude of the value indicates how far from equilibrium the reaction will go; a negative value close to zero indicates that the reaction is close to equilibrium, whereas a much more negative value indicates that the reactants of the system will react almost entirely to form the products [53].

The most basic way in which ΔG can be calculated is under standard conditions of temperature (25°C) and pressure (1 atm) using the standard Free energies of formation ($\Delta G^{\circ}{}_{f}$) of the relevant products and reactants. These values represent the Free energy change associated with the formation of a particular compound from its constituent elements in their most stable forms under standard conditions. For instance, the $\Delta G^{\circ}{}_{f}$ of carbon dioxide gas represents its formation from graphite ($C_{graphite}$) and di-oxygen (O_2) (values of $\Delta G^{\circ}{}_{f}$ are commonly listed for a range of compounds in the appendices of chemistry and thermodynamic text books). By subtracting the sum of the $\Delta G^{\circ}{}_{f}$ of the reactants from the sum of $\Delta G^{\circ}{}_{f}$ of the products, a value of ΔG° for the reaction is obtained:

$$\Delta G^{\circ}{}_{r} = \sum \Delta G^{\circ}{}_{products} - \sum \Delta G^{\circ}{}_{reactants} \tag{7.1}$$

This equation can be applied to any microbially-mediated redox reaction to determine its feasibility under standard conditions. For example, *Geobacter metallireducens* [54] reduces ferric iron, such as iron(III) hydroxide, by acquiring electrons from organic compounds, such as acetate:

$$\begin{gathered} 8\mathrm{Fe(OH)_3} + \mathrm{CH_3COO^-} + 15\mathrm{H^+} \rightleftharpoons 8\,\mathrm{Fe^{2+}} + 2\mathrm{HCo_3^-} + 20\mathrm{H_2O} \\ \Delta G^{\circ}{}_{r} = -112.94\ \mathrm{kJ/mol\ e^-} \end{gathered} \tag{R7.1}$$

The negative change in Gibbs energy for this reaction, obtained by applying $\Delta G^{\circ}{}_{f}$ values [55] to (7.1), indicates that under standard conditions the reaction will proceed spontaneously, and hence microbially-mediated iron-reduction in an equivalent environment is feasible. The following equations represent the other half of the microbial iron cycle, iron oxidation, in which ferrous iron is oxidised by oxygen ([3] e.g., *Acidiferrobacter thioxydans*; [56]) in aerobic conditions or nitrate ([4] e.g., *Desulfitobacterium frappieri*; [57]) in anaerobic conditions:

$$\begin{gathered} 4\mathrm{Fe^{2+}} + \mathrm{O_2} + 4\mathrm{H^+} \rightleftharpoons 4\mathrm{Fe^{3+}} + 2\mathrm{H_2O} \\ \Delta G^{\circ}{}_{r} = -44.32\ \mathrm{kJ/mol\ e^-} \end{gathered} \tag{R7.2}$$

$$\begin{gathered} 10\mathrm{Fe^{2+}} + 2\mathrm{NO_3^-} + 12\mathrm{H^+} \rightleftharpoons 10\mathrm{Fe^{3+}} + \mathrm{N_2} + 6\mathrm{H_2O} \\ \Delta G^{\circ}{}_{r} = -45.78\ \mathrm{kJ/mol\ e^-} \end{gathered} \tag{R7.3}$$

These reactions are also thermodynamically favourable under standard conditions, and the difference between utilizing oxygen or nitrate as the terminal electron acceptor is minimal. However, the larger difference in Gibbs energy change between the reduction and oxidation of iron indicates that reduction is more favourable. Indeed, by calculating $\Delta G^{\circ}{}_r$ for a range of microbially-mediated redox reactions and normalising for the number of electrons transferred, as in the equations above, one can get a sense of the hierarchy of favourability under the standard conditions specified.

Gibbs energy calculations are also useful for determining the difference between the utilisation of different electron donors and acceptors in the same metabolic process. For instance, *G. metallireducens* is capable of oxidising a wide range of organic compounds with the concurrent reduction in ferric iron. By calculating $\Delta G^{\circ}{}_r$ for iron reduction coupled to a variety of different electron donors, the relative favourability of each can be determined. The following equations represent the reduction of ferric iron in solution coupled to methanol [5], glucose [6], and toluene [7] oxidation, respectively:

$$6Fe^{3+} + CH_3OH + 2H_2O \rightleftharpoons 6Fe^{2+} + HCo_3^- + 7H^+ \quad \Delta G^{\circ}{}_r = -65.30\ \mathrm{kJ/mol\ e^-} \tag{R7.4}$$

$$24Fe^{3+} + C_6H_{12}O_6 + 12H_2O \rightleftharpoons 24Fe^{2+} + 6HCO_3^- + 3OH^+ \quad \Delta G^{\circ}{}_r = -64.17\ \mathrm{kJ/mol\ e^-} \tag{R7.5}$$

$$36Fe^{3+} + C_7H_8 + 21H_2O \rightleftharpoons 36Fe^{2+} + 7HCO_3^- + 3H^+ \quad \Delta G^{\circ}{}_r = -53.18\ \mathrm{kJ/mol\ e^-} \tag{R7.6}$$

The values are all negative, therefore each iron reduction reaction is thermodynamically favourable under standard conditions; however the relatively more negative value of methanol oxidation indicates that this is the most favourable combination of electron donor and acceptor for iron-reducing micro-organisms.

Compared with electron donors, the relative difference in Gibbs energy is much greater for a range of ferric iron electron acceptors in iron reduction, clearly illustrated by comparing the values of $\Delta G^{\circ}{}_r$ from (R7.4), (R7.5) and (R7.6) with the following reactions, in which the iron(III) hydroxide electron acceptor has been replaced with hematite [8] and goethite [9], respectively:

$$4Fe_2O_3 + CH_3COO^- + 15H^+ \rightleftharpoons 8Fe^{2+} + 2HCO_3^- + 8H_2O \quad \Delta G^{\circ}{}_r = -65.30\ \mathrm{kJ/mol\ e^-} \tag{R7.7}$$

$$8FeO(OH) + CH_3COO^- + 15H^+ \rightleftharpoons 8Fe^{2+} + 2HCO_3^- + 12H_2O \quad \Delta G^{\circ}{}_r = -5.82\ \mathrm{kJ/mol\ e^-} \tag{R7.8}$$

Again, all reactions are thermodynamically feasible, with the reduction of iron (III) hydroxide the most favourable. In contrast, goethite reduction is only just thermodynamically feasible. Such data are useful in considering the various redox pairs, and their relative favourability, available to iron-reducing micro-organisms.

It is worth highlighting that these data refer only to an idealised situation in which temperature and pressure remain constant at standard conditions. The following equation allows for the calculation of ΔG at temperatures other than 25°C:

$$\Delta G = \Delta H - T\Delta S \tag{7.2}$$

where:

$$\Delta H = \sum \Delta H^{\circ}{}_{products} - \sum \Delta H^{\circ}{}_{reactants} \tag{7.3}$$

$$\Delta S = \sum \Delta S^{\circ}{}_{products} - \sum \Delta S^{\circ}{}_{reactants} \tag{7.4}$$

Unlike (7.1), which incorporates only the standard Gibbs energies of formation for the relevant reactants and products, this (7.2) splits the Gibbs energy function into its constituent elements of enthalpy and energy changes, isolating the effect of temperature. In order to apply (7.2) data are required for the standard enthalpies of formation (change in heat content when a substance is formed from its elements in their most stable forms under standard conditions of 25°C and 1 atm) and entropies (degree of disorder) of each substance [58]. Such data are often listed alongside standard Gibbs energies of formation in data tables, and are applied to (7.5) and (7.6).

Equation (7.2) can be applied to microbially-mediated redox reactions in environments characterised by extremes in temperatures, where the calculation of standard changes in Gibbs energy (7.1) are misrepresentative of the system. Let us reconsider the reduction of ferric iron coupled to methanol oxidation. At 25°C this reaction has a Gibbs energy of −65.30 kJ per electron transferred. If we consider the same reaction at 100°C, a temperature representative of a hydrothermal vent system, the change in Gibbs energy is −76.30 kJ/mol e−. In contrast, at a temperature of 0°C representative of polar environments, the same reaction yields a less negative change in Gibbs energy of −61.43 kJ/mol e−. Thermodynamic considerations therefore indicate that this redox reaction is more favourable in environments with higher temperatures.

It is possible to calculate change in Gibbs energy for systems in which the pressure is other than 1 atm. However, the majority of microbially-mediated redox reactions occur in the aqueous phase, and hence pressure is not a major concern in this context. For reactions that do involve gaseous phases, such as methanogenesis from carbon dioxide and hydrogen, the pressure of the system and its changes through the course of the reaction will affect the Gibbs energy of the reaction. However, its calculation is beyond the scope of this chapter [58].

To assess the effect of different concentrations of redox constituents on the change in Gibbs energy of a reaction, the equilibrium constant (K_c) is required:

$$K_c = \frac{[\mathrm{Y}]^y [\mathrm{Z}]^z}{[\mathrm{B}]^b [\mathrm{D}]^d} \tag{7.5}$$

for the hypothetical reaction reaction: $b\mathrm{B} + d\mathrm{D} \rightleftharpoons y\mathrm{Y} + z\mathrm{Z}$.

The square brackets in (7.5) represent the concentration of reactants (B, D) and products (Y, Z; in moles per litre), and indices indicate stoichiometric coefficients. The change in Gibbs energy can then be calculated using the following equation:

$$\Delta G = -RT \ln K_c + RT \ln \frac{a_Y^y a_Z^z}{a_B^b a_D^d} \tag{7.6}$$

where R is the universal gas constant, T is temperature (Kelvin), and K_c is the equilibrium constant calculated from (7.5) above.

This equation is particular useful when assessing the effects of pH on a redox system, where the concentration of protons produced or reacted can be increased or decreased to represent relatively more acidic or alkali environments, respectively. It can also be applied to a particular environment where geochemical data are available in order to gain insight into the range of feasible microbially-mediated redox reactions that may occur over time and space.

It is important to point out that, whilst thermodynamic tools offer the means to deduce whether a given reaction will occur, the rate of such reactions require the application of chemical kinetics. That is to say, the Gibbs energy value may indicate that a reaction will proceed spontaneously for a given set of physico-chemical conditions, yet the rate of reaction may be slow. Additionally, that a reaction will proceed spontaneously does not alone indicate that this reaction is catalysed by life. It is therefore most powerful when used in conjunction with chemical kinetics, and complemented by microbiological data.

7.8.2 *Applications*

Thermodynamic considerations are increasingly being incorporated into discussions of redox-metabolising micro-organisms as tools for understanding metabolic activity in nature and laboratory studies [e.g., 59, 60]. Perhaps the most comprehensive example of the application of thermodynamics to life in extremes is the study of metabolic processes in the Vulcano shallow marine hydrothermal system in Italy [61]. In order to better understand the diverse lithotrophic and heterotrophic microbial assemblages found in this system, the authors used geochemical data (pH, temperature, conductivity, ionic and organic acid concentrations) collected from various locations to model reaction energetics of 145 organic and inorganic redox reactions under in situ conditions. To achieve this, they calculated individual activities for each aqueous compound, and obtained changes in Gibbs energy using (7.6). The resulting values were normalised for the number of electrons transferred to allow cross-comparison of different metabolic pathways. From these values they were able to deduce which redox reactions yielded the most and the least energy, with values ranging from 0 to 125 kJ/mol e−. Trends in most favourable terminal electron acceptor were also apparent. These data were complemented by a genetic survey of geothermal (56°C) fluids taken from the same locations. Additional geochemical data from a subsequent field campaign allowed for additional spatial coverage of the Vulcano hydrothermal system and for temporal comparisons to be drawn [62]; three additional locations were sampled, and six of the previous locations were re-sampled. Results from the subsequent energetic calculations revealed that spatial variation in iron redox reaction energy yields was significant, though temperature variations were moderate in their effects. Given that the microbial ecology of such an environment is contingent on geochemical composition, an understanding of the associated energetic potential available to life is a crucial aspect in understanding its limits.

Thermodynamic considerations have also been suggested as an approach to search for habitable environments elsewhere in our Solar System [e.g., 63–65]; a thermodynamic approach offers a universally-applicable constraint on habitability, given the universal energy demand across all life. This approach would entail defining potentially habitable environments, such as on Mars, in terms of energy sources, electron donors and electron acceptors, and characterising the extent of energy disequilibrium available across space and time [63]. This has been dubbed the "Follow the Energy" approach to detecting habitability, thought to build upon the well-known "Follow the Water" concept that has yielded so much valuable geochemical information on Mars to date. In addition to identifying areas in which life may reside, thermodynamic considerations will also serve to identify the suite of biomarkers that may exist, fuelling future life detection missions to Mars and beyond [65].

7.9 Many Biospaces?

The limits to life based on a water-carbon biochemistry elaborated earlier in the chapter give rise to a physical and chemical 'space' within which life can grow. We might refer to this as a 'biospace'. The biospace is an n-dimensional space bounded by the wide variety of physical and chemical extremes of life including pressure, high and low temperature, pH, radiation, salinity, etc. The absolute limits of life for any given physical and chemical extreme are in many cases not known, for example the upper temperature limit for life may not yet have been reached, so the biospace for terrestrial-like carbon-based life using water as a solvent is not yet properly defined.

Are the boundaries of the biospace universal? Many of the limits to life are set by basic biophysics. For example, the upper temperature limit for life is ultimately likely to be set by the damage done to biomolecules (i.e. the energy imparted to them by the high temperatures) and whether the energy required to repair molecules and synthesise new ones can be gathered by the organism from the environment. Although different evolutionary innovations might plausibly allow for more efficient energy gathering apparatus than in terrestrial organisms and novel types of molecular folding might make molecules even more resistant to high temperatures than those found in existing organisms, ultimately bond energies of carbon molecules are universal and one might hypothesise that the energies imparted by high temperatures will ultimately exceed any evolutionary innovations possible to counteract them. Terrestrial life has had over 3.5 billion years to evolve methods to cope with physical and chemical extremes and one might argue that the

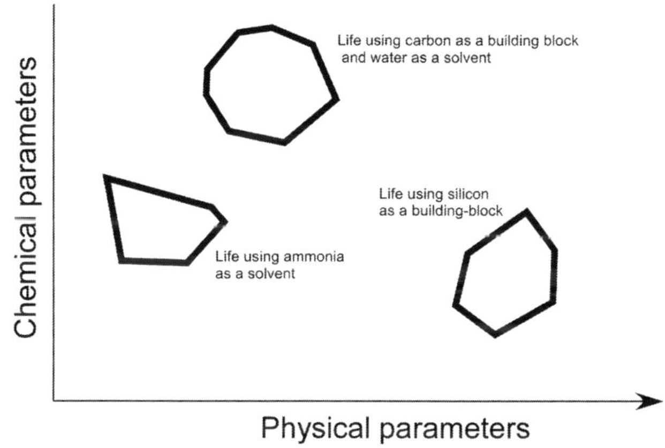

Fig. 7.4 A simple conceptual diagram of multiple 'biospaces' of life that use different physical chemistries and therefore are enclosed within different physical and chemical extremes. The boundaries of each biospace may, however, be universal

limits exhibited by terrestrial life, both today and in the past, represent universal boundaries of biophysical capabilities, although the boundaries may be slightly different depending on the exact architecture of key molecules that become incorporated into any particular carbon based life.

Regardless of whether the limits to terrestrial life displayed by lifeforms on the Earth are universal or whether the n-dimensional biospace of carbon and water based life can be radically different, another question is whether other biospaces exist in the Universe that are defined by the physical and chemical extremes of other types of biochemistry. Bains (2004) for example, presents an elaborate scheme for the biogeochemistry of a silane/silanol biochemistry in liquid nitrogen. In this scheme crustal CO and CO_2, water and ammonia react with silica in serpentinization reactions to produce methane, hydrogen and silanes/silanols. These silicon compounds are then transported to the oceans by geyser activity. These compounds react with unsaturated hydrocarbons to generate methane and other byproducts which can take part in feedback reactions to generate further catalytic silanols. Clearly, if biochemistries such as these exist and they were to lead to life, that life would occupy a very different biospace than life on the Earth with, for example, a lower temperature range.

The possibility of multiple biospaces which life can occupy with different biochemistries is conceptually illustrated in Fig. 7.4. As yet, there is no chemical evidence for life that uses an alternative elemental building block than carbon and an alternative solvent to water. Few experiments exist to test the possibility of the origin and evolution of life. As yet, we know of only the biospace occupied by carbon and water-based life exemplified by life on the Earth. Nevertheless, the alternative solvents and elemental building blocks reviewed in this chapter remind us that the physico-chemical limits of life are not unambiguously described by our knowledge of chemistry.

A future challenge of physical chemistry in astrobiology is to test the hypothesis that alternative biospaces exist on other planetary bodies and to test the plausibility of reaction schemes in the laboratory.

7.10 Some Concluding Remarks

The complexity of biology might superficially give the impression that every planetary 'experiment' in biological evolution would give rise to different outcomes. Although very few of the physical and chemical boundaries to terrestrial life that have been explored have been defined exactly, it is clear that for low temperature and water activity, two extensively studied extremes, physical chemistry does seem to define boundaries that are likely to be universal. These boundaries are further 'reinforced' by the thermodynamics of acquiring sufficient energy from biologically-available redox couples to allow organisms to repair damage and synthesise new biological material at these extremes. Clearly, however, our knowledge of the physical chemistry of the limits of life remains in its infancy. A further question that remains largely experimentally unexplored is whether the 'biospace' occupied by carbon and water based life is the only available biospace, or whether other biospaces exist in which life uses alternative solvents or elemental building blocks and which have quite different physical and chemical limits.

References

1. Gerday C, Glansdorff N (2007) Physiology and biochemistry of extremophiles. ASM Press, Washington, DC
2. Breezee J, Cady N, Staley JT (2004) Subfreezing growth of the sea ice bacterium *Psychromonas ingrahamii*. Microb Ecol 47:300–304
3. Wells LE, Deming JW (2006) Characterization of a cold-active bacteriophage on two psychrophilic marine hosts. Aquat Microb Ecol 45:15–29
4. Wells LE, Deming JW (2006) Modelled and measured dynamics of viruses in Arctic winter sea-ice brines. Environ Microbiol 8:1115–1121
5. Junge K, Deming JW, Hajo E (2001) A microscopic approach to investigate bacteria under in situ conditions in sea-ice samples. Ann Glaciol 33:304–310
6. Rivkina EM, Friedmann EI, McKay CP, Gilichinsky DA (2000) Metabolic activity of permafrost bacteria below the freezing point. Appl Environ Microbiol 66:3230–3233
7. Bakermans C, Tsapin AI, Souza-Egipsy V, Gilichinsky DA, Nealson KH (2003) Reproduction and metabolism at −10°C of bacteria isolated from Siberian permafrost. Environ Microbiol 5:321–326
8. Brock TD (1979) Biology of micro-organisms, 3rd edn. Prentice-Hall, Englewood Cliffs
9. Daniel RM, Dunn RV, Finney JL, Smith JC (2003) The role of dynamics in enzyme activity. Annu Rev Biophys Biomol Struct 32:69–92
10. Junge K, Eicken H, Swanson BD, Deming JW (2007) Bacterial incorporation of leucine into protein down to −20°C with evidence for potential activity in sub-eutectic saline ice formations. Cryobiology 52:417–429

11. Warren SG, Hudson SR (2003) Bacterial activity in South pole snow is questionable. Appl Environ Microbiol 69:6340–6341
12. Carpenter EJ, Lin S, Capone DG (2000) Bacterial activity in South pole snow. Appl Environ Microbiol 66:4514–4517
13. Panikov NS, Sizova MV (2007) Growth kinetics of microorganisms isolated from Alaskan soil and permafrost in solid media frozen down to −35°C. FEMS Microbiol Ecol 59:500–512
14. Junge K, Eicken H, Deming JW (2004) Bacterial activity at −2 to −20°C in Arctic wintertime sea ice. Appl Environ Microbiol 70:550–557
15. Rivkina EM, Laurinavichus KS, Gilichinsky DA, Scherbakova VA (2002) Methane generation in permafrost sediments. Dokl Biol Sci 383:179–181
16. Elberling B, Brandt KH (2003) Uncoupling of microbial CO_2 production and release in frozen soils and its implications for field studies of arctic C cycling. Soil Biol Biochem 35:263–272
17. Panikov NS, Flanagan PW, Oechel WC, Mastepanov MA, Christensen TR (2006) Microbial activity in soils frozen to below −39°C. Soil Biol Biochem 38:785–794
18. Campen RK, Sowers T, Alley RB (2003) Evidence of microbial consortia metabolizing within a low-latitude mountain glacier. Geology 31:231–234
19. Morita RY (1997) Bacteria in oligotrophic environments. Kluwer, Dordrecht
20. Price PB, Sowers T (2004) Temperature dependence of metabolic rates for microbial growth, maintenance and survival. Proc Natl Acad Sci USA 101:4631–4636
21. D'Amico S, Collins T, Marx J-C, Feller G, Gerday C (2006) Psychrophilic microorganisms: challenges for life. EMBO Rep 7:385–389
22. Kashefi K, Lovley DR (2003) Extending the upper temperature limit for life. Science 301:934
23. Daniel RM, Cowan DA (2000) Review: biomolecular stability and life at high temperatures. Cell Mol Life Sci 57(2):250–264
24. Cowan DA (2004) The upper temperature of life – how far can we go? Trends Microbiol 12:58–60
25. Phipps BM, Hoffmann A, Stetter KO, Baumeister W (1991) A novel ATPase complex selectively accumulated upon heat shock is a major cellular component of thermophilic archaebacteria. EMBO J 10:1711–1722
26. Carballeira NM, Reyes M, Sostre A, Huang H, Verhagen MFJM, Adams MWW (1997) Unusual fatty acid compositions of the hyperthermophilic archaeon *Pyrococcus furiosus* and the bacterium *Thermotoga maritime*. J Bacteriol 179:2766–2768
27. Bartlett DH (2002) Pressure effects on in vivo microbial processes. Biochem Biophys Acta 1595:367–381
28. Kato CL, Li Y, Nogi Y, Nakamura Y, Tamaoka J, Horikoshi K (1998) Extremely barophilic bacteria isolated from the Mariana Trench, Challenger Deep, at a depth of 11,000 meters. Appl Environ Microbiol 64:1510–1513
29. Sharma A, Scott JH, Cody GD, Fogel ML, Hazen RM, Hemley RJ, Huntress WT (2002) Microbial activity at gigapascal pressures. Science 295:1514–1516
30. Navarro-Gonzalez R, Rainey FA, Molina P, Bagaley DR, Hollen BJ, de la Rosa J, Small AM, Quinn RC, Grunthaner FJ, Caceres L, Gomez-Silva B, McKa CP (2003) Mars-like soils in the Atacama Desert, Chile, and the dry limit of microbial life. Science 302:1018–1021
31. Harris RF (1981) The effect of water potential on microbial growth and activity. In: Parr JF, Gardner WR (eds) Water potential relations in soil microbiology. Soil Science Society of America, Madison, pp 23–95
32. Grant WD (2004) Life at low water activity. Philos Trans R Soc Lond B Biol Sci 359:1249–1267
33. Brown AD (1990) Microbial water stress: physiology: principles and perspectives. Wiley, Chichester
34. Kis-Papo T, Oren A, Wasser SP, Nevo E (2003) Survival of filamentous fungi in hypersaline Dead Sea water. Microb Ecol 45:183–190
35. Hallsworth JE, Yakimov MM, Golyshin PN, Gillion JLM, D'Auria G, Alves FDL, La Cono V, Genovese M, Mckew BA, Hayes SL, Harris G, Giuliano L, Timmis KN, McGenity TJ (2007) Limits of life in $MgCl_2$-containing environments. Environ Microbiol 9:801–813

36. van der Wielen PWJJ, Bolhuis H, Borin S, Daffonchio D, Corselli C, Giuliano L, D'Auria G, de Lange GJ, Huebner A, Varnavas SP, Thomson J, Tamburini C, Marty D, McGenity TJ, Timmis KN, BioDeep Scientific Party (2005) The enigma of prokaryotic life in deep hypersaline anoxic basins. Science 307:121–123
37. Potts M (1994) Desiccation tolerance of prokaryotes. Microbiol Rev 58:735–805
38. Möhlmann D (2005) Adsorption of water-related potential chemical and biological processes in the upper martian surface. Astrobiology 5:770–777
39. Koop T (2002) The water activity of aqueous solutions in equilibrium with ice. Bull Chem Soc Jpn 75:2587–2588
40. Robbins EI, Rodgers TM, Alpers CN, Nordstrom DK (2000) Ecogeochemistry of the subsurface food web at pH 0–2.5 in Iron Mountain, California, USA. Hydrobiologia 433:15–23
41. Kelch BA, Eagen KP, Erciyas EP, Humphris EL, Thomason AR, Mitsuiki S, Agard DA (2007) Structural and mechanistic exploration of acid resistance: kinetic stability facilitates evolution of extremophilic behavior. J Mol Biol 368:870–883
42. Baross JA, Berner SA, Cody GD, Copley SD, Pace NR (2007) The limits of organic life in planetary systems. National Academies Press, Washington, DC
43. Nichols DS, Greenhill AR, Shadbolt CT, Ross T, McMeekin TA (1999) Physicochemical parameters for growth of the sea ice bacteria *Glaciecola punicea* ACAM 611^{T} and *Gelidibacter* sp. strain IC158. Appl Environ Microbiol 65:3757–3760
44. Gilichinsky D, Rivkina E, Shcherbakova V, Laurinavichuis K, Tiedje J (2003) Supercooled water brines within permafrost – an unknown ecological niche for microorganisms: a model for astrobiology. Astrobiology 3:331–341
45. Wilson JW, Ott CM, Höner zu Bentrup K, Ramamurthy R, Quick L, Porwollik S, Cheng P, McClelland M, Tsaprailis G, Radabaugh T, Hunt A, Fernandez D, Richter E, Shah M, Kilcoyne M, Joshi L, Nelman-Gonzalez M, Hing S, Parra M, Dumars P, Norwood K, Bober R, Devich J, Ruggles A, Goulart C, Rupert M, Stodieck L, Stafford P, Catella L, Schurr MJ, Buchanan K, Morici L, McCracken J, Allen P, Baker-Coleman C, Hammond T, Vogel J, Nelson R, Pierson DL, Stefanyshyn-Piper HM, Nickerson CA (2007) Space flight alters bacterial gene expression and virulence and reveals a role for global regulator Hfq. Proc Natl Acad Sci USA 104:16299–16304
46. Schulze-Makuch D, Irwin LN (2008) Life in the Universe. Springer, Heidelberg
47. Storey KB, Storey JM (1984) Biochemical adaption for freezing tolerance in the wood, *Rana sylvatica*. J Comp Physiol 155:29–36
48. Feinberg G, Shapiro R (1980) Life beyond Earth: the intelligent Earthling's guide to life in the Universe. William Morrow and Company, Inc., New York
49. Bains W (2004) Many chemistries could be used to build living systems. Astrobiology 4:137–167
50. Firsoff VA (1963) Life beyond the Earth. Basic Books, Inc., New York
51. Benner SA, Ricardo A, Carrigan MA (2004) Is there a common chemical model for life in the Universe? Curr Opin Chem Biol 8:672–689
52. Houtkooper JM, Schulze-Makuch D (2007) A possible biogenic origin for hydrogen peroxide on Mars: the Viking results re-interpreted. Int J Astrobiol 6:147–152
53. Krauskopf KB (1983) Introduction to geochemistry, 2nd edn. McGraw-Hill, London
54. Lovley DR, Lonergan DJ (1990) Anaerobic oxidation of toluene, phenol, and *p*-cresol by the dissimilatory iron-reducing organism, GS-15. Appl Environ Microbiol 56:1858–1864
55. Stumm W, Morgan JJ (1995) Aquatic chemistry – chemical equilibria and rates in natural waters, 3rd edn. Wiley-Blackwell, New York
56. Hallberg KB, Hedrich S, Johnson DB (2011) *Acidiferrobacter thiooxydans*, gen. nov. sp. nov.; an acidophilic, thermo-tolerant, facultatively anaerobic iron- and sulfur-oxidizier of the family *Ectothiorhodospiraceae*. Extremophiles 15:271–279
57. Shelobolina ES, VanPraagh CG, Lovley DR (2003) Use of ferric and ferrous iron containing minerals for respiration by *Desulfitobacterium frappieri*. Geomicrobiol J 20:143–156

58. Warn JRW, Peters APH (1996) Concise chemical thermodynamics, 2nd edn. CRC Press, Boca Raton/London
59. Blum JS, Bindi AB, Buzzelli J, Stolz JF, Oremland RS (1998) *Bacillus arsenicoselenatis*, sp. nov., and *Bacillus selenitireducens*, sp. nov.: two haloalkaliphiles from Mono Lake, California that respire oxyanions of selenium and arsenic. Arch Microbiol 171:19–30
60. Grinder-Vogel M, Criddle CS, Fendorf S (2006) Thermodynamic constraints on the oxidation of biogenic UO_2 by Fe(III) (hydr)oxides. Environ Sci Technol 40:3544–3550
61. Rogers KL, Amend JP (2005) Archaeal diversity and geochemical energy yields in a geothermal well on Vulcano Island, Italy. Geobiology 3:319–332
62. Rogers KL, Amend JP, Gurrieri S (2007) Temporal changes in fluid geochemistry and energy profiles in the Vulcano Island hydrothermal system. Astrobiology 7:905–932
63. Nealson KH, Tsapin A, Storrie-Lombardi M (2002) Searching for life in the Universe: unconventional methods for an unconventional problem. Int Microbiol 2:223–230
64. Hoehler TM (2007) An energy balance concept for habitability. Astrobiology 7:824–838
65. Hoehler TM, Amend JP, Shock EL (2007) A "follow the energy" approach to astrobiology. Astrobiology 7:819–823

Chapter 8
Life, Metabolism and Energy

Robert Pascal

Abstract The energy processes that support life are analysed with respect to their thermodynamic and kinetic requirements: (1) a flow of energy in order that self-organisation does not violate the 2nd Law of thermodynamics and (2) the fact that life must be regarded as a kinetic state of matter. Aside from anabolism consisting in the synthesis of metabolites, including the activated precursors of biopolymers, the need for energy flow coming from catabolism or physical sources of energy is emphasised. Quantitative conclusions are reached by considering the lifetime of side-reactions and the absolute temperature. This relationship is consistent with the fact that self-organisation involves covalent bonds and implies the contribution of energy sources with a high thermodynamic potential. These constraints lead to a definition of the conditions under which self-organisation is possible, contribute to determine the nature of the system, and bring about a new concept with regard to the habitability of exoplanets: the compatibility with the origin of life.

The issue of the metabolism of the first living organisms that emerged on the surface of the Earth is the object of several ongoing controversies [1]. Which energy sources were feeding its metabolism [2–6]? Was it autotrophic, synthesizing its own organic components from energy and mineral sources of carbon (carbon dioxide, for instance) [7–10], or heterotrophic [11–14], using abiotically formed organic matter found in the environment as starting materials? Between metabolism and genetic support, which character of living beings emerged first [15–17]? These issues are commonly addressed by considering the environment of the early Earth or the biological record by a molecular phylogeny approach. Considering the physicochemical laws underlying the emergence of life may be an alternative

R. Pascal (✉)
UMR 5247, Institut des Biomolécules Max Mousseron, CNRS-Université Montpellier 1 & 2, Place Eugène Bataillon, Montpellier Cedex 34095, France
e-mail: rpascal@univ-montp2.fr

I.W.M. Smith et al. (eds.), *Astrochemistry and Astrobiology*, Physical Chemistry in Action, 243
DOI 10.1007/978-3-642-31730-9_8,

approach that should be regarded when tackling the possibility that life emerged in several instances in the Universe as the result of general self-organisation principles. A difficulty in this approach is that there is no consensus on the mere definition of what is life and a simple formulation may be out of reach [18]. In contrast to inert things, living organisms cannot be isolated as a fixed category since dynamical features are clearly needed to consider them as alive. As a matter of fact it is not so easy to tell the difference between a dead body or a stuffed animal and a living one by looking at a single snapshot whereas motions immediately give information on their nature. The key differences lie in the dynamic nature of living beings and in considering that they are pieces of a genealogic (evolutionary) process, which has been expressed by Theodosius Dobzhansky as "*Nothing in biology makes sense except in the light of evolution*" [19]. However, considering the species living on Earth as the result of an evolutionary process governed by natural selection does not mean that life escapes the principles of physics or chemistry. Biology provides no indication that a single feature of life is in contradiction with the laws of physics even though the systemic nature of living beings or that of the whole biosphere are not so easy to explain in a simple way and are not possible to describe in a purely deterministic way. Before any proposition that new principles need to be added to account for the living world, it would be better, in a reductionist approach, to check if living organisms make use of sub-systems that are not explained by physical laws.

Actually, our difficulties lie in the need of incorporating the evolutionary process in a physicochemical description. Attempts to express Darwinian theory in physical terms have been made since the beginning of the twentieth century, and a noticeable one was proposed by Lotka who wrote: "*…a principle competent to extend our systematic knowledge in this field seems to be found in the principle of natural selection, the principle of the survival of the fittest, or, to speak in terms freed from biological implications, the principle of the persistence of stable forms*" [20]. It was also probably the first attempt to merge the special ability of autocatalytic processes with the irreversible thermodynamics of far from equilibrium systems [21].[1] This assumption means that metabolic features of life could be understood with available laws of physics and chemistry. Independently from this intuition, comprehensive studies of the kinetic specificities of self-replicating systems have been published [23–27] so that Addy Pross has recently expressed natural selection as the manifestation of *dynamic kinetic stability*, which has a scope that is beyond biology. Then, the transition from non-living to living (the emergence of life) can be considered as a transition[2] in a process governed by driving forces (including *dynamic kinetic*

[1] The question of why Lotka views have scarcely been considered by scientists (except in a few instances as for example ref. [22]) involved in origin of life studies during the twentieth century is open to debate by historians of science.

[2] It is worthy to note that considering the emergence of life as an event in a continuous process means that locating the emergence of life is arbitrary, as that of any transition in history, so that the mere definition of life cannot be purely objective but the result of a convention established and adopted by the scientific community.

stability) and not as two independent processes, the first one depending on physical laws and the second one on the Darwinian theory [28, 29].

When life is described as resulting from the dynamics of things that are able to replicate as developed in the preceding paragraph, there is apparently no contribution of metabolism to the systems (although most authors do not overlook its contribution). On the contrary, any system in which replicators are in the growth phase needs to be fed with energy, provided as activated species or activating agents, which must become a limiting factor because of exponential growth. Indeed, it has been demonstrated [30] that a model involving two replicators, the first living at the expense of activated building blocks and the second capable, in addition, of harvesting energy to build its own building blocks, demonstrated that the second one will predominate even in the case of a less efficient replication when activated monomers become rare. The issue of how energy can be provided to a self-replicating system is important to be addressed. The aim of this chapter is to deal with this question and to show that physicochemical principles have both qualitative and quantitative consequences on the metabolic features of living organisms and on the issue of the plausibility of the emergence of life in the Universe. It is not intended as a comprehensive account of the metabolisms found in biology and microbiology but a only as an analysis of the consequences of simple physical and chemical laws on the proto-metabolisms that played a role in the emergence of life and on the driving forces that ruled their evolution towards modern enzymatic systems.

8.1 Self-Organisation and the Nature of Matter and Energy

One of the main reasons that probably led Erwin Schrödinger to write his booklet "What is life" [31] was related to his view that the description of our physical world by quantum theory had consequences for understanding how living organisms are functioning. In fact, macroscopic physics usually considers matter and energy as physical quantities that can be varied continuously, which is not right when considering the microscopic scale in which the atomic and subatomic descriptions of matter have to be considered and in which energy is changed by discontinuous amounts (quanta).

8.1.1 Matter

The view that solids, liquids and gases around us are made of atoms or molecules that can also be organised in crystal or other supramolecular assemblies seems unmistakable because of the progress of physics during the nineteenth and twentieth centuries. But it is worth noting that Greek philosophers considered that the structures observed in our world at the macroscopic scale result from the existence of the assembly of discrete particles. In many instances, the description of our world requires to consider that matter is not continuous; it is made of particles, atoms and molecules. Organized systems of molecules exceeding the atomic or molecular scale can be built from these molecular building blocks by non-covalent interactions leading to crystals, micelles, vesicles or other supramolecular architectures. The formation of these structures corresponds to the evolution that takes place provided that they are thermodynamically more stable than a disordered state. Their formation is usually thermodynamically driven. In some cases, they are capable of leading to the formation of biomimetic structures [32].

But kinetically driven processes can also lead to structures that have consequences at the macroscopic scale, which is meaningful with regards to life and its origin. This specificity is found for things (molecules, cells, organisms constituting replicators) that are capable of reproducing themselves leading in favourable cases to exponential growth. As a matter of fact, the existence of a finite set of atomic constituents is the source of the possibility of self-reproducing systems (molecules or assemblies), the dynamic behaviour of which has been considered as one of the main features responsible for the specificity of life and of its *kinetic stability* [28, 29]. The non-sustainable power of exponential growth may be illustrated by the possibility of a single grain of wheat, considering that it can be multiplied by a factor of let us say 10 every year (one generation), which could potentially lead to 10^{80} grains after 80 year (a life span), which would exhaust much more than the possibilities of the whole known Universe (*ca.* 10^{80} nucleons). This growth efficiency is the source of *dynamic kinetic stability* [28, 29] observable for systems held away from equilibrium and behaving as a kinetic state of matter [33–35]. In fact, a genuine exponential growth is a key for the possibility of selection of the most efficient replicators [25], those that have the more numerous offspring. This need of self-replicating or self-reproducing systems is taken into account in most scenarios for the origins of life since the replication of information bearing polymers, as achieved with nucleic acids thanks to Watson-Crick pairing, has been considered as a requirement for an unlimited variability required for an open-ended evolutionary process [36].

8.1.2 Energy

Energy changes in atomic or molecular systems occur through discrete amounts rather than being varied continuously or by any arbitrary amount. Then any collection of atoms, molecules or any other physical systems, which can only exist in discrete energy states considered individually, is populated at the equilibrium state according to statistical laws. Except at 0 K where the system can only occupy the lowest possible energy microstates, energy is distributed according to the Maxwell–Boltzmann law (see Chap. 1). This distribution of energy in a population of objects is at the origin of the macroscopic quantity called entropy. Following Schrödinger [31], it is essential to take it into account when dealing with living systems. In any system, entropy is

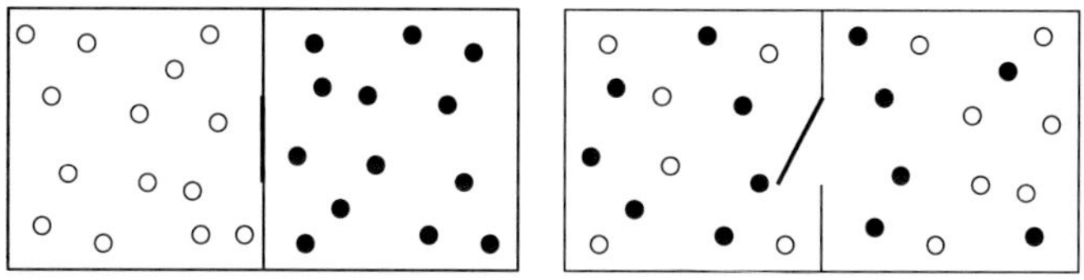

Fig. 8.1 Gas molecules, as do any other particles, tend to occupy the whole volume available to their motion. Mixing two gases leads to an increase in entropy

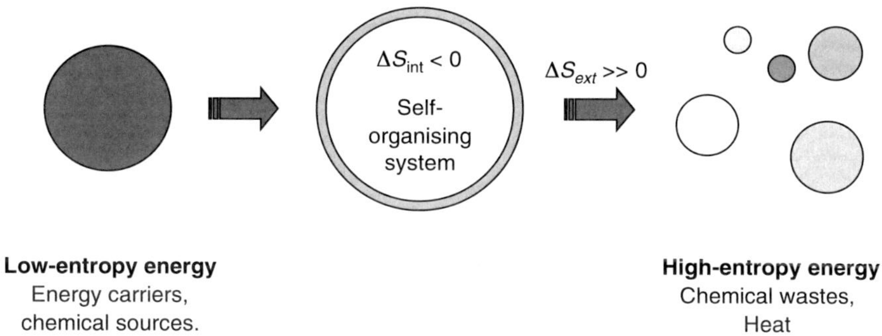

Fig. 8.2 A self-organising system needs a flow of energy. To continuously overcome the loss of entropy due to irreversible processes, it must be coupled to a flow delivering energy in a low entropy form and dissipate it under the form of heat or inactivated chemical derivatives

considered as having its lower value (S = 0 by definition) at absolute zero and increases with temperature as the energy of the system of interest is distributed in an increasing number of microstates with higher energy levels. Organised systems, which correspond to states in which energy is not distributed according to these laws, are characterised by lower entropy values than the equilibrium state and their evolution can only lead to an increase in disorder and then in entropy in agreement with the second law of thermodynamics as for example molecules of gas that tend to occupy the entire volume available (Fig. 8.1).

As living organisms are intrinsically working out of equilibrium, the physical nature of energy has consequences on the description of the living state and on its origin, which was pointed out by Schrödinger [31] and developed in several other investigations on the early development of metabolism [14, 37–39]. Considering metabolism, this rule applies to living systems, and means that they have to continuously collect energy from physicochemical sources or nutrients in a low entropy state – corresponding to a non-equilibrium distribution and high energy levels – from an external source and to release products in a high entropy state – close to the equilibrium distribution at the temperature of the system (Fig. 8.2) [14]. The overall result of this transformation is capable of compensating for the inescapable increase in entropy or, in other words, would be helpful in maintaining (or reproducing) the organised structure of living organisms. These views are consistent with the formulation of the second law that the entropy of an isolated system

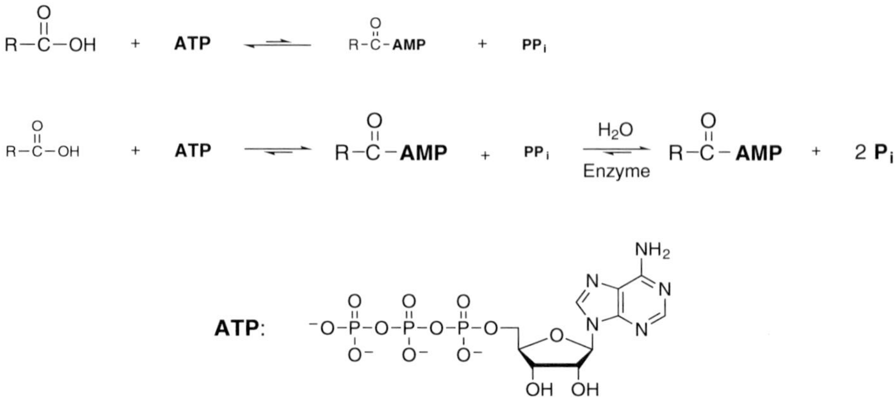

Fig. 8.3 The chemical artifice used for extracting more energy than the potential of a single phosphate anhydride linkage in ATP through coupling acylation with the hydrolysis of pyrophosphate (**PP**$_i$) into inorganic phosphate (**P**$_i$). The addition of a pyrophosphatase enzyme is capable of shifting the equilibrium to the right side thus allowing the system to reach significant concentration of acylated AMP intermediate

cannot decrease so that a self-organising system must not be isolated (exchanging neither matter nor energy with its environment), but open (exchanging matter and energy) or closed (exchanging energy only) [40]. As chemical energy corresponds to a thermodynamic potential, the realisation of chemical potential energy by an irreversible transformation of chemical energy carriers (capable of releasing a certain amount of energy in an individual chemical event [37, 38, 41]) through reactions with components of the system would lead to chemical intermediates with a lower potential accompanied by the dissipation of heat.

It is important to realise that a chemical intermediate with a higher chemical potential than the energy source cannot accumulate in high concentration. Either it would be formed in equilibrium from reactants (with population of microstates consistent with Maxwell-Boltzmann distribution) or its formation has to be additionally driven by coupling with an energy-releasing or entropy-producing process. An example of this latter possibility is found in biology by some reactions of ATP, the universal biochemical energy currency [42]. Some biochemical processes, such as acyl group activation, require an amount of energy that is above the potential of the hydrolysis of a single phosphate anhydride in ATP. In these cases, biological organisms use the following solution: the equilibrium is shifted towards the products by coupling the acylation process with the release of the free energy of the pyrophosphate side product of the acylation step in the presence of an efficient inorganic pyrophosphatase enzyme (Fig. 8.3). This example demonstrates that highly specific catalysis is needed to take advantage of this possibility, which is unlikely for very early processes.

8.2 Self-Organisation and Kinetics

As indicated above, self-organisation brings about states that are far from equilibrium. Thermodynamics predicts that these states must spontaneously evolve towards the equilibrium state but provides information neither about the time needed for this evolution nor about the corresponding chemical path. A kinetic analysis is therefore indispensable to understand why living systems are capable of remaining constantly away from equilibrium. Both a self-organising system (possibly in the course of its evolution towards life) and a living system (able to reproduce itself in a way that is in principle open ended) need to be protected from the spontaneous evolution towards the thermodynamic equilibrium. This condition has quantitative consequences on the rate of spontaneous chemical reactions leading to the deactivation of metabolites: they have to be slower than the course of the metabolic reactions otherwise the metabolic network would rapidly vanish. Because they correspond to a kinetic state rather than structures governed by thermodynamics it would be logical that quantitative information about life or metabolism could be derived from by kinetic laws rather than from thermodynamic principles.

8.2.1 *Kinetic Barriers*

The requirement that spontaneous chemical reactions leading to the deactivation of metabolites into inert waste in a metabolic network have to be much slower than the rates at which the metabolic or protometabolic reaction network performs its task in any metabolism needs to be studied in more detail. Albert Eschenmoser expressed this idea in a convenient way: "*Emergence of an evolving system in Nature according to such a model requires a chemical environment contained far from thermodynamic equilibrium by kinetic barriers*" [43]. As a result, metabolites and especially those bearing a high content in energy have to be protected from spontaneous degradation by chemical barriers [43–45]. This assumption is counter-intuitive since it is generally considered that reactions needed to trigger the emergence of life had to be as fast as most reactions involved in biological systems [46]. But it is consistent for example with the selection of relatively non-reactive phosphate anhydrides in which the ionised character inhibits nucleophilic attack at phosphate centres (as needed for hydrolysis). Then the selection of ATP as an energy currency in biology [47] could be described as resulting from its lack of kinetic reactivity easily compensated in the active site of enzymes. Therefore, processes involving metabolites that are not spontaneously subject to a fast decomposition is as important as their high reactivity through pathways useful to self-organisation, which means that kinetic and catalytic processes responsible for the origin of life had to be selective. However, by themselves, free energy barriers protecting a system from its evolution towards equilibrium are not sufficient to explain self-organisation. The development of a metabolism is needed for this aim and its consequence would be that species that are absent in the environment have to be produced and need to be re-produced by the whole system so that the whole set of reactions works as an autocatalytic network. Then all metabolites have to be synthesized at the expense of precursors available in the environment and the free energy of high-energy species has to be directed to this aim instead of being spontaneously dissipated. The emergence and selection of reacting systems capable of performing these tasks was probably essential in this process, or as stated by Eschenmoser "*What has been paramount to the origin of life with respect to the dichotomy of thermodynamic versus kinetic control is the central role of catalysis in imposing kinetic control on structural changes of a chemical environment held far from equilibrium by kinetic barriers*" [44]. Another important question with respect to the origin of metabolism is to determine how a metabolic network can be coupled with the reproduction of a support for heredity so that the entire system constitutes a unit of evolution [48]. Finally, it is worth noting that the importance of kinetic barriers is a general feature in organised systems that has obvious consequences in biology with many networks of metabolites or biopolymers connected to each others by very specific enzymatic catalysts that work at rates much faster than the spontaneous decay of most components. More surprisingly a similar complex behaviour can also be recognized in the description of nuclear reactions.[3]

8.2.2 Architecture of Metabolisms

Contemplating the whole metabolism of extant living organisms illustrates their complexity and the interconnection of metabolic pathways. However, it can be disconnected into different modules that are easier to analyse, which helps in understanding how metabolism could emerge from chemical systems. In this chapter, the concept of metabolism or proto-metabolism is regarded as that of networks of reactions that allow the energy coming from a source of energy or a chemical carrier to perform chemical work in the system in order to maintain or

[3] Interestingly, the height of barriers inhibiting spontaneous nuclear reactions and related to the strong repulsion existing between nuclei could be depicted in the same way as responsible for the possibilities of reaction cycles presenting all the attributes of catalytic or autocatalytic cycles in chemistry. For instance, the direct formation of ^{4}He helium nucleus from four protons is hardly possible by nuclear fusion of four protons but takes place through the C N O cycle in the core of massive stars (with ^{12}C, ^{13}N, ^{13}C, ^{14}N, ^{15}O, and ^{15}N as components of the reacting loop in which four protons are captured stepwise and an α particle is produced). Eigen and Schuster already pointed out this analogy with chemical catalytic cycles [24]. In the same way, nuclear fission reactions may also be described as autocatalytic replications of neutrons starting from unstable heavy nuclei. However, the main difference in the self-organisation processes that take place through nuclear reactions compared to chemistry probably lies in the limited set of atoms presenting a sufficient stability that can be obtained, whereas there is no limit to the number of structures accessible to organic chemistry.

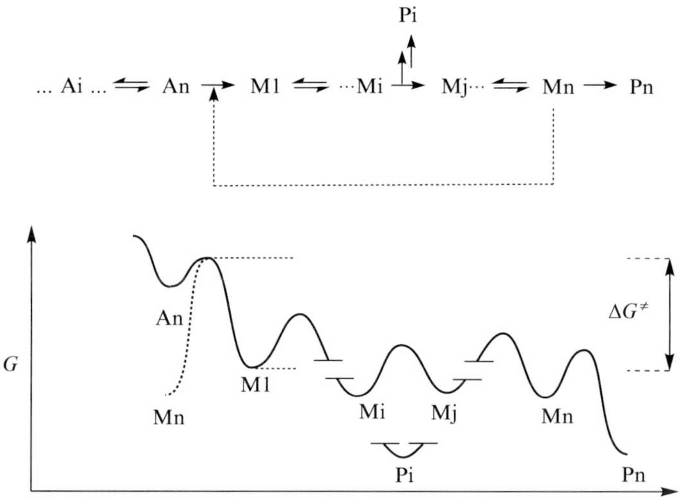

Fig. 8.4 Chemical path and free energy changes as a function of the progress along the reaction pathway in a proto-metabolic system. A metabolism or proto-metabolism can be divided into different parts: the activation part (energy carriers Ai, An) in relation with energy sources, the metabolic part (metabolites M1, Mi, Mj… Mn), and the product part (Pi, Pn). At least one of the reactions in the metabolic part must be irreversible on the timescale of the progress of the whole system (for instance if the value of $\Delta G^{\neq}$ is sufficient to avoid the backward reaction from M1 to An). This condition allows the system to operate as a one-way process (provided that energy is continuously brought about to the system as Ai, An). Products Pi, Pn are considered here as inactive *i.e.* as metabolic wastes. Recycling a metabolite (Mn, *dashed arrow*) converts the consecutive set of reactions into a metabolic cycle, which is actually collectively equivalent to a catalyst since an increase in the concentrations of any of the metabolites M1, Mi, Mj, or Mn increases the kinetic rates through which activated species Ai…An are consumed. A catalytic step is not explicitly involved because any chemical network with a cyclic topology results in catalysis. An autocatalytic network could be formed if anyone of the products of a downstream process is identical to one of the metabolites (for instance if Pi = Mn): one round of the cycle would then produce two molecules of Mn; note that all the intermediates involved in the loop (M1, Mi, Mj… Mn) are stoichiometrically reproduced every time the whole cycle is repeated

increase its degree of chemical organisation. Such systems correspond to sets of metabolites (reactants, intermediates or products) that include the reactions connecting them and that can result in a local decrease in entropy, which requires the metabolic or proto-metabolic[4] pathway to be irreversible, as any biochemical metabolic pathway [42]. Then, the sequence of reactions (some of which can be reversible) must involve at least an irreversible step (Fig. 8.4). It can be constituted

[4] In this chapter we use the term proto-metabolism for specifying networks of reactions capable of performing chemical transformations and inducing self-organisation features in a chemical system. From this definition, there is fundamentally no difference between metabolic and proto-metabolic pathways except that enzyme catalysis makes metabolism much more efficient in achieving its function and generates multiple possibilities of feedback control almost without any limitations.

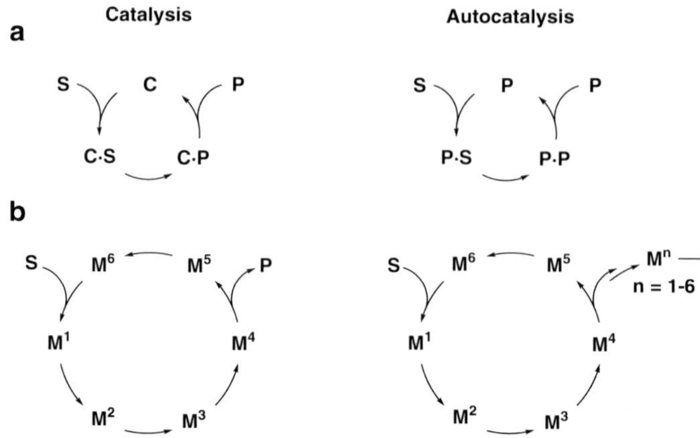

Fig. 8.5 Comparison between (**a**) the usual representations of catalysis and autocatalysis and (**b**) a more general version resulting from considering the cyclic architecture of reaction networks. The usual representation of enzyme catalysis deduced from Michaelis-Menten kinetics with two non-covalently bound complexes C·S and C·P fits the general description of a cycle by including the three states of the enzyme (free, bound to substrate, and bound to product). Genuine autocatalysis in its simplest version without covalent intermediate (*up right*) may be much more demanding than network autocatalysis because efficient autocatalysis requires that strong transient non-covalent interactions are present at the transition state whereas the reactant and product are stable in a monomer state. Moreover, the possibility that products or intermediates of downstream processes could be identical to intermediates of the metabolic cycle (M^1 to M^6) is statistically increased

of consecutive reactions with a linear or cyclic topology or involve a more complex organisation. In this discussion the metabolism is divided into different parts (1) activation in which energy is provided to the system as energy carriers or as energy source (photochemistry for instance), (2) the metabolic part which has to include an irreversible step, and (3) the product part corresponding to the formation species devoid of a sufficient chemical potential in the environment so that they can be considered as metabolic waste, though they can play a biochemical role independently of the metabolism. The architecture of the metabolic part has important consequences that are worth emphasising (Fig. 8.5):

- Any set of reactions with a cyclic arrangement will collectively behave as a catalyst since an increase in the concentration of one of the metabolites constituting the loop will result in an increase in the rates of consumption of the energy source,
- Any set of reactions with a cyclic arrangement that additionally includes a downstream process that reproduces one of the metabolites involved in the loop collectively behaves as an autocatalytic network in which all the components of the loop are reproduced when the cycle is running [23, 24].

Many biochemical properties are then simply the result of feedback control in a metabolism either in a positive way as mentioned above for autocatalysis or in a negative way by inhibition of a preceding step. It is important to notice that these properties can emerge from the topology of the system as a whole and not only from the behaviour of a single metabolite or chemical step.

8.2.3 Catalysis and Autocatalysis

The importance of kinetic barriers to maintain a chemical system out of equilibrium and to bring about possibilities of self-organisation governed by kinetic control [43, 44] has the important consequence that factors that increase the kinetic rates in a non-selective way are usually not favourable to this complex behaviour. High temperatures and non-selective catalysis (possibly acting on deactivation steps) are then to be avoided. Actually, catalysis by itself is unimportant for self-organisation, it is just another factor to increase rates, it is only when catalysis is *involved in the rate-determining step* of a selected pathway in a metabolic network that it becomes important (and subject to a selective pressure when, for example, the catalyst is constituted of a reproducible sequence of building blocks). Then, the emergence of selective biochemical catalysts (ribozymes or enzymes) for a specific step is, in principle, capable of orientating the reactant flux towards a path that is spontaneously not favoured, which illustrates the fact that the emergence and development of life needed to be based on non-robust chemical processes [49], in which finely tuned conditions or catalysis are capable of leading to different outcomes. Efficient genetically encoded catalysts are also likely to induce a change of rate-determining step in a stepwise metabolic pathway so that the selective pressure will affect the upstream or downstream step that has become limiting. It is therefore the emergence of early biochemical catalysts that was the driving force for the emergence of further enzymes explaining why almost every biochemical reaction is facilitated by an enzyme. These catalytic processes are likely to increase the overall *dynamic kinetic stability* [28, 29] of the whole replicating system. Genetically encoded catalysts have additionally the capacity of increasing the complexity of the network because the cycle of reproduction of the catalysts constitutes a supplementary feedback process by itself. The origin of life and the emergence of catalysts based on non-limited possibilities of variations must have led rapidly to a complex and interconnected network since any enzyme or ribozyme is produced by an independent metabolism so that any biochemical catalyst can be part of two metabolic loops, the catalytic process and its own replication.

Catalysis is usually described by an interaction of a catalyst with the transition state of a reaction lessening the activation barrier and implies that the catalyst is recovered in an unchanged form at the end of the reaction [50] (see Chap. 1). However, interactions at the reactant state, useless in an ideal description of catalysis, are capable of promoting an increase in rate because they induce a close proximity of reacting groups [51, 52]. This means that any catalytic process must involve at least three different steps (1) a non-covalent (or covalent) association of the catalyst and reactants, (2) the chemical step in which bonding changes take place and (3) the diffusion away of the catalyst and the products (Fig. 8.5).

Actually, adding several chemical steps (Fig. 8.5) to constitute a catalytic cycle does not change the chemical topology of the system, except that the catalyst loses its ideal unchanged character and interacts with reacting species for a time much longer than the lifetime of a transition state.[5] The advantage of this process is that covalent bonds can be involved in catalysis in addition to non-covalent interactions, a single interaction is then likely to stabilise the transition state much better than a weak interaction.[6] The ideal view of catalysis, based on a thermodynamic analysis, that catalysts only stabilise transition states by weak interactions can then be amended by a view encompassing a more general kinetic view that catalysis requires the catalyst to be regenerated after an undetermined number of steps, which allows the integration of other kinds of bonding interactions such as covalent bonds or bonding with metallic ions. This means that catalytic reactions can proceed through pathways very different from the non-catalysed reaction mechanism. Organocatalysis [54, 55] and metallic ion catalysis acting through strong interactions with substrates and transition states may then be viewed as early biocatalysts that have been improved latter by the action of enzymes [51]. Acid–base catalysis is also likely to have played an important role since most reactions of biomolecules involve proton transfers.

Autocatalysis corresponds to processes involving a product as a catalyst in a chemical step leading to its own formation. As noticed above for catalysis, a direct interaction of the autocatalyst with the transition state of the non-catalysed upstream reaction is formally not needed since autocatalysis can result from the architecture of a reaction network in which a product or intermediate is involved as a component of a metabolic cycle (catalytic cycle) [24] (Fig. 8.5). This observation on metabolic cycles is important since any metabolism requires that every component is reproduced.

8.2.4 Relationship Between Chemistry and Timescale in a Protometabolic System

Albert Eschenmoser proposed a "*kinetic version of Le Chatelier principle*" expressed as "*a chemical environment constrained by kinetic barriers will react to the stress of being kept far from equilibrium in such a way as to seek maximization of its equilibration rate*" [43]. This expression is equivalent to the observation

---

[5] Efficient catalysis requires that the interaction of a catalyst with a transition state is preceded by an interaction with reactants that has been analysed to impose constraints on enzymatic catalysis [52].

[6] This limitation does not apply to folded genetically encoded biocatalysts in which a well-defined three-dimensional structure gives rise to multiple non-covalent interactions specific of the transition state of the reaction and allows the binding energy with non-reacting portions of the substrate to substantially contribute to catalysis [53].

that any chemical process resulting from autocatalysis or replication will be kinetically preferred in such systems merely because they are faster than other non-catalysed processes. This picture of *dynamic kinetic stability* [28, 29] is shown here to yield to quantitative developments. Transition state theory supposes that the evolution to products takes place through a transient state corresponding to a maximum of free energy and that remain in quasi-equilibrium with reactants, which allows the definition of the corresponding equilibrium constant $K^{\neq}$. The free energy of the transition state will then correspond to:

$$\Delta G^{\neq} = -RT \ \ln \ K^{\neq} \tag{8.1}$$

The second tenet of this theory is that the breakdown from the transition state to products is the result of a vibration at the frequency ν between two moieties in a locally flat energy profile at the saddle point corresponding to the transition from the reactant part to the product part of the free energy landscape.

$$\nu = (k_{\mathrm{B}}T/h) \tag{8.2}$$

In other words it expresses than no reaction can progress at rates faster than a vibration frequency. Then the rates of the reaction will be proportional to the product of this frequency and the ratio of reactants in this instable state. The Eyring equation deduced in this way from the transition state theory provides a relationship between the height of the barrier ($\Delta G^{\neq}$ the free energy of activation), absolute temperature T, and the rate constant k of the reaction.

$$k = \kappa \frac{k_B T}{h} e^{-\Delta G^{\neq}/RT} \tag{8.3}$$

This relationship does not depend on another variable than these three parameters and will be used as a tool to detect conditions in which self-organisation based on proto-metabolic fluxes of energy can take place. Selecting a value of 1 for the transmission coefficient κ (meaning that there is no possibility of reverting to the reactants after the system has crossed the transition state), the value of the free energy of activation can then be deduced as a function of the half-life of a first-order (or pseudo-first-order) reaction at different values of temperature (Fig. 8.6).

$$\Delta G^{\neq} = R\,T \ \ln(\frac{k_B\,T\,t_{1/2}}{h\ \ln(2)}) \tag{8.4}$$

At a moderate temperature (300 K) and at half-lives spanning from 1 s to 100 year (a factor of 3×10^9), the kinetic barriers correspond to a free energy range of 74–129 kJ mol^{-1}, an order of magnitude remaining quite close to 100 kJ mol^{-1}. This value for a kinetic barrier represents a significant fraction of the free energy of a covalent bond (ca. 350 kJ mol^{-1} for a C–C bond). Then at the

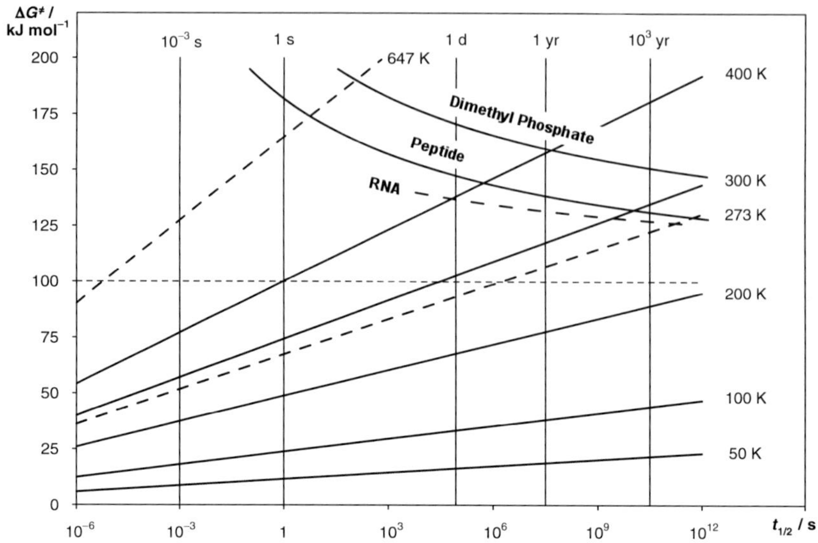

Fig. 8.6 Relationship between the free energy of activation and the half-lives ($t_{1/2} = \ln(2)/k$, logarithmic scale) of metabolites at different temperatures (in Kelvin) calculated using (8.4). Supposing that activating agents (energy carriers) or metabolites with a sufficient lifetime need to resist for periods of time of 1 s to 100 year leads to the conclusion that they must be protected from spontaneous degradation by barriers corresponding to an order of magnitude of 100 kJ mol^{-1} (74–129 kJ mol^{-1}) at 300 K. Solid and dashed lines represent the variation of $\Delta G^{\neq}$ as a function of $t_{1/2}$ at different values of absolute temperatures including the temperature of fusion of ice (273 K) and the critical temperature of water (647 K). Additional curves, calculated using published values of $\Delta H^{\neq}$ and $\Delta S^{\neq}$ are displayed for different specific reactions of biological interest (**a**) the cleavage of dimethyl phosphate, chosen for representing the stability of the phosphodiester bond in DNA [56], (**b**) acetylglycine *N*-methylamide as a representative example of peptide bond [56], and (**c**) the curve deduced from the published values for RNA cleavage at 25 °C and 92 °C [57]

timescale of the diffusion of chemical species on the surface of the Earth and at moderate temperature a self-organisation process is likely to involve covalent bonds. Alternatively, this condition can be expressed as: a living system based on covalent bonds (and then with a large predilection for carbon chemistry giving covalent bonds easily) is more resistant when the temperature is not sufficient to overcome a free energy barrier representing an important fraction of the bond dissociation energy of covalent bonds and in particular that of C–C bonds (350 kJ mol^{-1}). This means that the more favourable temperatures for life (and probably its emergence) are close to the lower temperatures compatible with liquid water (273 K under a pressure of 10^2 kPa – 1 atm – and 251 K at a pressure of 210 MPa – 2,070 atm). Structures based on covalent bonds are then likely to constitute the building blocks of self-organisation processes driven by the search of a more favourable kinetic stability in our physical world. Although covalent bonds match the requirement for kinetic barriers, systems remaining far from the equilibrium state and based on multiple interactions involving weak bonds cannot be excluded. But reaching kinetic barriers having a similar height would need the multiple weak interactions to be broken simultaneously for the individual barriers to be additive, which seems unlikely since this would be true for all the metabolites involved in the system and in addition requires perfectly rigid structures.

The influence of temperature needs to be examined more deeply. An increase in temperature from 300 K to 400 K has a strong influence on the barrier needed for protection of the chemical intermediates (an increase of 30–45 kJ mol^{-1} is needed to maintain similar half-lives). Enzyme reactions are usually much less sensitive to temperature than spontaneous reactions [58, 59]. This difference is more crucial than expected by the fact that the common "rule" that reactions double in rates by raising the temperature by 10 °C is more or less verified for reactions taking place with lifetimes spanning from minutes to days but not for long-lived reaction systems as those found in most non-catalysed processes of degradation of biomolecules [58, 60]. The consequence of this observation is that processes leading to the spontaneous decomposition of energy carriers or metabolites would have a much sharper temperature dependence. The rates of non-catalysed processes increase faster than those of catalytic ones so that high temperatures are harmful to biomolecules and to complex processes, which strongly supports the belief that they are not favourable to the origins and development of life [61]. The domain of existence of liquid water covers the range of temperature 251–647 K including the presence of liquid at high pressure below 0 °C. It is then likely that complex networks of reactions are more easily obtained in the lower range of temperatures. Actually, high temperatures raise the question of the stability of covalent bonds found in biopolymers. There is a limit of stability for life related to the lifetimes of active metabolic polymers (proteins) and information carriers (DNA and RNA) that depend on the rate of cleavage of a large number of repetitive linkages. Indeed, the data displayed in Fig. 8.6 show that the lifetime of a peptide bond, of a RNA internucleotidic linkage, or of a model of the corresponding one in DNA (dimethyl phosphate), estimated from published data, decrease sharply above 400 K in a way consistent with the fact that there are no living organisms capable of growing above this temperature. Among these three biopolymers, the most sensitive to high temperatures is RNA, with a lifetime that does not exceed 1 day for a single bond at 400 K so that the size of a genome based on RNA sequences (RNA world) would need to be short at this temperature to ensure the transmission of information from generation to generation. The hypothesis of an origin of life in a world based on RNA as an information support needs a lifetime sufficient for RNA to be replicated before cleavage without loss of information leading to an error catastrophe [62], except if the environment provided conditions to preserve RNA from degradation without affecting its replication rate. Cooling the system to a temperature of 0 °C or temperatures of eutectic phases compatible with liquid water [63–65] increases the lifetime by a factor exceeding five orders of magnitude, compatible with sequences of much greater lengths.

Revealing that there is a quantitative relationship between the timescale of a proto-metabolism (this timescale may be related to the turnover frequency for a catalytic cycle [66]) and the height of the kinetic barriers that must be present for protecting the metabolites or carriers that are involved in exchanges with the environment, gives rise to the possibility of identifying factors influencing the probability for a complex chemical system to survive and then the more favourable conditions for its evolution. Understanding these conditions will lead to a prediction of the most favourable energy sources capable of eliciting life by introducing a further assumption.

8.2.5 The Nature of Energy Sources

Many kinds of sources of energy have been proposed as capable of feeding early organisms with energy on the primitive Earth [67, 68]. The idea that a living or self-organising system must continuously tend to reduce its internal entropy is closely linked to the requirement that metabolic pathways work irreversibly and that the process must be maintained far from equilibrium. The fact that the whole system must work irreversibly means that energy has then to be delivered as a one-way process so that the system works as a dissipative structure constantly generating entropy in the environment to maintain its own organization [31]. The second condition is related to the kinetic barriers that are required to avoid spontaneous deactivation leading to side-products that may not be coupled with the metabolic process. These barriers must protect every metabolite and especially species presenting the highest free energy potential, the energy carriers. We showed above that the constraint on kinetic barriers could be assessed quite precisely using transition state theory and making a conservative assumption on the timescale of the whole system. This evaluation was a first step in the determination of the requirement with respect to energy sources that could feed a proto-metabolism. In the specific case of activating agents (energy carriers that are formed from non-chemical energy sources from processes distinct or included in the metabolism, Fig. 8.7), these remarks mean that they must be formed through an irreversible process, which raises the quantitative condition on the free energy of their formation that the energy source must release an amount of energy exceeding that of the transition state separating them from inactive reactants, which is a specifically acute condition for the energy carriers that have to move from the location of their formation to an environment favourable to the development of a metabolism. The free energy source feeding the system in energy by activating inert species into chemical carriers must then provide a chemical potential exceeding the amount needed for the whole metabolism to work irreversibly and allowing a sufficient lifetime for energy carriers or high energy metabolites. This analysis (depicted in Fig. 8.7) leads to the conclusion that the free energy source must provide at least a free energy potential corresponding to the sum of that for the kinetic barrier $\Delta G^{\neq}$ and that of the carrier ΔG relative to inactive reactants. Many high energy metabolites found in biochemistry have free energies of hydrolysis reaching or exceeding 50 kJ mol^{-1} and many of them are likely to have been needed for essential processes leading to the emergence of life [14, 39]. A value close to or exceeding 150 kJ mol^{-1} can then be considered as a likely potential for an energy

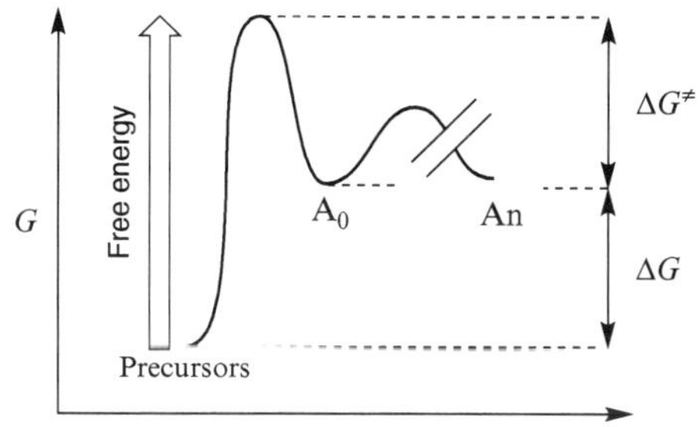

Fig. 8.7 Formation of an energy carrier from an energy source and non-activated chemical precursors (free energy change along the reaction path specifying the activation part of Fig. 8.4). Energy carriers formed in the environment from energy sources must survive a lifetime sufficient to allow their migration to locations where they will be used. Considering that activation must be irreversible (which is equivalent to the fact that the energy carrier A_0 must be protected from spontaneous degradation as other metabolites), it follows that the free energy potential brought about by the energy source in the activation step must exceed the sum of the free energy of activation of the reverse reaction and of that of the carrier

source capable of leading to the origin of life at temperatures compatible with liquid water. Using this conclusion, we can then examine the potential sources of energy for early life *i.e.* light, heat or chemical gradients.

For photochemistry, the correspondence of energy and frequency is given by the Plank relation ($E = h\nu$) so that a chemical potential reaching or exceeding 150 kJ mol^{-1} can be easily attributed to photons with $\leq$0.8 μm wavelengths corresponding to the visible domain.

The thermodynamic potential of heat sources interacting with a chemical system corresponds to the quanta of energy that can be obtained through spontaneous irreversible processes such as conduction or radiation.[7] Conduction cannot constitute a progress towards the conversion of heat into chemical energy since it results in an irreversible diffusion process. By contrast, radiation can constitute a first step leading to photochemistry, so that the radiation of a black body provides a correspondence between the properties of heat sources with respect to the potential needed to induce a proto-metabolism determined above for photochemistry. A thermodynamic potential corresponding to a free energy $\geq$150 kJ mol^{-1}, attributed to photons of $\leq$0.8 μm wavelengths can thus be obtained as the maximum of emission of a black body heated to 3,600 K. Then only heat sources with temperatures of several thousand Kelvin can induce self-organisation through proto-metabolisms at temperatures of *ca.* 300 K. The preceding conclusions are

[7] It is important to notice that the conversion of heat into other forms of energy by other processes requires heat engines that use a cold sink (entropy sink) to get energy with higher thermodynamic potentials. Natural processes such as storms are capable of doing so by giving rise to lightning with local temperatures transiently exceeding 10,000 K, but considering lightning as an alternative source of energy will be preferred here in order to avoid any misleading statement about the potential of heat sources.

consistent with the properties of the Sun defined as a black body with a surface temperature of *ca.* 6,000 K that is capable of inducing the formation of high-energy metabolites through radiation. With regards to heat, lightning and shock waves provoked by the impact of meteorites are the only other obvious possibilities of providing the required thermodynamic potential on the early Earth.

8.2.6 The Question of Redox Energy Sources

The existence of chemical barriers, developed here through a quantitative approach as a condition for the emergence of self-organisation, leads to the conclusion that only covalent bonds can maintain far from equilibrium situations prone to self-organisation in temperate environments. The transfer of an electron from a reducer to an oxidized species usually cannot provide a similar kinetic barrier as, for instance, during a transfer between transition metal ions. But the situation is different when the redox reaction (see Chap. 1) involves species in which covalent bonds have to be altered for the redox reaction to take place and that usually need the transfer of two electrons as for instance when the oxidation state of a carbon atom is changed. Translating directly the condition for irreversibility that long-lived metabolites based on covalent bonds made in this way must have a sufficient lifetime compared to the progress of the whole system, and must be sufficiently rich in energy to drive early biochemistry, would mean that the redox reaction must correspond to a difference of potential of at least 0.75 V $[-(\Delta G^{\neq} + \Delta G)/n\mathcal{F}]$ for a two electron redox reaction. These views are confirmed by the analysis of the specific case of the reduction of CO_2 by the $FeS\text{-}H_2S/FeS_2$ redox couple that involves high activation energy making the generation of energy through this process unlikely [69]. The presence of redox gradients on the early Earth is the result of geodynamics that could bring into contact reservoirs with a different state of oxidation [4–6]. The question of the possibility that life could, or not, have emerged through the use of redox sources of energy is generally combined with the need of reducing power for synthesising biomolecules from inorganic carbon sources (mainly CO_2, which corresponds to the oxidized state of carbon) [7, 8]. But these two questions can be disconnected, even though extant life provides examples of organisms that are capable, depending on redox potential between inorganic species, of generating energy and that use inorganic sources of carbon to synthesise their own components. With regard to the question of redox sources of energy, generating a proto-metabolism from these sources would require that a system in which species are contained far from equilibrium by kinetic barriers (and that is considered above to be dependant on covalent bonds) could be formed from the conversion of a redox potential. Possibilities that the chemistry of sulphur and especially the fact that energy-rich thioesters could be obtained by redox chemistry have been proposed [70–72]. Alternatively, biochemistry provides a different solution [73] in which the dissipation of redox potential through electron transfers is coupled to the generation of a proton concentration gradient between two compartments. The persistence of the concentration gradient requires an efficient barrier between the compartments so that it can be used in a second stage for the generation of an energy-rich metabolite – i.e. ATP in cells – through the use of a molecular engine capable of coupling the translocation of several protons. Alternatively, alkali metal ions may replace protons to perform this task and sodium is likely to have preceded protons in primitive ATP synthases [74, 75]. It is worth noting that the utility of the phospholipid membranes of cells in the generation of energy through chemiosmosis can be related to the need of kinetic barriers in the description proposed by Eschenmoser and developed here (the need of holding the system far from equilibrium). An observation of importance if the evidences that membranes made of fatty acids were very probably leaky to protons and alkali metal ions are considered [76, 77]. The possibility that redox energy sources could have fed energy to emerging life has been analysed elsewhere through its capacity of providing species having a thermodynamic potential sufficient for sustaining essential biochemical pathways (translation) [14]. However, the fact that evolution has found a system involving very sophisticated molecular machines indicates that the conversion of redox gradients into energy available to the metabolism is far from being trivial and that its role in emerging life remains questionable [14].

8.3 Emergence and Evolution of Metabolism

8.3.1 Free Energy of Organic Matter

In contrast with the composition of the present day atmosphere, the environment of the primitive Earth was poor in oxygen so that organic matter did not constitute a nutrient capable of providing amounts of energy similar to that presently available through the respiratory metabolism, although other electron acceptors may have replaced oxygen (sulphate for example).[8] This means that a catabolism may not have constituted a source of energy sufficient to drive an irreversible metabolism. However, this statement has to be mitigated by the fact that the formation of organic matter required less energy provided that starting materials in a less oxidized state were present, which is consistent with the fact that aerobic organisms require more energy than anaerobes to synthesize the same biomass [80]. Anyway, the degree of oxidation of the primitive Earth was an important constraint on the emergence and development of life [81, 82]. Self-organisation, as for example the formation of biopolymers, needed the additional availability of a source of energy capable of inducing the irreversible processes defined above. The fact that the barriers needed

[8] Carbon at the carbonyl state of oxidation (sugars) represents the most efficient source of organic matter capable of feeding early heterotrophic organisms in energy in the absence of a respiratory metabolism [14, 78, 79].

to hold the chemical environment hosting a far from equilibrium proto-metabolism could only be available from covalent bonds constitutes a strong rationalisation explaining why organic chemistry (defined as the chemistry of carbon) has a preponderant role in biochemistry.

8.3.2 Anabolism, Catabolism, and Supply in Energy

In living organisms, metabolism fulfils two different tasks. The first one corresponds to the need of maintaining the cell in a far from equilibrium state. This task, in which energy is collected from the environment and irreversibly transformed into useless inactive products and heat, serves in a coupled way to maintain and develop the degree of organisation of the system for instance by the formation of biopolymers or other structures in cells or through their reproduction. The second task is the formation of organic matter from inorganic precursors (*anabolism*). These roles may have developed simultaneously in a coupled manner, but there is a possibility that they were initially disconnected since many organic derivatives are not so unstable (in a far from equilibrium state) in environments depleted in oxygen. This observation is confirmed by the fact that abiotic organic matter is formed spontaneously in different extraterrestrial environments and does not correspond to a free energy state that is far from equilibrium (incapable of inducing energetically useful transformations in very primitive life forms devoid of a complex energy collecting apparatus [14]). As a result, the reducing power of minerals may have been used for the formation of low-energy organic derivatives from inorganic sources of carbon under conditions that are not far from equilibrium (in which organic matter may be both formed and destroyed), in hydrothermal vents for example, in a way that does not contradict the heterotrophic origin of life hypothesis [83]. These sources may have contributed to the pool of organic matter present on the early Earth together with other sources: the delivery of exogenic organic matter from impacts of meteorites and the chemistry in the atmosphere. However the emergence of a metabolism corresponding to the *self-organisation of a system governed by dynamic kinetic stability* [28, 29] which constitutes the distinctive feature of the living state, required the availability of sources of energy corresponding to the conditions defined above (with a free energy potential exceeding 150 kJ mol^{-1}). Many misleading statements about metabolism may result from the confusion of the two different roles of metabolism, which is increased by the fact that they are closely connected in the biochemistry of many evolved living organisms (e.g. aerobic respiration) that is dominated by the importance of catabolism in energy production because the presence of a high content in oxygen in the environment gives the organic matter a high chemical potential [14].

8.3.3 Scenarios for the Emergence of Metabolism

As far as the emergence of life is concerned, the main role of metabolism is its contribution to self-organisation by locally reducing the internal entropy of the system. The emergence of this essential property is then closely related to the availability of energy sources capable of providing quanta of energy sufficient to ensure an irreversible flow of metabolites and then to drive self-organisation. Then only certain forms of organic matter present in the environment may have carried enough energy to induce an irreversible metabolism leading to steady states ruled by *dynamic kinetic stability*, simply because the chemical potential of the other forms is reduced in an anoxic environment. Generating states presenting common properties with the living state requires therefore the direct coupling to physical sources of energy (photochemistry[9] for instance) or the availability of high-energy molecules by chemistry in the atmosphere. These processes would lead to metabolites capable of releasing energy by chemical processes such as hydrolysis. In this perspective, most low-molecular weight organic molecules with CC or CN triple bonds are, in principle, capable of releasing an important amount of free energy potentially coupled to the formation of the most activated biochemical intermediates [14]. In this view, processes leading to more or less complex inactivated organic matter were additionally needed for the formation of simple building blocks, but their contribution to the origin of life may not be considered as more important than other abiotic pathways leading to carbon derivatives in the interstellar medium or in the atmosphere. These processes may have included the reducing power associated with minerals and/or catalysts and even catalytic cycles involving small molecule interactions [16]. But *systems chemistry* [84] defined by the specific behaviour emerging from *dynamic kinetic stability* [29] could only have emerged through irreversible processes functioning as a one-way flow of reactants and in which kinetic barriers tend to suppress both side-reactions and the spontaneous breakdown of activated species into starting materials. In other words, the classification of metabolisms in autotrophic and heterotrophic and anabolism or catabolism may have a limited interest with respect to the origin of life since they omit the most important property which is the need of coupling irreversibility with the production of organisation. Any scenario for the emergence of metabolism should then include, on the one hand, pathways for the formation of building blocks and simple biomolecules that may be reversible or irreversible, and, on the other hand, a mechanism for coupling the supply in energy to the formation of structures or complex processes, which may be either associated with or independent of the

[9] The difference between photosynthesis and the photochemical generation of activated molecules must be emphasised. The latter corresponds to the chemistry occurring after bringing a chemical system into a highly activated state that can generate intermediates with a chemical potential in a determined environment. Oxygenic photosynthesis is a complex process coupling three different actions: the collection of energy, the oxidation of water, and the generation of reducing power exploitable for the synthesis of organic matter.

previous ones. This observation means that independently of the actual source of organic matter, the early organisation of life depended on the presence of energy sources from lightning or from the primitive photochemical conversion of light into activated intermediates that may be directly connected to the metabolism or disconnected as in the case of a generation of activated species in the atmosphere.

8.3.4 Different Conditions for the Origin of Life and for Evolved Organisms

Natural selection is likely to have increased the kinetic stability of living organisms through many additional means that could not be available from abiotic processes, such as catalysis by enzymes or ribozymes. Considering indeed that life emerged in a chemical environment allowing complexity to organise in the presence of kinetic barriers through a limited number of processes, the activity of polymers with defined structures is a powerful tool that may have selectively oriented the reactivity of unstable metabolites, stabilized high-energy intermediates or opened connexions with unrelated processes. It is important to emphasise that these processes are active by binding unstable intermediates or transition states but does not alter the thermodynamic properties of metabolites that remain in a free unbound state in the cell. This increase of complexity brought about by the realisation of the binding energy of biopolymers with unstable intermediates and transition states is likely to have increased the kinetic stability of living organisms but also to have opened adaptability to environments different from that in which life emerged. As already shown in the discussion of the importance of redox processes as energy sources, natural selection led to the development of highly complex machineries working as thermodynamic engines capable of taking advantage of processes that would not be capable of initiating life because the utilisation of sources of chemical potential with less strict free energy conditions became possible. It has, for instance, been proposed that life required the availability of free energy corresponding to a minimal free energy quantum [38, 41] and evolution has developed highly complex cooperative processes to reduce this minimum value in order to settle in new environments [85, 86]. However, the fact that life has found solutions allowing its development in environments that are poor in low-entropy energy does not mean that life could emerge under these conditions and more generally that extremophiles may be relevant to the origin of life [87].

8.4 Implications for Astrobiology

The concept of habitability in astrobiology is associated with the idea that life can exist in environments where life is found on Earth. A rough definition of the habitability zone that has been selected by astronomers in their search for exoplanets [88] is the zone around a star in which water is stable in the liquid state at the surface. The constraints defined here for sustaining life based on covalent bonds are likely to lead to more precise descriptions of planets that may allow life to emerge on their surface with conditions on the temperature and the presence of free energy from visible light or lightning. But it may also extend our vision of the emergence of complex systems considered as the outcome of environments held far from equilibrium by kinetic barriers and in which the processes that are the object of study of *systems chemistry* [84, 89] can emerge. It is striking that this definition encompasses catalytic cycles responsible for nuclear reactions at very high temperature [24], but in this case the finite number of nuclei with a significant lifetime strongly limits complexity. The representation of (8.4) displayed on Fig. 8.6 suggests that weak bonds such as hydrogen bonds or van der Waals interactions may give rise to such a kind of chemistry at temperatures of a few tens of Kelvin, but indications on the physical reality of such processes are not available.

8.5 Conclusion

Science cannot endorse vitalist views to account for the very specificity of the living state. Therefore any description of the driving forces sustaining life must be rooted in the physicochemical properties of matter and energy. The physical nature of matter (made of particles, atoms, molecules...) is taken into account in most scenarios proposed to account for the origin of life. As a matter of fact, the existence of a finite set of atomic constituents is the source of the possibility of self-reproducing systems (molecules or assemblies), the dynamic behaviour of which has been considered as one of the main features responsible for the specificity of life and of its kinetic stability [28]. The physical nature of energy has, in a similar way, important consequences on the description of the living state and on its origin, although this approach has only been used except in a limited number of attempts [14, 37–39]. More generally, living systems do not escape thermodynamic laws and the constraints due to the Second Law have been considered in the description of life [20, 21, 31]. But thermodynamic laws are not, by themselves, sufficient to define driving forces supporting life. This issue was addressed by considering the consequences of the physical nature of energy on the description of the living state and of the driving forces that may have led to its emergence. The quantum description of our world, implying that energy is usually changed by discrete amounts (or quanta rather than being varied continuously or by arbitrary amounts) is essential in the description of the living state as well as the range of free energies of both the high-energy quanta (energy source) and low-energy quanta (entropy sink). Then the free energy available from physical sources and the thermodynamic potential of high-energy chemical species into which they can be converted (including the range of free energy quanta exchanged by these systems) constitute key components determining the nature of self-replicating systems accessible under conditions determined by a specific environment. The fact that this approach uses kinetic laws is consistent with the description of life as *a kinetic state of matter* [33, 34]. Hypotheses proposed to account for the origin of life were examined in relation with the constraints defined here. These views may introduce additional requirements for habitability in the search for exoplanets capable of harbouring life as we know it, or conversely may help in predicting what specific form of life (including free energies of biochemical intermediates and energy carriers) could emerge in environments different from that of the Earth. This approach does not mean that metabolism emerged first but that the introduction of a genetic support and that of selection between variants must have proceeded together with a contribution of a metabolism obeying specific rules. Thus the emergence of life could have been less unlikely provided that suitable conditions could be realised, but the contribution of contingency in this historical process is unavoidable. Determining these conditions is a genuine goal for scientific investigation.

References

1. Pereto J (2005) Controversies on the origin of life. Int Microbiol 8:23–31
2. Lazcano A, Miller SL (1999) On the origin of metabolic pathways. J Mol Evol 49:424–431
3. Delaye L, Lazcano A (2005) Prebiological evolution and the physics of the origin of life. Phys Life Rev 2:47–64
4. Martin W, Russell MJ (2003) On the origins of cells: a hypothesis for the evolutionary transitions from abiotic geochemistry to chemoautotrophic prokaryotes, and from prokaryotes to nucleated cells. Philos Trans R Soc Lond B Biol Sci 358:59–85
5. Lane N, Allen JF, Martin W (2010) How did LUCA make a living? Chemiosmosis in the origin of life. Bioessays 32:271–280
6. Russell MJ, Hall AJ, Martin W (2010) Serpentinization as a source of energy at the origin of life. Geobiology 8:355–371
7. Wächtershäuser G (1988) Before enzymes and templates: theory of surface metabolism. Microbiol Rev 52:452–484
8. Wächtershäuser G (2006) From volcanic origins of chemoautotrophic life to bacteria, archaea and eukarya. Philos Trans R Soc Lond B Biol Sci 361:1787–1808
9. Wächtershäuser G (2006) Evolution of the first metabolic cycles. Proc Natl Acad Sci USA 87:200–204
10. Wächtershäuser G (2007) On the chemistry and evolution of the pioneer organism. Chem Biodivers 4:584–602
11. Oparin AI (1924) The origin of life. Proiskhodenie Zhini. Reprinted and translated in Bernal JD (1967) The origin of life. Weidenfeld and Nicolson, London
12. Haldane JBS (1929) The origin of life. Rationalist Annu 148:3–10
13. Urey HC (1951) On the early chemical history of the earth and the origin of life. Proc Natl Acad Sci USA 38:351–363
14. Pascal R, Boiteau L (2011) Energy flows, metabolism, and translation. Philos Trans R Soc Lond B Biol Sci 366:2949–2958
15. Pross A (2004) Causation and the origin of life metabolism or replication first? Orig Life Evol Biosph 34:307–321, 1017
16. Shapiro R (2006) Small molecule interactions were central to the origin of life. Q Rev Biol 81:106–125
17. Fry I (2011) The role of natural selection in the origin of life. Orig Life Evol Biosph 1:3–16

18. Bruylants G, Bartik K, Reisse J (2011) Prebiotic chemistry: a fuzzy field. C R Chimie 14:388–391
19. Dobzhansky T (1973) Nothing in biology makes sense except in the light of evolution. Am Biol Teach 35:125–129
20. Lotka AJ (1922) Natural selection as a physical principle. Proc Natl Acad Sci USA 8:151–154
21. Lotka AJ (1922) Contribution to the energetics of evolution. Proc Natl Acad Sci USA 8:147–151
22. Cavalier-Smith T (2001) Obcells as proto-organisms: membrane heredity, lithophosphorylation, and the origin of the genetic code, the first cells and photosynthesis. J Mol Evol 53:555–595
23. Eigen M (1971) Selforganisation of matter and the evolution of biological macromolecules. Naturwissenschaften 58:465–523
24. Eigen M, Schuster P (1977) The hypercycle. A principle of natural self-organization. Part A. The emergence of the hypercycle. Naturwissenschaften 64:541–565
25. Szathmáry E, Gladkih I (1989) Sub-exponential growth and coexistence of non-enzymatically replicating templates. J Theor Biol 138:55–58
26. Bag BG, von Kiedrowski G (1996) Templates, autocatalysis and molecular replication. Pure Appl Chem 68:2145–2152
27. Gánti T (2003) The principles of life. Oxford University Press, Oxford
28. Pross A (2009) Seeking the chemical roots of Darwinism: bridging between chemistry and biology. Chem Eur J 15:8374–8381
29. Pross A (2011) Toward a general theory of evolution: extending Darwinian theory to inanimate matter. J Syst Chem 2:1
30. Wagner N, Pross A, Tannenbaum E (2009) Selective advantage of metabolic over non-metabolic replicators: a kinetic analysis. Biosystems 99:126–129
31. Schrödinger E (1946) What is life. McMillan, New York
32. Nassif N, Livage J (2011) From diatoms to silica-based biohybrids. Chem Soc Rev 40:849–859
33. Pross A (2005) Stability in chemistry and biology: life as a kinetic state of matter. Pure Appl Chem 77:1905–1921
34. Pross A (2005) On the emergence of biological complexity: life as a kinetic state of matter. Orig Life Evol Biosph 35:151–166
35. Pross A (2005) On the chemical nature and origin of teleonomy. Orig Life Evol Biosph 35:383–394
36. Ruiz-Mirazo K, Pereto J, Moreno A (2004) A universal definition of life: autonomy and open-ended evolution. Orig Life Evol Biosph 34:323–346
37. Hoehler TM (2007) An energy balance concept for habitability. Astrobiology 7:824–838
38. Hoehler TM, Amend JP, Shock EL (2007) A "follow the energy" approach for astrobiology. Astrobiology 6:819–823
39. Boiteau L, Pascal R (2011) Energy sources, self-organization, and the origin of life. Orig Life Evol Biosph 41:23–33
40. Kondepudi D, Prigogine I (1998) Modern thermodynamics—from heat engines to dissipative structures. Wiley, Chichester
41. Schink B (1997) Energetics of syntrophic cooperation in methanogenic degradation. Microbiol Mol Biol Rev 61:262–280
42. Voet D, Voet JG, Pratt CW (2006) Fundamentals of biochemistry, 2nd edn. Wiley, New York
43. Eschenmoser A (1994) Chemistry of potentially prebiological natural products. Orig Life Evol Biosph 24:389–423
44. Eschenmoser A (2007) Question 1: commentary referring to the statement "The origin of life can be traced back to the origin of kinetic control" and the question "Do you agree with this statement; and how would you envisage the prebiotic evolutionary bridge between thermodynamic and kinetic control?" Stated in section 1.1. Orig Life Evol Biosph 37:309–314

45. Eschenmoser A (2011) Etiology of potentially primordial biomolecular structures: from vitamin B12 to the nucleic acids and an inquiry into the chemistry of life's origin – a retrospective. Angew Chem Int Ed 50:12412–12472
46. Stockbridge RB, Lewis CE Jr, Yang Y, Wolfenden R (2010) Impact of temperature on the time required for the establishment of primordial biochemistry, and for the evolution of enzymes. Proc Natl Acad Sci USA 38:22102–22105
47. Westheimer FH (1987) Why nature chose phosphate. Science 235:1173–1178
48. Vasas V, Szathmáry E, Santos M (2010) Lack of evolvability in self-sustaining autocatalytic networks constraints metabolism-first scenarios for the origin of life. Proc Natl Acad Sci USA 107:1470–1475
49. Eschenmoser A (2007) The search for the chemistry of life's origin. Tetrahedron 63:12821–12844
50. Schowen RL (1978) Catalytic power and transition-state stabilization. In: Gandour RD, Schowen RL (eds) Transition states of biochemical processes. Plenum, New York, pp 77–114
51. Pascal R (2003) Catalysis through induced intramolecularity: what can be learned by mimicking enzymes with carbonyl compounds that covalently bind substrates? Eur J Org Chem 2003:1813–1824
52. Pascal R (2003) Do enzymes bind their substrates in the ground state because of a physicochemical requirement? Bioorg Chem 31:485–493
53. Jencks WP (1975) Binding energy, specificity, and enzymic catalysis: the circe effect. Adv Enzymol Relat Areas Mol Biol 43:219–410
54. List B, Yang JW (2006) The organic approach to asymmetric catalysis. Science 313:1584–1587
55. Barbas CF (2008) Organocatalysis lost: modern chemistry, ancient chemistry, and an unseen biosynthetic apparatus. Angew Chem Int Ed 47:42–47
56. Wolfenden R (2006) Degrees of difficulty of water-consuming reactions in the absence of enzymes. Chem Rev 106:3379–3396
57. Wolfenden R (2011) Benchmark reaction rates, the stability of biological molecules in water, and the evolution of catalytic power in enzymes. Annu Rev Biochem 80:645–667
58. Wolfenden R, Snider M, Ridgway C, Miller B (1999) The temperature dependence of enzyme rate enhancements. J Am Chem Soc 121:7419–7420
59. Snider MJ, Gaunitz S, Ridgway C, Short SA, Wolfenden R (2000) Temperature effects on the catalytic efficiency, rate enhancement, and transition state affinity of cytidine deaminase, and the thermodynamic consequences for catalysis of removing a substrate "Anchor". Biochemistry 38:9746–9753
60. Wolfenden R, Snider MJ (2001) The depth of chemical time and the power of enzymes as catalysts. Acc Chem Res 34:938–945
61. Bada JL, Lazcano A (2002) Some like it hot, but not the first biomolecules. Science 296:1982–1983
62. Eigen M, McCaskill J, Schuster P (1988) Molecular quasi-species. J Phys Chem 92:6881–6891
63. Monnard PA, Kanavarioti A, Deamer DW (2003) Eutectic phase polymerization of activated ribonucleotide mixtures yields quasi-equimolar incorporation of purine and pyrimidine nucleobases. J Am Chem Soc 125:13734–13740
64. Vlassov AV, Johnston BH, Landweber LF, Kazakov SA (2004) Ligation activity of fragmented ribozymes in frozen solution: implications for the RNA world. Nucleic Acids Res 32:2966–2974
65. Monnard PA, Ziok H (2008) Eutectic phase in water-ice: a self-assembled environment conducive to metal-catalyzed non-enzymatic RNA polymerization. Chem Biodiv 5:1521–1539
66. Kozuch S, Shaik S (2011) How to conceptualize catalytic cycles? The energetic span model. Acc Chem Res 44:101–110
67. Deamer DW (1997) The first living systems: a bioenergetic perspective. Microbiol Mol Biol Rev 61:239–261
68. Deamer D, Weber AL (2010) Bioenergetics and life's origins. Cold Spring Harb Perspect Biol 2:a004929

69. Schoonen MAA, Xu Y, Bebie J (1999) Energetics and kinetics of the prebiotic synthesis of simple organic acids and amino acids with the FeS-H2S/FeS2 redox couple as reductant. Orig Life Evol Biosph 29:5–32
70. De Duve C (1988) Prebiotic syntheses and the mechanism of early chemical evolution. In: Kleinkauf H, Von Dohren H, Jaenicke L (eds) The roots of modern biochemistry – Fritz Lipmann's Squiggle and its consequences. De Gruyte, Berlin, pp 881–894
71. De Duve C (1992) The thioester world. In: Tran Thanh Van J, Tran Thanh Van K, Mounolou JC, Schneider J, McKay C (eds) Frontiers of life. Frontières, Gif-sur-Yvette, pp 1–20
72. De Duve C (2003) A research proposal on the origin of life. Orig Life Evol Biosph 33:559–574
73. Mitchell P (1961) Coupling of phosphorylation to electron and hydrogen transfer by a chemi-osmotic type of mechanism. Nature 191:144–148
74. Mulkidjanian AY, Galperin MY, Makarova KS, Wolf YI, Koonin EV (2008) Evolutionary primacy of sodium bioenergetics. Biol Direct 3:13–32
75. Mulkidjanian AY, Galperin MY, Koonin EV (2009) Co-evolution of primordial membranes and membrane proteins. Trends Biochem Sci 34:206–215
76. Deamer D (2008) How leaky were primitive cells? Nature 454:37–38
77. Mansy SS (2010) Membrane transport in primitive cells. Cold Spring Harb Perspect Biol 2:a002188
78. Weber AL (2000) Sugars as the optimal biosynthetic carbon substrate of aqueous life throughout the universe. Orig Life Evol Biosph 30:33–43
79. Weber AL (2002) Chemical constraints governing the origin of metabolism: the thermodynamic landscape of carbon group transformations under mild aqueous conditions. Orig Life Evol Biosph 32:333–357
80. McCollom TM, Amend JP (2005) A thermodynamic assessment of energy requirements for biomass synthesis by chemolithoautotrophic micro-organisms in oxic and anoxic environments. Geobiology 3:135–144
81. Cleaves HJ, Chalmers JH, Lazcano A, Miller SL, Bada JL (2008) A reassessment of prebiotic organic synthesis in neutral planetary atmospheres. Orig Life Evol Biosph 38:105–115
82. Chyba C, Sagan C (1992) Endogenous production, exogenous delivery and impact-shock synthesis of organic molecules: an inventory for the origins of life. Nature 355:125–132
83. Rivas M, Becerra A, Peretó J, Bada JL, Lazcano A (2011) Metalloproteins and the pyrite-based origin of life: a critical assessment. Orig Life Evol Biosph 41:347–356
84. Kindermann M, Stahl I, Reimold M, Pankau WM, von Kiedrowski G (2005) Systems chemistry: kinetic and computational analysis of a nearly exponential organic replicator. Angew Chem Int Ed 44:6750–6755
85. McInerney MJ, Rohlin L, Mouttaki H, Kim U, Krupp RS, Rios-Hernandez L, Sieber J, Struchtemeyer CG, Bhattacharyy A, Campbell JW, Gunsalus RP (2007) The genome of Syntrophus aciditrophicus: life at the thermodynamic limit of microbial growth. Proc Natl Acad Sci USA 104:7600–7605
86. Delong EF (2007) Life on the thermodynamic edge. Science 317:327–328
87. Cleaves HJ, Chalmers JH (2004) Extremophiles may be irrelevant to the origin of life. Astrobiology 4:1–9
88. Lammer H, Selsis F, Chassefière E, Breuer D et al (2010) Geophysical and atmospheric evolution of habitable planets. Astrobiology 10:45–68
89. Ludlow RF, Otto S (2008) Systems chemistry. Chem Soc Rev 37:101–108

Chapter 9
Life: The Physical Underpinnings of Replication

Rebecca Turk-MacLeod, Ulrich Gerland, and Irene Chen

Abstract Replication is a fundamental process that is critical to life as we know it. While replication today is carried out by complex biochemical machineries that have been evolving for billions of years, it must have originated with relatively small molecules in simple systems. Here we explore this concept, focusing on the physicochemical characteristics and prebiotic potential of two classes of biological macromolecules: nucleic acids and lipids. We discuss the informational and catalytic capabilities of DNA and RNA, the thermodynamic limits of information transfer, the structure and function of lipid membranes, and the formation and maintenance of primitive 'protocells'.

9.1 Introduction

The origin of life was a special point in our history when the principles of physics and chemistry first blossomed into the complex interactions that characterise living organisms. Biological phenomena, like replication, can be thought of as emerging from deeper microscopic structural and dynamic properties, in the same way that the physical phenomenon of friction emerges from microscopic interactions among materials. Although living organisms today are often so sophisticated that it can be difficult to see the roots of physical chemistry in their everyday operation, the very first organisms and transitional forms would have been quite close to those roots.

R. Turk-MacLeod
FAS Center for Systems Biology, Harvard University, 52 Oxford St, Cambridge, MA 02138, USA

U. Gerland
Department of Physics, Arnold Sommerfeld Center for Theoretical Physics and Center for NanoScience, Ludwig-Maximilians Universitat, Theresienstr. 37, Munich D-80333, Germany

I. Chen (✉)
Department of Chemistry and Biochemistry, University of California, Santa Barbara
e-mail: ichen@post.harvard.edu

I.W.M. Smith et al. (eds.), *Astrochemistry and Astrobiology*, Physical Chemistry in Action, 271
DOI 10.1007/978-3-642-31730-9_9, © Springer-Verlag Berlin Heidelberg 2013

We do not know, and it is probably impossible to know, the details of how life arose on Earth. What we can study, however, are experimental and theoretical models that are inspired by a combination of historical inferences and reasonable suppositions. One might say that we are not studying *the* origin of life, but instead we are studying *possible* origins of life. There is a great deal of fascinating debate among those who study origins of life regarding the sequence of events, such as whether metabolic cycles arose before or after genetic information [1–3]. It may well be that there are many ways to 'skin' this cat, or many possible roads to life. For our purpose of demonstrating physico-chemical principles, however, we take the practical approach of sidestepping these riddles and focusing on molecular structures that almost all would agree were important at some point. In this chapter, we examine two types of molecular structures that are thought to have particular general importance during an origin of life similar to our own:

1. Carriers of information, particularly nucleic acids, and
2. Boundaries that define self and non-self, particularly lipid membranes.

9.2 Nucleic Acids

9.2.1 What Are Nucleic Acids?

Deoxyribonucleic acid (DNA) (Fig. 9.1a), and its chemical relative, ribonucleic acid (RNA) (Fig. 9.1b), are often said to be the instruction manuals for building organisms. These 'manuals' contain long sequences of letters that are translated into specific proteins by a molecular machine that is itself composed of RNA and protein (the ribosome). The proteins do the real work of the organism, such as catalysing reactions and joining together to build scaffolds. The alphabet of DNA has four canonical nucleobases that are analogous to letters: adenine (A), guanine (G), cytosine (C), and thymine (T). The alphabet of RNA is similar with the exception that T is replaced by its chemical relative uracil (U). While the letters carry the information, a backbone made of phosphorylated sugars holds the letters in the proper order, fulfilling the same function as the paper that the instruction manual is printed upon. To copy the information, a strand of nucleic acid acts as a template for a newly synthesised strand, which will contain A opposite T or U, and G opposite C; these pairings are known as Watson-Crick base pairs (Fig. 9.2). It therefore takes two rounds of copying to reproduce the original single strand.

The conformation of the sugar largely determines the overall structure of the nucleic acid polymer. By convention, the carbons in the pentose sugar are numbered as C1′, C2′, … C5′, beginning with the carbon bearing the nitrogenous base. In DNA, the 2′-deoxyribose nucleotides tend to form the classic B-form double helix (Fig. 9.3a) with antiparallel strands, whose stability is determined primarily by the overlap of orbitals from adjacent aromatic nucleobases (π-π stacking interactions). In the B-form, each base pair is rotated by about 36° relative to the next, such that a

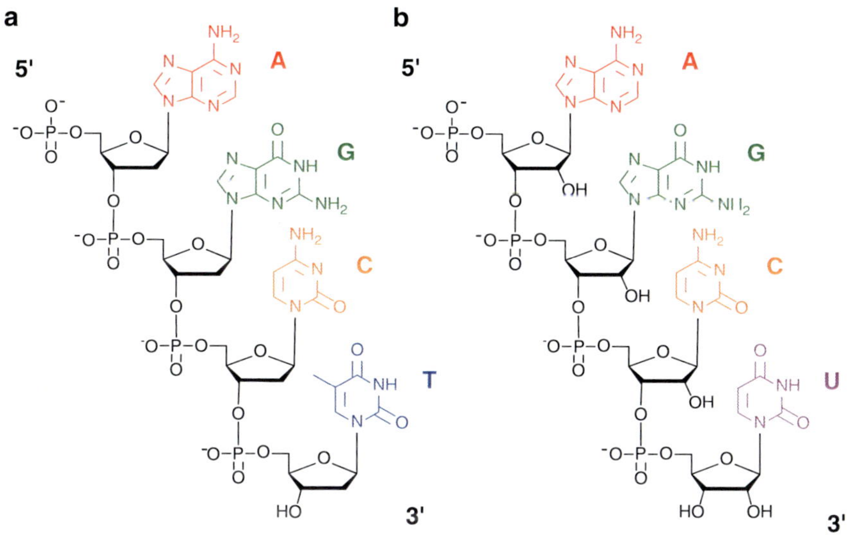

Fig. 9.1 Chemical structures of nucleic acids. (**a**) Deoxyribonucleic acid (DNA). (**b**) Ribonucleic acid (RNA). *A* adenine, *G* guanine, *C* cytosine, *T* thymine, *U* uracil

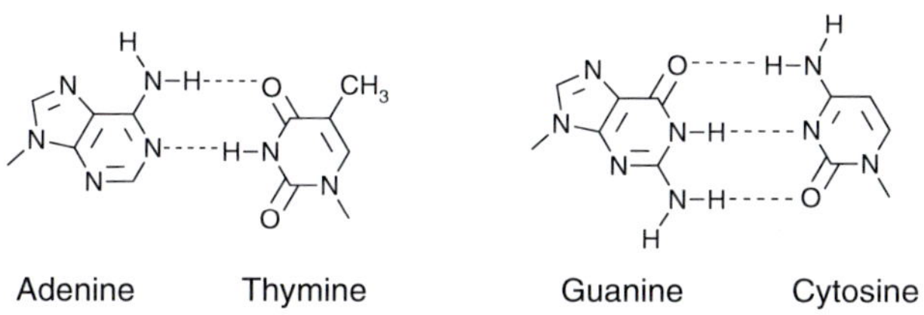

Fig. 9.2 Watson-Crick base pairs. *Dashed lines* indicate hydrogen bonds

full turn of the helix takes about ten base pairs. The plane of each base pair lies nearly orthogonal to the helical axis, so B-form DNA somewhat resembles an upright helical staircase. This structure is determined in large part by the conformation of the 2′-deoxyribose sugar. If all five atoms of the sugar ring lie in a single plane, their substituents will clash with one another because they are lined up in the eclipsed conformation. To relieve the steric strain, the five-membered ring can pucker such that one atom of the ring moves out of the plane. In DNA and RNA, this puckered atom is usually either C2′ or C3′, and the preferred direction of the pucker is *endo* (i.e., on the same face of the ring as C5′). Whether C2′ or C3′ puckers out of the plane is determined by the steric bulk of the ring's substituents. For DNA, the substituents of the 2′ carbon are both hydrogen, allowing the 3′-phosphate (and adjoining nucleotide subunits) to adopt its preferred orientation as far as possible from C5′ and its adjoining subunits (the C2′-*endo* conformation). However, the relatively bulky 2′-hydroxyl of RNA pushes for the C3′-*endo* pucker, resulting in RNA

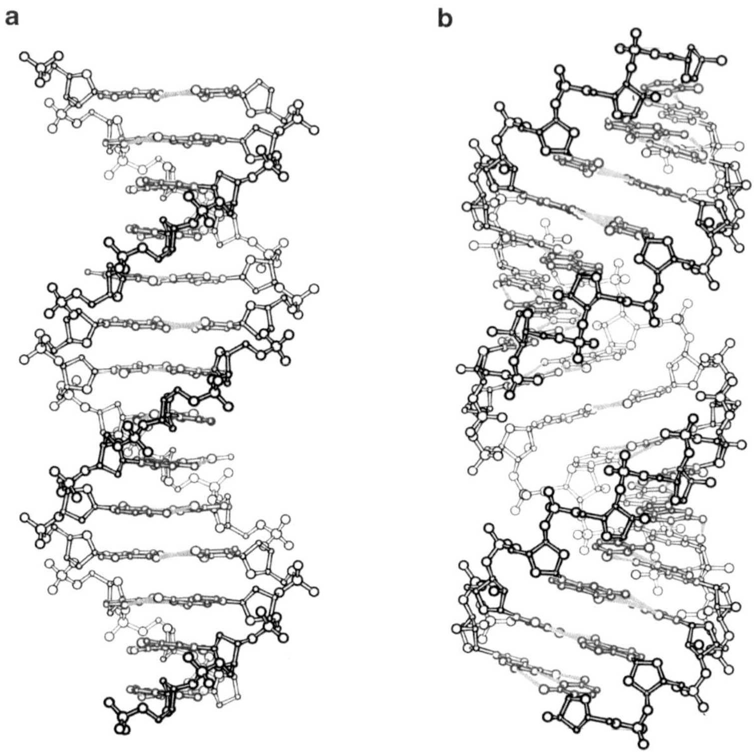

Fig. 9.3 Three-dimensional structures of nucleic acid double helices. (**a**) DNA B-form double helix. (**b**) RNA A-form double helix (Reproduced with permission from Saenger W (1984) Principles of nucleic acid structure. Springer-Verlag, New York. Figs. 10.1 and 11.3, pp 244, 262)

typically adopting a different helical structure, the A-form helix (Fig. 9.3b). In contrast to the regular, upright staircase of the B-form helix, RNA resembles a crazy funhouse staircase: the bases in RNA are strongly tilted relative to the plane orthogonal to the helical axis (by 20°), and the bases are closer to one another (2.3 Å apart compared to 3.3 Å in B-form DNA) while the sugar-phosphate backbone swings out more widely (helical diameter of 26 Å vs. 20 Å for B-form DNA). DNA can also adopt the A-form helix under certain conditions, particularly when dehydrated. The structural repertoire of RNA appears to be quite diverse, including non-canonical base pairs (commonly G:U wobbles) and hydrogen-bonding with the 2′-OH, and combining single-stranded and double-stranded regions into a functional molecule.

9.2.2 *Why Nucleic Acids: The Polyelectrolyte Theory of the Gene*

All molecules carry information about themselves in the precise organisation of their constituent atoms. But not all molecules can replicate and ‘communicate’ that information. There are two fundamental requirements for evolution by Darwinian natural selection: heritable variation and differential reproduction or survival based on the variations. Although there may be many possible ways to generate heritable variation at a molecular level, varying the sequence of a polymer chain (the genome) using a small set of monomers has proven to be a highly successful way to explore functional space.

In extant life on Earth, genetic information is stored in DNA and translated into useable material through RNA. The nitrogenous bases in DNA and RNA recognise each other specifically through hydrogen bonds, forming base pairs, while the phosphate groups give nucleic acids a more or less uniformly distributed negative charge. DNA and RNA are therefore polyanions, or more generally polyelectrolytes.

This feature has been proposed as a defining characteristic of genetic polymers, as described in the polyelectrolyte theory of the gene [4]. The theory states that the following characteristics make nucleic acids particularly good at storing genetic information:

1. The polyanionic charge makes DNA and RNA easily soluble in water, the universal solvent of life on Earth.
2. Nucleic acid backbones in close proximity will repel each other, allowing the nitrogenous bases to come close together, which facilitates base pairing.
3. A repeating polyanionic charge causes nucleic acids to exist in an extended, rather than folded, conformation (at least in the absence of buffering positive charges). This allows them to act as efficient templates for replication/polymerization.

These characteristics make DNA and RNA adept as genetic macromolecules. An important point arises in the final criterion, however. In an environment with enough positive charges (such as divalent cations) to counteract the negative charges, intramolecular interactions, and consequently folding, can occur. This will particularly happen in the case of RNA, which exists in a single-stranded form in nature, whereas DNA is always double-stranded except when it is being replicated (though exceptions to both these rules occur in some viruses).

9.2.3 *The RNA World*

Although the major function of nucleic acids in modern organisms is to carry and transfer information, both RNA and DNA can also fold into complex structures, much like proteins. In the 1960s, this observation caused several scientists to raise the possibility that nucleic acids might also be able to catalyse reactions and bind specific molecular targets [5–7]. If this hypothesis were correct, then one could imagine a simplified living system. Instead of information in DNA being copied into messenger RNA and then translated into protein by a complex protein-RNA machine (the ribosome in conjunction with tRNAs, aminoacyl-tRNA synthetases, and a host of other proteins), RNA might have simply encoded the information, and

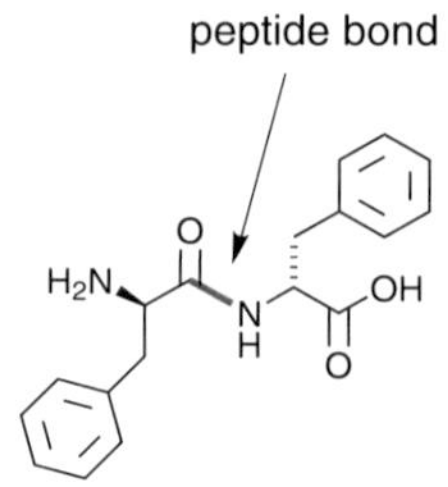

Fig. 9.4 Chemical structure of a peptide bond. Amino acids (here two phenylalanines) are joined through peptide bonds to form proteins

RNA copies of the genome would have directly acted as the chemical workhorses of this primitive life. This concept is known as the RNA world theory.

Evidence supporting this theory first came from Tom Cech and Sidney Altman's groups, which showed that extant RNAs can function as enzymes, termed ribozymes [8, 9]. Both synthetic and naturally-occurring ribozymes have been shown to make (ligate) and break phosphodiester bonds (the polyanionic backbone of nucleic acids) [10–12], and a derivative of the Class I ligase ribozyme [13] can polymerize up to 95 nucleotides on an RNA primer [14]. In addition, RNAs can fold into structures that bind to specific targets. Such RNAs are known as aptamers (from the Latin *aptus*, meaning 'to fit'). Many people believe that RNA should be able to direct its own replication.

Although modern protein enzymes typically mediate rate accelerations of 10^6–10^{13} [15], an early genome would benefit greatly from even a small rate acceleration over non-enzymatic, templated polymerization (see Sect. 9.2.4). A mutant genome whose fitness increased by only a small amount would accumulate tremendous replicative advantage over many generations:

$$p_t/q_t = w^t(p_0/q_0), \tag{9.1}$$

where p is the frequency of genotype A, q is the frequency of genotype B, w is the fitness of A relative to B, and t is the number of generations. Small improvements may be rather common in a random pool of sequences. One indication for this is the finding that, on average, the affinity of RNA aptamers for GTP increases twofold when fixing one position (2 bits of information) in a nucleotide sequence. A similar relationship also appears to hold for ribozyme ligases [16] although a different set of sequences might show a different dependence of function on information content.

RNA can also react with amino acids to form biologically relevant products, suggesting a role for RNA in the origin of protein synthesis. Ribozymes can activate amino acids as both mixed phosphate anhydrides [17] and aminoacyl-RNA esters [18–20]. These are the same substrates that extant biology uses in protein synthesis: mixed phosphate anhydride amino acids are transferred to tRNAs, forming esters, before they are strung together through peptide bonds (Fig. 9.4). In addition, synthetic RNA aptamers have been found that specifically bind amino acids [21]. This implies that current RNA-amino acid associations (i.e., the genetic code) could have evolved from direct stereochemical interactions. Finally, the crystal structure of the modern ribosome shows that its peptide bond-forming centre is completely composed of RNA, with no proteins within 18 Å [22, 23]. All these observations suggest that early protein synthesis (translation) may have been directed by RNA enzymes.

This vision for an early stage of life, the RNA world theory, elegantly outlines an intermediate on the pathway toward living systems. Only a single biopolymer (RNA) would have been necessary, thereby avoiding the difficulty of imagining the wholesale emergence of DNA, RNA, and the protein translation apparatus. Evidence for an RNA world includes at least three major lines of reasoning. First, several metabolic cofactors closely resemble nucleotides; the central role of ATP in our metabolism underlines this fact. Second, RNA can, in principle, replicate without protein intervention and is subject to mutation, and thus Darwinian evolution. Third, all the reactions necessary for protein synthesis can be catalysed by RNA. These fundamental features of our biology appear to be fossils from the RNA world.

9.2.4 *Nucleic Acid-Based Catalysis: Reaction Geometry and Effective Concentration*

One of the defining features of life as we know it is the ability to replicate. This starts with the duplication of an organism's entire collection of genetic information (genome) in the form of DNA. Today this is a highly complex process mediated by an impressive suite of proteins that functions to make sure each strand of DNA is copied efficiently and accurately. While we may never be sure if there truly was an RNA world, logic demands a mechanism for early genome replication without the intervention of complex, genome-encoded enzymes. But is such a mechanism chemically plausible?

One possible answer lies in the fact that positioning and proximity are important factors in chemical reactions [24]. Enzymes can 'find' reactants even in relatively dilute solutions; they accelerate chemistry, in part, by bringing reactants close to each other, increasing their effective concentration. Enzyme-catalysed reactions can occur at rates that would normally require very high concentrations of reactants in solution. Enzymes often use tricks such as acid or base catalysis to achieve huge rate enhancements above spontaneous reactions, but substrate positioning is a part of what makes the chemistry work.

Nucleic acids can use this phenomenon of substrate positioning in the absence of enzymes to build complementary strands. When pieces of DNA and RNA (oligonucleotides) are incubated with activated nucleotides (such as nucleoside 5′-phosphorimidazolides), the activated monomers align themselves opposite the 'template' strands and spontaneously form new phosphodiester bonds (reviewed in

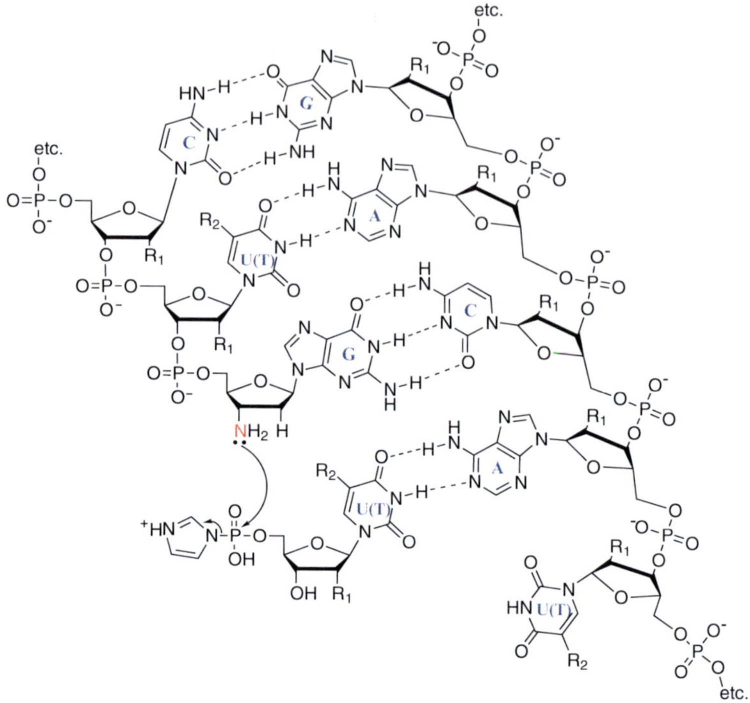

Fig. 9.5 Spontaneous template-directed polymerization. 5′-phosphorimidazolide activated nucleotide forms a new bond with the terminal 3′ nucleophile of the primer (here an amino group for rapid reaction). $R_1 = H$ for DNA, OH for RNA; $R_2 = CH_3$ for DNA, H for RNA (Reproduced with permission from [27])

[25]) (Fig. 9.5). This can lead to the synthesis of complementary oligonucleotides several to tens of bases long, depending on experimental conditions. Further, if a template strand hybridizes with a shorter 'primer' strand and is incubated with appropriate activated monomers, the primer can be extended by several nucleotides relatively efficiently; similarly, two adjacent activated oligonucleotides will bond to each other, or ligate, when annealed to a complementary sequence [26]. In principle, then, a DNA or RNA strand can be copied in the complete absence of enzymes, under the right conditions and provided the sequence is not too long.

Spontaneous nucleic acid polymerization is not perfect, however; several obstacles needed to be overcome in evolution, likely with the aid of early enzymes. Firstly, this reaction is not particularly efficient. Nucleoside 5′-phosphorimidazolides are not used in nature; their highly activated leaving groups make spontaneous polymerization reactions measurable on a laboratory time scale. Biology uses nucleoside 5′-triphosphates, which react more slowly. Secondly, all activated nucleotides do not react at the same rate. G, for example, is incorporated across a template C much more quickly than T or U across A [27]. Finally, spontaneous polymerization is prone to error; Watson-Crick base pairing is not always conserved.

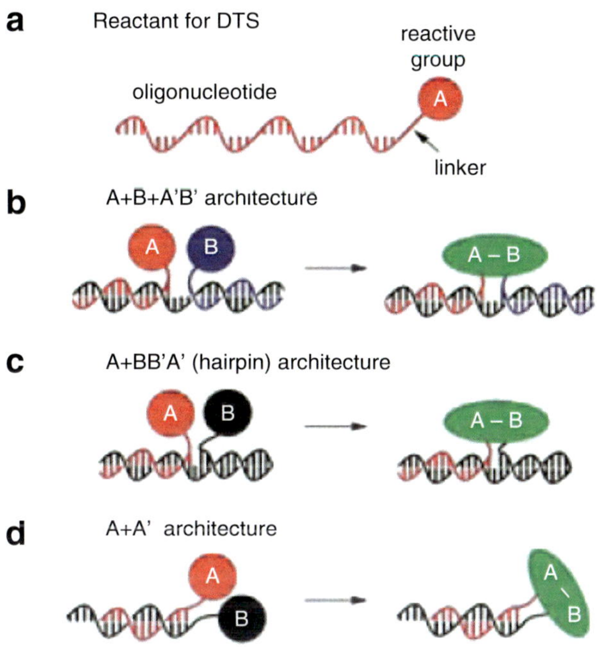

Fig. 9.6 DNA-templated synthesis. (**a**) The three components of a reactant for DTS. (**b–d**) Template architectures for DTS. A/B and A′/B′ refer to reactants containing complementary oligonucleotides, and + symbols indicate separate molecules (Reproduced with permission from [28])

G frequently pairs across from U, forming a G:U wobble, which means that information is often lost. Some level of mutation is necessary for evolution to occur, but too many mutations will result in an error catastrophe, the loss of all information (see Sect. 9.3). Still, it is important to note the potential for information transfer in the complete absence of enzymes, and the importance of chemical reactions based on proximity and positioning.

Nucleic acids can also function as scaffolds for the synthesis of products other than nucleic acids. DNA-templated organic synthesis (DTS) has been used to catalyse S_N2 reactions, conjugate addition, reductive amination, amine acylation, oxazolidine formation, nitro-aldol addition, Wittig olefination, Heck coupling, and 1,3-nitrone cycloaddition, among other reactions (reviewed in [28]). These syntheses utilise reactants that are conjugated to the 5′- and 3′-ends of DNA strands, which hybridize sequence-specifically to proximally orient the reactants (Fig. 9.6). Molecules joined to the ends of hybridized DNA strands in this way can approach each other to a distance of <10 Å, which corresponds to an effective molarity of >1 M. Double-stranded DNA can also be used to direct the synthesis of DNA-like molecules by binding reactants to the major or minor groove, the 'outer face' of DNA.

Template-directed synthesis allows the formation of certain products that are difficult to synthesise chemically. Not only do nucleic acid templates increase the effective concentration of conjugated reactants and thus accelerate reaction rates, but they also hinder the formation of unwanted byproducts that often emerge in solution-based chemical synthesis. Multistep syntheses can also be carried out in DTS systems, provided the intermediate products can be separated from the DNA templates before subsequent reactions [28]. The principle of template-directed synthesis might be applicable to almost any reaction, if the substrates are conjugated to nucleic acids or their relatives.

Nucleic acid-directed synthesis may have also been important in early peptide formation. Weber and Orgel showed that when the amino acid glycine is esterified to derivatives of adenosine (in the same manner that amino acids are bonded to tRNA in extant protein synthesis), the amino acids will form peptide bonds, resulting in cyclic Gly-Gly dipeptides [29]. Further, when poly-uracil (poly(U)) is added to the mixture, the amount of cyclic Gly-Gly formed increases about 3 times [30]. The temperature and concentration effects of the reactions suggested dependence on formation of a poly(U) helix; specifically, a triple helix of two strands of poly(U) complexed with the glycine-esterified adenosine derivatives. While the exact mechanism of peptide bond formation in this case has not been established, the increased yield of dipeptide could be due to increased local concentration/ optimal orientation of the glycine derivatives based on specific interactions between poly(U) and adenosine.

Amino acids covalently bonded to short oligonucleotide sequences also function as substrates for peptide bond formation in prebiotic experiments. Tamura and Schimmel [31] showed that a peptide bond will form between puromycin (an amino acid analogue) and alanine, when both amino acids are covalently attached to complementary oligonucleotides. Only four base pairs were sufficient to bring the two substrates together for the reaction to occur. Interestingly, the alanyl-oligonucleotide was a tRNA analogue, which suggests that such a reaction could have been used by early biology. This reaction, however, required an imidazole catalyst, which most likely functions as a general base (imidazole was also used in the reaction buffer for the aforementioned poly(U)-stimulated reactions). Therefore, while substrate positioning is important, it is not always sufficient to accelerate reactions; additional catalysis is sometimes necessary.

It was later shown, however, that similar peptide-bond-forming reactions can occur in the absence of added catalyst, based on similar Watson-Crick base pair positioning and highly reactive aminoacyl substrates. When phenylalanine is bonded to an oligonucleotide through a mixed phosphoric anhydride linkage, it is a ready substrate for nucleophilic attack by an oligonucleotide-esterified alanine [32], when the two amino acids are again proximally oriented by Watson-Crick base pairing. This reaction results in Phe-Ala dipeptide, as well as small amounts of Phe-Phe-Ala tripeptide, if the reactant concentrations are increased. It is important to note, however, that these peptides are assembled at random (their sequence is not dictated by a genetic code), and come from different starting materials than those used in extant biology, though similar reactions could have occurred to form short, non-coded peptides in the past.

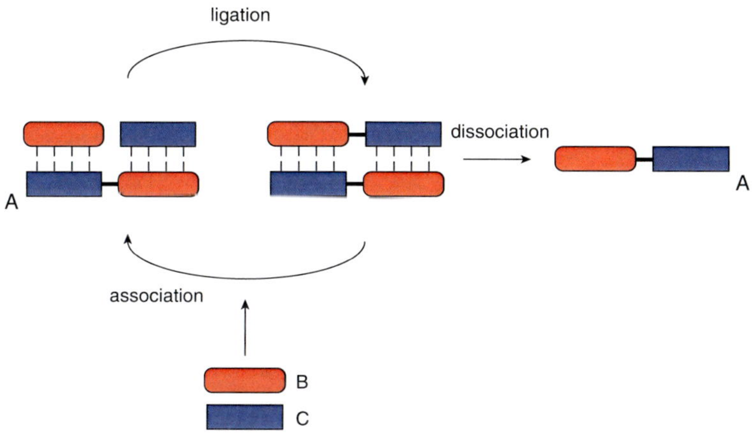

Fig. 9.7 Autocatalytic system. Autocatalyst A associates with components B and C through noncovalent interactions (*dashed lines*) to create a new autocatalyst A through ligation (here used as a general term for bond formation). After product dissociation, the cycle continues. This theoretically leads to exponential growth of the system

Compounds other than nucleic acids can also serve as templates for polymerization to an extent. Tjivikua et al. [33] showed that an adenosine derivative will couple with a pentafluorophenyl ester derivative to form an amide bond. The product is self-complementary; hydrogen bonds form between two products in a manner analogous to Watson-Crick base pairing, though the products are not nucleotides. The formation of one product enhances the formation of future products; therefore the system is autocatalytic, and may be a model for self-replicating systems using alternate chemistry (Fig. 9.7).

Similarly, peptides can form self-replicating systems. Certain short proteins (32 amino acids long) that form three-dimensional structures known as α-helices can serve as scaffolds for two halves of the same protein, enhancing ligation of the two fragments and thus self-propagation [34]. The mode of interaction between the peptides, however, is not as specific as base pairing; presumably a number of different peptides can be ligated together, as long as they are α-helices. Still, the processes of templated synthesis and self-replication can be carried out by a range of molecules, in addition to nucleic acids.

9.2.5 *Non-templated Catalysis*

Some nucleic acid templates that direct synthesis could be considered enzymes, since they facilitate chemical reactions while remaining themselves unchanged. Most of the previously described reactions are dependent on base pairing between nucleic acid strands to facilitate proximity of reactants. Based on these observations

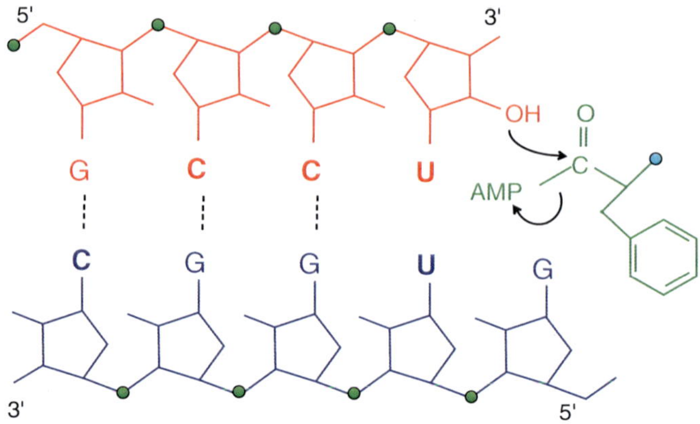

Fig. 9.8 Schematic of aminoacylation reaction catalysed by a 5-nucleotide ribozyme. An ester bond forms between the 4-nt substrate (*red*) 2′-OH and the carboxyl group of Phe-AMP (*green*)

alone, one might conclude that early enzymatic polymerization reactions of any kind were highly dependent on primary nucleic acid sequences.

Long paired regions would have been difficult to synthesise in early biology, however. A simple, general, and processive RNA polymerase ribozyme (i.e., capable of proceeding base-by-base to replicate an entire RNA sequence) has not yet been found. Early RNAs were likely short, perhaps synthesised relatively slowly on mineral surfaces [35]. Since most known ribozymes are tens or hundreds of nucleotides long, and since most protein enzymes are composed of chains of hundreds of amino acids, it may seem that short nucleic acids are incapable of biologically relevant catalysis, beyond simple template-driven reactions.

This is not the case, however. A 5-nucleotide-long RNA can react with a 4-nt-long substrate and activated phenylalanine (in the form of Phe-AMP, the biological substrate) to form an aminoacyl-RNA ester [20] (Fig. 9.8). This ribozyme is a true enzyme, as it exhibits a rate enhancement 25 times that of the spontaneous (no-ribozyme) reaction, and remains itself unmodified once the reaction has taken place. Here base pairing plays a part in the interaction between the ribozyme and the 4-nt RNA substrate, but Phe-AMP likely associates through hydrogen bonding interactions (rather than being forced into place by covalent bonding) with the enzyme-substrate complex.

The tiny ribozyme is promiscuous; it functions not only with Phe-AMP, but also with Phe-UMP and Met-AMP (and possibly other amino acids as well). This may explain why it does not display as great a rate enhancement as an evolved, complex enzyme. A less complex enzyme means fewer points of contact with substrates, and thus less efficient catalysis. These kinds of reactions, however, may have been important in early biology; some degree of promiscuous catalytic activity may have been necessary to generate a diverse repertoire of biological reactants.

Furthermore, the initially formed acyl-RNA acts as a substrate for further polymerization from activated phenylalanine. Peptide bonds form between phenylalanine molecules, resulting in RNA-peptides [20, 36]. These peptides range in size

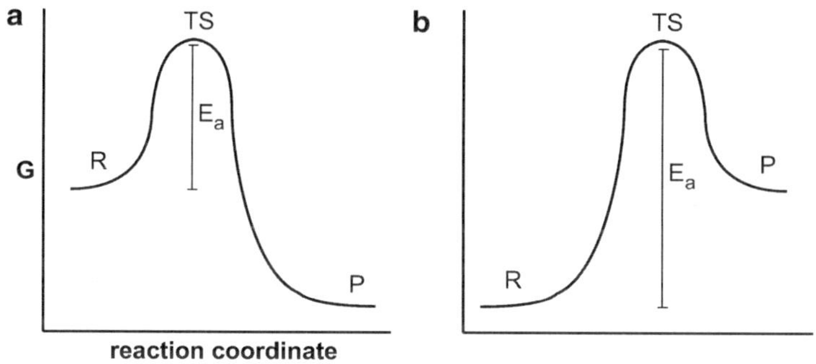

Fig. 9.9 Reaction coordinates for (**a**) exergonic, and (**b**) endergonic, reactions. The transition state resembles whichever state (reactants or products) it is closest to in free energy. *R* reactants, *P* products, *TS* transition state, *G* free energy, E_a activation energy

up to RNA-Phe$_5$, and include a complex array of RNA-peptide and RNA-peptide diester products (some products contain phenylalanines on both hydroxyl groups of the terminal ribose). Peptide formation is not *catalysed* by RNA in this case, but arises when the pK$_a$ value of the amino group of the first phenylalanine decreases upon esterification, making the amino group a stronger nucleophile. While these peptides are, again, not coded, a system such as this could have been important in early peptide formation.

The tiny ribozyme puts a revealing constraint on the amount of information that may have been required to generate relevant materials in early biology. The three G-C base pairs between the ribozyme and RNA substrate are likely not sequence-constrained; in fact, the catalytic parts of the molecule are the unpaired G and U of the ribozyme, and the U of the substrate. Therefore, there are only three nucleotides necessary for the reaction to proceed. Considering this, any number of nine-nucleotide complexes may have functioned as early systems of aminoacyl-ester formation. It is perhaps not hard to imagine, then, similar systems arising on other worlds, perhaps made of different chemical components, but based on similar chemical principles.

At some point, primitive enzymes had to evolve to become more specific and more efficient. In order to increase the rate of a reaction, an enzyme must decrease the reaction's activation energy, or the free energy required to convert reactants to a transition state (Fig. 9.9, see also Chap. 1). One way of doing this is to stabilise (e.g. by binding) the transition state of a reaction. The Hammond postulate says that for a highly exergonic reaction, the transition state will structurally resemble the reactants; conversely, for a highly endergonic reaction, the transition state will resemble the products [37] (Fig. 9.9). One might speculate that enzymes may have started as aptamers, or structures that bind particularly oriented reactants or products, and evolved through mutation to stabilise transition states for the chemical reaction.

De novo discovery of catalytic sequences has been investigated experimentally using transition state analogues. Researchers have constructed molecules that resemble proposed transition states (using covalent bonds to phosphorus, for example, to mimic the temporary tetrahedral intermediate of peptide bond formation) of reactions. This method has been used to design novel enzymes that catalyse many reactions, such as the Claisen rearrangement, a carbon-carbon bond-forming reaction normally catalysed by chorismate mutase [38], and peptide bond formation [39]. These enzymes come from antibodies, which arise from exposing a model organism's immune system to a relevant transition state analogue. This process of selecting molecules that bind reactants may mimic what occurred in evolution, if binding to a small molecule somehow increased the fitness of an evolving enzyme.

While selection for RNA aptamers that bind transition state analogues have not yet yielded catalysts, researchers have found many RNAs that specifically bind amino acids [21], as well as other small molecules. It seems possible that RNAs that bind amino acids, for example, may have evolved to catalyse peptide bond formation. Subtle mutations that conferred less binding to reactants or products may have led to transition state stabilisation and subsequent product formation. As enzymes (nucleic acid or otherwise) developed diverse catalytic strategies, biological complexity would have increased.

9.3 Copying Fidelity and Base-Pairing Thermodynamics

9.3.1 Maintenance of Molecular Function

There are two complementary fundamental questions about the possible origins of life from a physico-chemical point of view: the question of how catalytic functions can spontaneously arise and the question of how such functions, once discovered, can be maintained. These same two aspects of emergence and maintenance are fundamental questions in the evolutionary theory of living organisms, and are often referred to as evolvability [40] and evolutionary robustness. In this section, we focus on the maintenance aspect, which appears particularly challenging in a prebiotic world without sophisticated polymerases and repair enzymes: Is the copying fidelity of primitive replication processes sufficient to maintain molecular function after it has spontaneously arisen? Which types of replication processes and conditions favour maintenance of function?

9.3.2 The Importance of Fidelity: The Error Threshold

Low copying fidelities can lead to an 'error catastrophe' where a replicative process cannot maintain the sequence information [41–45]. The 'error threshold' marks the onset of this catastrophe: it refers to a critical value in the ratio μ/s of the mutation rate μ to the selection strength s. Beyond the threshold, the mutants with compromised function dominate over the original 'master' sequence, because the replicative advantage of the master is not sufficient to overcome the accumulation of mutants. The evolutionary process then becomes one of random diffusion through sequence space. This concept is well-established in evolutionary theory and is applied to the evolution of organisms, in particular viruses which appear to live close to the error threshold [46–48], as well as molecular evolution. An example of the latter kind is the evolution of transcription factor binding sites on the genomic DNA, which are under selection to match the binding preferences of the protein while being exposed to point mutations. The balance of evolutionary forces naturally leads to 'fuzzy' binding sites, while a mutational pressure that is too high or a selection pressure that is too low may lead to the loss of the binding site [49].

It is important to note, however, that the concept of an error threshold can acquire a somewhat different meaning in the prebiotic context. In particular, if we consider the molecules that form the core of a primitive replication process, then the most fundamental maintenance requirement is that the process generates at least one functional copy per core replication molecule before the template is destroyed. This situation differs from the standard error threshold scenario: In the latter case, both master and mutants replicate and an error catastrophe results from their competition, whereas in the former case the catastrophe is no replication at all. However, this replication breakdown can also be induced by a low copying fidelity.

9.3.3 Thermodynamic Bounds on Fidelity

The polymerase enzymes of living organisms can exploit the chemical energy of nucleoside triphosphates to obtain copying accuracies as high as one error in ten billion. Kinetic proofreading [50, 51] is one important mechanism, which can in principle achieve arbitrarily high fidelity by 'burning' nucleotides, i.e. consuming more nucleoside triphosphates than are finally integrated with the help of an exonuclease that preferentially excises erroneously incorporated nucleotides. Polymerases lacking exonuclease activity can still achieve accuracies up to about 1 error in a million using a conformational coupling mechanism that relies only on the chemical energy of the nucleoside triphosphate being incorporated [52]. However, all of these mechanisms require large, highly sophisticated protein enzymes which likely emerged from a long evolutionary optimisation process and were not part of the first replicators. What is the fidelity that could have been achieved under prebiotic conditions? If template-directed polymerization was non-enzymatic or catalysed only by primitive enzymes that merely speed up the process, then the thermodynamics of base-pairing puts a limit on the achievable fidelity. To see explicitly how such a limit arises, it is instructive to consider a simple Michaelis-Menten-type reaction model (see Chap. 1) for template-directed polymerization, where the incorporation of a new base is a two-step process,

$$
\begin{aligned}
&p_j + n \underset{k_{\text{off}}}{\overset{k_{\text{on}}}{\rightleftharpoons}} p_j \cdot n \overset{W}{\rightarrow} p_{j+1} \\
&p_j + \tilde{n} \underset{\tilde{k}_{\text{off}}}{\overset{\tilde{k}_{\text{on}}}{\rightleftharpoons}} p_j \cdot \tilde{n} \overset{W}{\rightarrow} \tilde{p}_{j+1}
\end{aligned}
\tag{9.2}
$$

Here, p_j denotes a polynucleotide of length j, which is growing along a template. The correct next nucleotide n competes with a mismatched nucleotide $\tilde{n}$ for incorporation. Incorporation of either nucleotide is assumed to be a two-step process, with a reversible first step that subsumes the initial binding and any conformational rearrangement required prior to the formation of the chemical bond in the sugar-phosphate backbone. The second step corresponds to the bond formation, which is assumed to be irreversible and to occur at the same rate W for the correct and the mismatched nucleotide. Thus, all processes contributing to the discrimination between correct and mismatched nucleotides are described, on a coarse-grained level, by the effective rate constants for the first step, k_{on} and k_{off} for n and analogous ones for $\tilde{n}$.

For the above reaction scheme, the error ratio Φ, i.e. the rate of generating the incorrect product relative to the rate of generating the correct product, is simply the relative flux through the two competing Michaelis-Menten reaction 'channels',

$$
\Phi = \frac{\tilde{k}_{\text{on}}}{\tilde{k}_{\text{off}} + W} \frac{k_{\text{off}} + W}{k_{\text{on}}} = \frac{K}{\tilde{K}} \cdot \frac{1 + W/k_{\text{off}}}{1 + W/\tilde{k}_{\text{off}}},
\tag{9.3}
$$

where we assume equal concentrations of the free nucleotides. In the last expression, where $K = k_{\text{off}}/k_{\text{on}}$ denotes the dissociation constant for the correct nucleotide (and a tilde again denotes the same quantity for a mismatched nucleotide), the error ratio is expressed as a product of the thermodynamic error ratio

$$
\Phi_0 = \frac{K}{\tilde{K}} = \exp(-\Delta G/k_B T)
\tag{9.4}
$$

and a kinetic ratio

$$
\gamma = \frac{1 + W/k_{\text{off}}}{1 + W/\tilde{k}_{\text{off}}}
\tag{9.5}
$$

The thermodynamic ratio $\Phi_0 < 1$ is connected to the free energy of discrimination, $\Delta G > 0$, between the correct and incorrect nucleotides ($k_B T$ is the thermal energy unit, with $k_B T \approx 2.5$ kJ mol^{-1} at room temperature). The kinetic ratio γ depends on the bond formation rate W and the effective off-rates. It approaches the value 1 when the bond formation rate W is negligibly slow compared to both off-rates. Furthermore, as long as the off-rate for the correct nucleotide is not larger than that for the mismatched one (a reasonable assumption), the kinetic ratio cannot

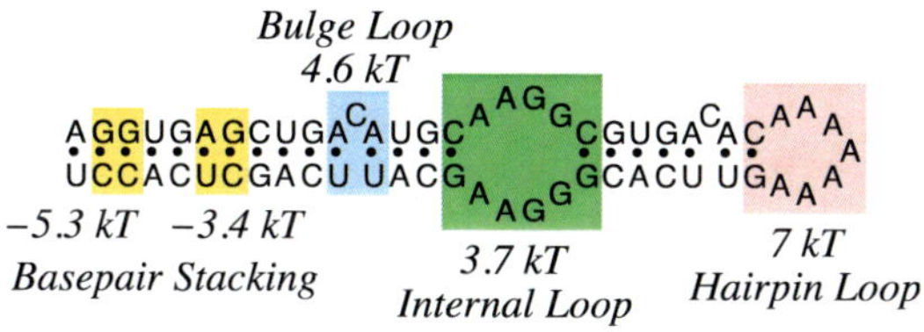

Fig. 9.10 Illustration of the free energy rules for secondary structure formation. Different structural elements of a small RNA molecule are highlighted in *colour* and the corresponding free energy reward (negative) or penalty (positive) are indicated for a temperature of 37 °C and standard ionic conditions of 1 M salt

be smaller than one, or $\gamma \geq 1$, such that fast bond formation can only increase the error ratio Φ.

In summary, the above argument leads to a thermodynamic limit Φ_0 for the error ratio of a non-enzymatic template-directed polymerization process, where the discrimination must primarily stem from the off-rates. How can we obtain estimates for the value of Φ_0? We should, in fact, expect that Φ_0 not only depends on the identity of the template base and the mismatched base under question, but also on the local sequence context and, more fundamentally, on the type of the polynucleotide. Extensive experimental studies of the thermodynamics of base-pairing for RNA [53, 54] and DNA [55] have led to so-called 'nearest-neighbour' models for the base-pairing free energies. These models assign sequence-dependent free energy rewards or penalties for local structural elements such as stacked base pairs, internal loops or bulge loops, as illustrated in Fig. 9.10 for a small RNA molecule that exhibits different structural elements. The free energy parameters also display a fairly well characterized dependence on the local temperature and ion concentrations. Generally, these free energy rules for RNA and DNA molecules lead to fairly accurate predictions of the folding free energies and melting temperatures of small structures, and they have also been used to quantitatively describe mechanical single-molecule experiments, such as unfolding by translocation through a nanopore [56]. With the help of the free energy rules it is straightforward to calculate sequence-specific estimates for the thermodynamic limit Φ_0 on the error ratio [27]. Specifically, the discrimination free energy ΔG can at most be equal to the difference in the free energy of the fully polymerized duplexes, with one duplex containing the mismatched bases and the perfectly paired duplex as reference.

9.3.4 Correlation with Experiment Hints at Universal Trends

How do the resulting Φ_0 values compare to experimental error ratios Φ for non-enzymatic template-directed polymerization? Using chemically activated nucleotides, ref. [27] reported the measured errors incurred during non-enzymatic template-directed polymerization for all combinations of DNA and RNA templates and primers. The resulting mis-incorporation probabilities are plotted in Fig. 9.11

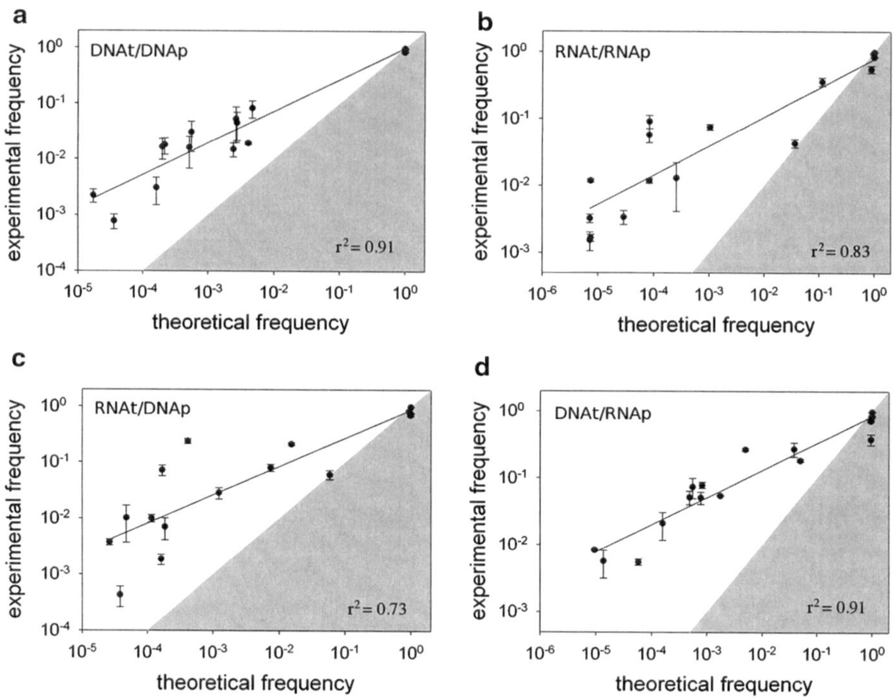

Fig. 9.11 Correlation between experimental incorporation and mis-incorporation frequencies vs. thermodynamic predictions for copying (**a**) DNA into DNA, (**b**) RNA into RNA, (**c**) RNA into DNA, (**d**) DNA into RNA (Reproduced with permission from [27])

against the corresponding calculated thermodynamic values. The thermodynamic values do indeed bound the experimentally observed rates from below. The thermodynamically predicted error rates are typically smaller by one to three orders of magnitude compared to the experimental error rates. This is not unexpected since the kinetically activated experimental reaction is far from equilibrium. Furthermore, the simple template-directed polymerization process likely cannot make use of the entire discrimination free energy of the fully polymerized duplex. Interestingly, however, the two sets of error rates display a substantial correlation. This correlation suggests that the underlying thermodynamics of the eventual product of replication influences the kinetics of the experimental system. Notably, both the experimental and the thermodynamic values follow the same general trends in the comparison of DNA to RNA, and hybrid systems:

1. DNA replication is intrinsically more faithful than RNA replication,
2. Copying RNA into DNA is about as faithful as RNA replication, and
3. Copying DNA into RNA is error-prone compared to pure DNA replication.

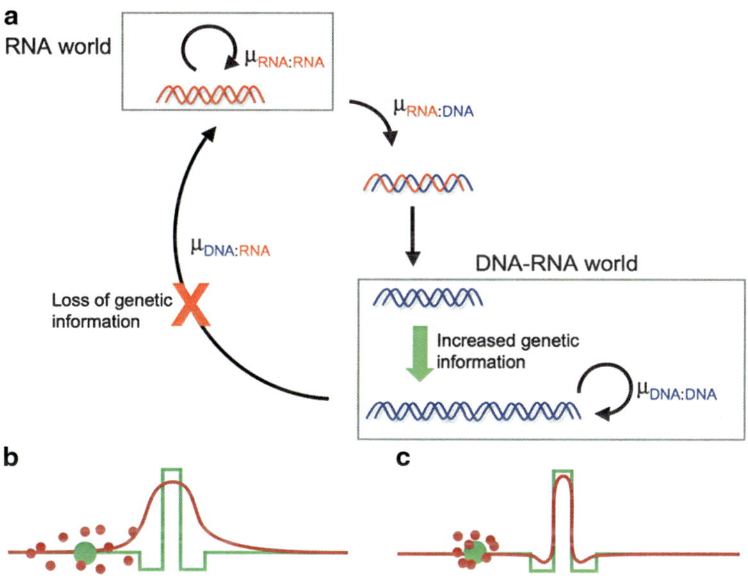

Fig. 9.12 (**a**) Scheme for the genetic takeover (Reproduced with permission from [27]). (**b**, **c**) Illustration of the 'look-ahead' effect for a particular genotypic fitness landscape (*green line*; height corresponds to fitness and horizontal position indicates a position in sequence space). The phenotypic landscape (*red line*) is shown at high phenotypic variability (**b**) and low phenotypic variability (**c**). The genotype (*green dot*) spins off many different phenotypes (*red dots*) due to transcription errors, resulting in a smoothened phenotypic landscape

9.3.5 The Genetic Takeover

At some point during the origin of life, the RNA world gave way to the DNA-RNA-protein world that dominates biology today. During this transition, DNA must have taken over the role as genetic material. This step presumably involved copying the genomic information from RNA into DNA, a process known as the 'genetic takeover'. What do the comparisons of error ratios of RNA, DNA, and the RNA-DNA hybrid systems tell us about the genetic takeover? They suggest that such a transition would have been accompanied by the advantage of a higher intrinsic copying fidelity, since non-enzymatic RNA polymerization had about twice the mis-incorporation rate of DNA polymerization, suggesting that more information could be stably encoded after the switch to DNA as the genetic material (Fig. 9.12a). Furthermore, the observation that copying RNA into DNA occurs with a mutation rate similar to RNA replication suggests that the genetic takeover itself would not cause much loss of information. In contrast, copying DNA back into RNA appears to be a highly error-prone process, suggesting that an organism attempting to switch from DNA back to RNA would be at an immediate disadvantage caused by the corruption of genetic information.

9.3.6 The Evolutionary 'Look-Ahead' Effect

The observed hierarchy of experimental and thermodynamic mutation rates also suggests that after the genetic takeover, non-enzymatic RNA synthesis, or transcription (e.g., of ribozymes encoded on the DNA genome) would be significantly more error-prone than genome duplication. Error-prone transcription implies high phenotypic variability, which could lead to an evolutionary 'look-ahead' effect: while each genotype specifies a particular ribozyme sequence, it also leads to a cloud of transcripts nearby in RNA sequence space. A given genotype may thus exhibit an overall phenotype (function) influenced by its neighbours, resulting in a locally smoothened phenotypic fitness landscape (Fig. 9.12b, c). The smoothing due to phenotypic variability may enhance evolvability by producing a selective benefit from relatively distant optima and facilitating evolutionary paths across low-fitness regions.

9.4 Lipid Membranes

9.4.1 The Hydrophobic Effect

Everyday experiences show us that non-polar and polar compounds, such as oil and water, respectively, do not mix. On a macroscopic scale, clear but oily soups (e.g., French onion soup) separate readily into their components, with the lower density liquid rising to the top and forming a bulk phase of oil. This separation occurs because the oil–water interface is highly ordered at a molecular scale. In contrast, in the bulk water and bulk oil phases the molecules can adopt many possible configurations with respect to one another, and in bulk water many polar interactions can be satisfied. The entropy of the system is therefore highest when the interface is minimised, causing phase separation between oil and water. This is the essence of the hydrophobic effect. Although the word 'hydrophobic' is derived from Greek for 'water-fearing', this is somewhat misleading with regard to its etiology because non-polar molecules do not have a particular repulsion to water. Instead, the effect is better thought of as being caused by a preference of water molecules to interact with one another, which has the consequence of excluding the non-polar substances. One might say that although oil does not fear water, water loves itself to the exclusion of oil.

9.4.2 Self-Assembly of Amphiphiles

What happens if a single molecule contains a substantial polar and nonpolar region? Such molecules are called amphiphiles (see Chap. 1), from Greek for 'loving both' (see Fig. 9.13 for examples). These molecules dissolve in water to a limited extent, or they can form a bulk oil phase. But if they are present in a high enough

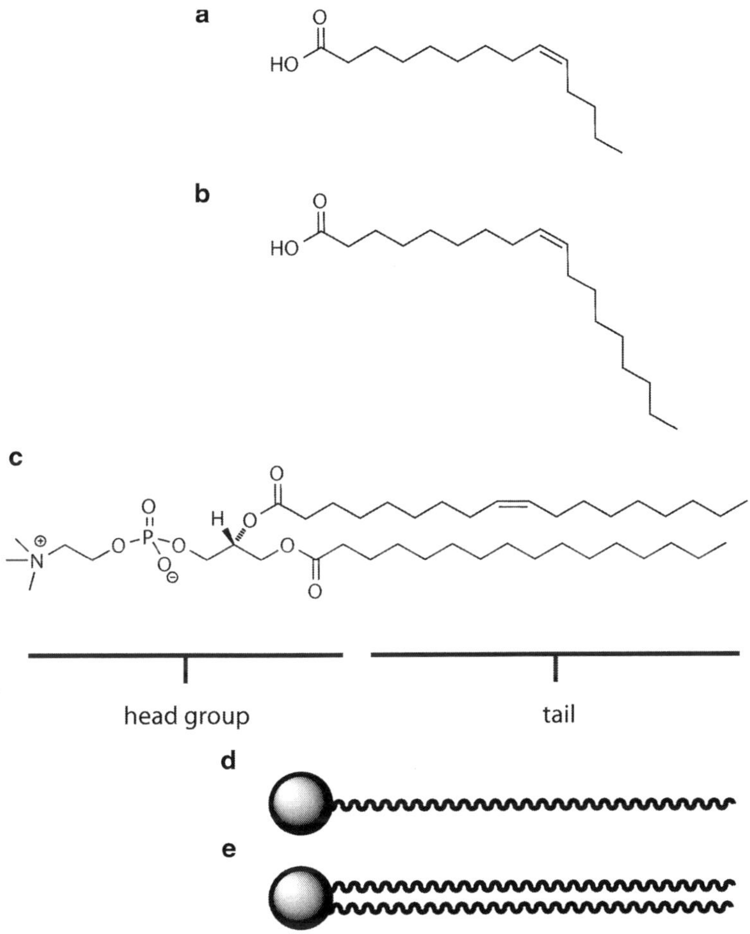

Fig. 9.13 Chemical structures of amphiphiles. (**a**) Myristoleic acid; (**b**) oleic acid; (**c**) 1-Palmitoyl-2-oleoylphosphatidylcholine (POPC), a phospholipid; (**d**) schematic representation of a fatty acid; (**e**) schematic representation of a phospholipid

concentration in aqueous solution under the right conditions, a remarkable phase transition can occur. The hydrophobic effect causes the nonpolar regions of the molecules to aggregate together, while at the same time the polar regions prefer to interact with water. This molecular tug-of-war results in the self-assembly of large, quasi-ordered structures, such as micelles and vesicles (Fig. 9.14, see also Chap. 1).

The details of the structures formed depend on multiple factors, including temperature, salt concentration, and molecular geometry. Large, electrostatically repulsive head groups tend to cause formation of micelles, with a hydrophobic interior. For example, in a spherical micelle, the nonpolar components are locally aggregated and the polar components become the interface with the bulk water phase, like a herd of musk oxen assembling in a circle with their horns facing

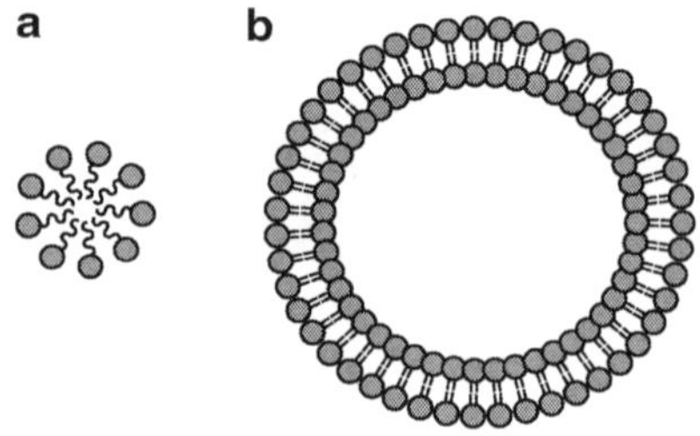

Fig. 9.14 Supramolecular structures of amphiphiles. (**a**) Micelle; (**b**) phospholipid vesicle. Note that the micelle has a hydrophobic interior, while the vesicle encloses an aqueous compartment. Structures are drawn here in a two-dimensional cross-section. The structures themselves are three-dimensional, with the micelle resembling a solid sphere (or cylinder, depending on the amphiphile and buffer conditions) and the vesicle resembling a hollow, flexible shell

outward when confronted by predatory wolves. Smaller, less repulsive head groups can cause formation of vesicles, which are locally planar membranes that form large shells enclosing a volume of water. Each vesicle has an aqueous interior compartment that is topologically distinct from the bulk water. This turns out to be an important evolutionary property.

9.4.3 Why Lipid Membranes: Competition and Cooperation

Living systems are characterised by cooperation at many levels, from the complex interweavings of biochemical pathways to the coordinated action of many cells and up to the formation of tribes and nations. Was cooperation important at the origin of life? Consider a ribozyme that copies genetic information. The enzymatic activity might be ligating short pieces of RNA together using a template, or copying sequence information from a template as polymerases do today. To survive the process of natural selection, the enzyme must derive some benefit from its own activity.

In order for the genome to accommodate variation and thereby evolve, the replicating genomic enzyme (the ‘replicase’) should not be limited to copying a single sequence (e.g., its original sequence), but should be able to copy an arbitrary template sequence. This constrains the replicase to copying physically distinct molecules, being unable to serve simultaneously as the enzyme and the template. A replicase would thus improve the fitness of other sequences while having no effect on its own absolute fitness and actually decreasing its own relative fitness. Such traits are sometimes called altruistic traits and are disfavoured by natural selection among individuals at this level (replicase molecules). Even if a replicase happened to copy a sequence to produce another replicase, the newly formed replicase would simply diffuse away, having no effect on the fitness of the first one. In free solution, where individual sequences interact randomly with one another, natural selection would favour the evolution of very good template sequences (i.e., unfolded and flexible) that parasitise the enzymes, but good replicase enzymes (i.e., well-folded and active) would be disfavoured. The enzyme in free solution is essentially an altruist because it confers a benefit on the template sequence without receiving any benefit itself; the best template sequences grow in number while the enzyme itself does not.

One way this problem could be avoided would be to have enzymes preferentially copy one another. Apparently altruistic traits, such as the replicase trait, can evolve when interactions among individuals are non-random.[1] For example, an individual may display altruistic behavior preferentially toward a relative. The selection of these traits is based on including the fitness of genetically similar relatives with the fitness of the individual (inclusive fitness). According to W.D. Hamilton's rule [57], an altruistic trait is evolutionarily favoured if the cost (c) of the behavior is less than the benefit (b) to the relative, weighted by the coefficient of relatedness (r), summed over all the affected relatives (i), or

$$c < \sum_i b_i r_i. \tag{9.6}$$

Inclusive fitness takes into account selection at the level of genes in addition to individuals. A classic example of an altruistic behaviour selected through inclusive fitness is the predator alarm call of female Belding's ground squirrels. Predators kill individuals who give the alarm call at a higher rate than non-callers, indicating a direct fitness cost. Females with relatives nearby, including females without offspring, are more likely to call than females without relatives nearby, implicating inclusive fitness (male Belding's ground squirrels disperse from their relatives) [58, 59].

A general way to create non-random interactions among individuals is to group them, creating a higher level of selection. Indeed, selection through inclusive fitness is sometimes considered to be a particularly effective type of grouping, in which groups consist of relatives [60, 61]. If the groups vary with respect to the composition of altruists and non-altruists, and the groups containing altruists out-compete the groups lacking altruists, then the proportion of altruists can increase in the total population (even as the proportion of altruists decreases within any given group) [62, 63]. This can be stated as Price's equation [64]:

$$\Delta P = \mathrm{cov}_n(s,p) + \mathrm{ave}_{n'}(\Delta p), \tag{9.7}$$

where ΔP is the change in global allele frequency, n is the population of a group and n' is the population of the group in the next generation, p is the allele frequency within a group, s is the mean relative fitness of a group (normalised by the global relative fitness), and Δp is the change of p in the next generation. The first term is the covariance of allele frequency and fitness of a group, weighted by the

[1] One important mode of altruism, reciprocal altruism (e.g., in the Prisoner's dilemma), can evolve under certain conditions, such as repeated interactions. However, the first replicases were unlikely to have a mechanism to decide whether or not to cooperate, so reciprocal altruism will not be considered further here.

population of the group, and therefore represents the change of P due to competition between groups (a positive value if the allele is altruistic). The second term is the average change in the allele frequency within groups, weighted by the group population in the next generation, and therefore represents the change in P due to competition within groups (a negative value if the allele is altruistic). Thus, the frequency of the altruistic allele can increase if between-group competition is sufficiently strong.

Intense competition between colonies is believed to explain the altruistic behavior of foraging specialists in leaf-cutter ants (*Acromyrmex versicolor*) [65, 66]. Colonies of leaf-cutter ants raid other nests, capturing broods to be raised as workers in their own colonies, and raided colonies disappear. This creates strong selection between groups. Multiple unrelated females found a single colony, and a female who becomes a forager takes on increased predation risk and lays fewer eggs than non-foragers. The forager role appears to be assigned without conflict, and a forager shares food with the other queens in her colony. Although foragers experience decreased fitness within their colonies, colonies with foragers produce more new workers, leading to greater success at raiding other colonies. Competition between groups thus selects for the foraging behavior. The effects of intra- and inter-group selection have also been investigated experimentally, particularly with regard to population size in the flour beetle *Tribolium castaneum* [67, 68].

The presence of multiple levels of selection can therefore lead to the evolution of traits that appear altruistic at the individual level, as selective pressure at one level may be offset by an opposing selective pressure at another level. Indeed, multilevel selection is thought to have contributed to major evolutionary innovations in cooperation, including the formation of cell-like systems and multicellular organisms [69–71].

Several mechanisms for preferential interactions might be envisioned, but one easy solution is to spatially segregate small numbers of sequences into cellular compartments. A compartment that happens to contain multiple enzymes could begin a cycle of cooperation, with each enzyme copying others and being copied in turn. Cells that contain more enzymes could grow faster, so natural selection among cells would lead to the evolution of greater enzymatic activity. During the origin of life, physical grouping is a plausible way for replicases to interact non-randomly. For example, replicases encapsulated in membrane vesicles would preferentially copy sequences related to themselves, even in the absence of the ability to recognise kin. This would lead to a particularly strong form of group selection with relatives grouped together. Because the vesicles separated different genomes from each other, poor replicases would not have access to active replicases and could not parasitise them. But mutants with improved replicase activity would benefit directly themselves, as their descendants would remain in the same vesicle and copy each other.[2] As the vesicles grew and divided, they would continue to segregate the

[2] It actually takes a pair of replicases to start an autocatalytic cycle of replication. The second replicase might have been encapsulated in the same vesicle by chance, or may have been generated via templated non-enzymatic polymerization.

encapsulated molecules. An occasional parasitic sequence would be separated from most of the active polymerases during vesicle division and thus could not poison the entire system (the 'stochastic corrector' model) [72, 73]. Thus, a higher level of population organisation, the cell, is necessary for the evolution of more efficient replicases [73–77]. Therefore, in this section, we delve into the physical chemistry of vesicle membranes as an important, although not necessarily essential, ingredient that would support the evolution of 'altruistic' enzymatic activity.

Membrane vesicles are not the only way to segregate different genomes. The attachment of molecules onto surfaces also creates a heterogeneous distribution of interactions based on spatial proximity, a scenario that has been investigated theoretically using cellular automaton models [78]. However, membrane vesicles are of particular interest because they also serve as semi-permeable barriers that mediate the chemical fluxes of molecular species, creating a relatively protected environment for replicating genomes. Although they may not have been the initial means of achieving genomic segregation during the origin of life, membranes are the dominant means of separating cells today. Membranes presumably assumed this function very long ago, at least 3–4 billion years ago, at some time prior to the diversification from the last common ancestor.

9.5 Protocell Competition

9.5.1 Model Membranes for the Origin of Life

Today biological membranes have complex compositions including phospholipids, sphingolipids, sterols, and proteins. However, membrane vesicles are also formed in an aqueous suspension of relatively simple, prebiotically plausible amphiphiles, such as fatty acids [79]. Fatty acids can be synthesized abiotically in several ways. For example, Miller-Urey-type electrical discharge reactions in a solution of ammonia under a nitrogen and methane atmosphere yield fatty acids with a chain length up to C12 [80, 81]. Abiotic syntheses generally yield decreasing amounts of fatty acids of longer chain lengths. A synthesis simulating hydrothermal vents yielded fatty acids up to C33, from an aqueous Fischer-Tropsch-type synthesis using a heated solution of oxalic acid (which disproportionates into H_2, CO_2, and CO) [82, 83].

Direct evidence for the abiotic presence of fatty acids comes from the detection of fatty acids in the interior of the Murray and Murchison carbonaceous chondrite meteorites from Australia (up to C8), as well as an Asuka carbonaceous chondrite meteorite (A-881458) from Antarctica (up to C12) [84–87]. Fatty acids are relatively abundant in these meteorites, being 20 times more abundant than amino acids in the organic extract of A-881458. Indeed, organic extracts from the Murchison meteorite form boundary membranes when rehydrated [88, 89]. The presence of fatty acids is particularly suggestive because the chemical composition of these meteorites is believed to resemble that of the early solar system.

Depending on the solution pH, fatty acids self-assemble into different structures [90, 91]. At low pH, the molecules are protonated and uncharged, resulting in an oil phase. At high pH, they are deprotonated and negatively charged, resulting in the formation of micelles with a hydrophobic core and surface-exposed carboxylates that repel one another (Fig. 9.14a). Although the pK_a of a carboxylic acid is typically 4–5, the self-assembly of fatty acids leads to a cooperative effect that increases the pK_a. For example, oleic acid monomers have a pK_a of 4.5, but oleic acid assembled into a bilayer membrane has a pK_a of 8.5. Medium- and long-chain fatty acids incorporated into membranes have pK_as in the general range of 7–9 [90]. When the solution pH is near the pK_a, fatty acids assemble into bilayer membrane vesicles (Fig. 9.14b) that are capable of entrapping solutes [92, 93]. These vesicles have a net negative charge, with a formal surface charge density close to half of the molecular density of the membrane. As a result, cations are also associated with the surface, forming an electrical double layer [94, 95].

At a given pH, the structures formed (free molecules, micelles, or vesicles) also depend on the concentration. Because the number of molecules required to form a micelle or vesicle is quite high, the equilibrium can be treated as a phase transition:

$$\Delta G^0{}_{\text{transition}} = m\text{RT}ln(\text{critical concentration}), \tag{9.8}$$

where m accounts for the entropic loss from the association of counterions with the aggregate (for fatty acids, $m = 1.5$ for partially ionised vesicles) [96–98]. Free molecules aggregate into micelles when the concentration reaches a certain value, the critical micelle concentration (cmc). Similarly, membrane vesicles form above a critical aggregate concentration (cac). The cac decreases as the fatty acyl chain length increases. For example, the cac of myristoleic acid is 4 mM, while the cac of oleic acid is 80 μM. The temperature and hydration of the suspension also determine the physical phase of the fatty acid, which may be one of several ordered gel phases or a liquid-like phase in which the hydrocarbon chains are relatively disordered. At room temperature and low mole fraction (e.g., <0.2 M in water), oleic acid forms bilayers with liquid-like hydrocarbon chains [99].

If fatty acid micelles are added to a solution buffered at a pH close to the pK_a, they are thermodynamically unstable and will eventually aggregate into vesicles in an autocatalytic process [100]. The initial formation of vesicles accelerates the formation of more vesicles, giving rise to sigmoidal kinetics, and the addition of preformed vesicles shortens the lag phase and accelerates the reaction [101]. There is also some evidence for an effect on the size of newly formed vesicles, in which the size distribution of the new vesicles may resemble that of the preformed vesicles [101, 102]. The autocatalysis is probably caused by electrostatic interactions between a preformed vesicle surface and a bilayer in the process of forming. Although both surfaces are negatively charged, correlations among counterions can reduce the electrostatic repulsion and even mediate an attractive interaction [103–108]. In addition, a depletion effect may favour the interaction of two negatively charged surfaces due to the entropic gain from the release of displaced anions into the buffer [109, 110]. Catalysis of the assembly of fatty acid vesicles also occurs on a variety of ionic surfaces that adsorb fatty acid, including montmorillonite clay and silicates, but not on surfaces that do not adsorb fatty acid, such as Teflon [111, 112]. However, Teflon particles coated with fluorinated fatty acid do catalyse vesicle assembly, suggesting that the fatty acid adsorbed to the surface mediates vesicle assembly, similar to autocatalysis and catalysis by preformed vesicles [112].

In the presence of preformed vesicles, fatty acid from the micelles forms new vesicles, but some fatty acid is also incorporated into the preformed vesicles, with a yield determined by the particular method of addition [111, 113]. The slow and steady addition of micelles to preformed vesicles leads to 90–100 % incorporation of fatty acid into preformed vesicles [111]. Other methods of adding free fatty acid to the system also cause preformed vesicles to grow larger [101, 114]. For example, in a heterogeneous two-phase system with oleic anhydride layered on top of the aqueous phase containing vesicles, hydrolysis of the anhydride to oleic acid causes both *de novo* vesicle formation and growth of preformed vesicles [101]. Previous studies have demonstrated the feasibility of several cycles of growth and division (i.e., at least five division cycles can be induced by extrusion of the vesicles through small pores or by gentle agitation of large floppy vesicles), mimicking a life cycle for protocellular vesicles [111, 115].

9.5.2 The Second Law of Thermodynamics

Concentration differences are out of equilibrium. This is because there are more ways to distribute items evenly than to distribute them unevenly. Suppose we have a pool table with several balls, and we designate a right and left side. We jostle the balls around in a random fashion, much like thermal energy would. Using your intuition from the natural world, you would be quite surprised if all of the balls wound up on one side. The most likely outcome is for approximately half of the balls to end up on the right, and half to end up on the left, because (if there is no particular reason to favour one direction over the other) there are many more ways to spread the balls evenly than ways that would generate a noticeable imbalance. If the pool table started with all the balls on the right, random jostling would quickly spread them roughly evenly on both sides. This statistical phenomenon is the root of the second law of thermodynamics, which states that the entropy of a system tends to increase.

9.5.3 Osmotic Pressure and the Gibbs-Donnan Equilibrium

Like atoms in an ideal gas, water molecules also tend to move down their own concentration gradient because there are more ways to distribute the molecules evenly than ways to maintain the gradient. Suppose we have a spherical membrane (vesicle) in water, and water is able to pass through the membrane but a large polymer, like RNA, is trapped inside the vesicle. The RNA, being negatively charged, also traps counterions like Na^+ to maintain approximate electrical neutrality (more on this later). Therefore, the bulk exterior solution has a higher water activity than the interior (i.e., the bulk water is hypotonic relative to the vesicle interior), creating a driving force for water to enter the vesicle. The resulting pressure difference (the osmotic pressure Π) is equal to $i\Delta c RT$, where Δc is the solute concentration difference and i is the van't Hoff factor of the solute (e.g., i is approximately 1 for a non-ionic solute). Therefore, a charged solute generates more osmotic pressure, and the osmotic pressure of a charged polymer, such as RNA, will be due primarily to the counterions associated with it.

The encapsulation of the impermeable solute and its counterions creates an ion gradient across the membrane, causing ions to flow out of the hypertonic volume if both cations and anions are permeable. This effect reduces the osmotic pressure exerted by the impermeable solute. The result is a transmembrane potential and an altered distribution of ions, known as the Gibbs-Donnan equilibrium [116, 117]. For a permeable cation, the ratio of its interior to exterior concentration (r) is equal to

$$-Z_M c_M/(2c) + \left[(Z_M c_M/(2c))^2 + 1\right]^{1/2}, \tag{9.9}$$

where Z_M and c_M are the charge and concentration of the impermeable solute and c is the external concentration of the salt. The ratio of concentrations for the permeable anion is $1/r$. However, if either ion is impermeable, the requirement for electroneutrality prevents the equilibration of both ions, so the osmotic pressure difference ΔΠ is maximal and equal to $(Z_M + 1)\Delta c_M RT$. The real value of ΔΠ is also affected if the impermeable solute is very large and excludes significant volume.

This pressure difference is counterbalanced by membrane tension, as the membrane holds together via the hydrophobic effect, limiting the volume expansion of the vesicle. During osmotic swelling, a vesicle membrane rounds up to a spherical shape (if it was not already spherical), and then becomes tense due to the osmotic pressure. This balance gives red blood cells a well-known bloated appearance when put into hypotonic solution. If swelling continues, the strain on the membrane increases until a critical areal strain is reached, at which point the membrane ruptures and releases contents into the exterior. This rupture does not result in full equilibration of the solutes because the membrane reseals as soon as a tolerable (non-zero) level of tension is reached [118, 119]. The tensile strength (τ^*) of the membrane depends somewhat on its composition, but is generally 3–40 dyn/cm (0.003–0.04 N m^{-1}) for phospholipid bilayer membranes [119–121]. The maximum tolerable osmotic pressure depends on both the tensile strength and the size of the vesicle, in accordance with Laplace's law ($\tau = \Delta\Pi r/2$, where r is the internal radius of the vesicle) [119]. The concentration gradient required to achieve the rupture tension is inversely proportional to the radius of the vesicle, so small vesicles can actually withstand quite large pressure gradients (several atm for a vesicle with diameter of a few hundred nanometres).

9.5.4 Relieving Membrane Tension

In principle, there are at least two ways to relieve the tension in the membrane caused by osmotic pressure. First, osmolytes might leak, as happens if the tension is too great and the membrane ruptures. Second, the volume of the vesicle could increase, diluting the internal osmolyte concentration and thus reducing the gradient. Leakage is a sudden event that only occurs above the rupture tension, but vesicle volume could increase in a more gradual fashion. For example, if a vesicle containing a certain amount of osmolytes has an elongated shape, it may round up to a sphere to increase its volume while keeping the surface area constant. How would a vesicle begin in a non-spherical shape? This might happen as a result of the method of preparation; for example, vesicles extruded through small pores are elongated as they are squeezed through the pore. Once extruded, the volume of the vesicle is determined by the osmolarity; if no gradient is applied, the vesicle remains elongated. But if placed in a hypotonic solution, such a vesicle could round up to relieve membrane tension. Alternatively, the surface area of a spherical, tense vesicle might increase, allowing the volume to increase. How might the surface area increase? If a source of amphiphiles is present, fresh amphiphiles could insert into the existing vesicle membrane, effectively growing the vesicle while keeping the osmolyte number constant and thereby decreasing the osmolyte concentration inside the vesicle.

Membrane growth can actually be observed experimentally under the right conditions with highly dynamic amphiphiles that insert, flip-flop, and exit the membrane at observable timescales. While phospholipid membranes, consisting mostly of double-chain amphiphiles, are fairly static because they are so stable within the membrane, single-chain amphiphiles are much more easily removed and flipped in the membrane. A particularly interesting setup is as follows: imagine a vesicle, 'Bob', that encapsulates RNA and is therefore osmotically tensed in a hypotonic solution. Another vesicle, 'Joe', by chance, does not encapsulate RNA and therefore its membrane is relaxed; perhaps its membrane is also elongated from extrusion through a pore. What happens? The system can minimise tension by transferring amphiphiles from Joe to Bob, increasing the surface area and volume of Bob while decreasing the surface area of Joe. At some point, Joe will also round up as the loss of surface area continues. This may lead to an increase in membrane tension for Joe, stopping the transfer of amphiphiles. But as a protocell, Bob will have grown larger at the expense of Joe. By the mere act of encapsulating an osmolyte, Bob 'wins' a competition with Joe [122].

9.5.5 Kinetics and Thermodynamics of Protocell Competition

In this scenario, there are two important facets that enable a replicator's properties to translate into fitness during the protocell competition. First, there must be a thermodynamic driving force that creates an asymmetry to be exploited during the competition. The driving force in this case is the tension of the membrane caused by osmotic pressure, itself induced by the chance encapsulation of osmolytes. A lower energy state could be achieved by increasing the surface area of the membrane. Second, there must be a kinetic pathway that approaches the lower energy state. In this case, the fact that the amphiphiles can transfer from one vesicle to another by exiting a membrane, diffusing through solution, and inserting into another membrane, provides a pathway for movement of the system toward equilibrium.

9.5.6 The Evolution of Membranes and the 'Red Queen'

In Lewis Carroll's famous tale "Through the Looking-Glass," young Alice encounters the Red Queen, who shows her the rules of live Chess in a fanciful alternate universe. Alice finds herself running as fast as she can with the Red Queen, but the landscape never changes. When Alice expresses surprise that they find themselves in the exact same spot, the Red Queen explains, "Now, *here*, you see, it takes all the running you can do, to keep in the same place." In evolutionary biology, rapid evolution is sometimes observed and appears to be the result of natural selection for change. For example, hosts and parasites seem to be engaged in an evolutionary 'arms race', with the host evolving a defence (e.g., mutated receptor protein) and the parasite countering (mutant attachment protein). Indeed, genes involved in host defence and parasite offence often exhibit a greater ratio (dN/dS) of non-synonymous (resulting in an amino acid change) to synonymous changes (resulting in no change at the amino acid level) in the DNA sequence compared to the rest of the genome. This acceleration in evolution by natural selection for change is called the "Red Queen" effect; an evolutionarily static organism would not survive, and instead the organisms need to evolve quickly just to survive in a dynamic environment.

The competition among protocells for membrane components could drive rapid evolution of sophisticated membranes. While fatty acids were likely to be available through prebiotic synthesis, the synthesis of phospholipids may have required enzymatic activity, such as a ribozyme that could catalyse condensation of fatty acids with phosphate head groups. If such an activity arose, protocells containing phospholipids would be able to 'hold on' to their fatty acids more tightly, in the sense that the rate constant for dissociation of the fatty acids from the bulk phase is reduced by the presence of the bipartite lipid. In addition, the dissociation rate for bipartite lipids themselves is also much lower than that of fatty acids. Both effects result in overall stabilisation of the membrane in the competition among protocells, causing net transfer of lipid toward the population of protocells with phospholipids [123]. However, a consequence of a stabilised membrane is reduced permeability, as the fluidity of the membrane decreases. Therefore, a protocell that won the competition for membrane components would need an additional mechanism to preserve its ability to utilise nutrients (as the Red Queen would say, "to keep in the same place"). One may imagine how further ribozyme and membrane evolution, sometimes in response to competition, sometimes in response to prior changes, could snowball into a great deal of biochemical complexity.

References

1. Copley SD, Smith E, Morowitz HJ (2007) The origin of the RNA world: co-evolution of genes and metabolism. Bioorg Chem 35:430–443. doi:10.1016/j.bioorg.2007.08.001
2. Orgel LE (2008) The implausibility of metabolic cycles on the prebiotic Earth. PLoS Biol 6:e18. doi:10.1371/journal.pbio.0060018
3. Shapiro R (2007) A simpler origin of life. Scientific American. http://www.scientificamerican.com/article.cfm?id=a-simpler-origin-for-life. Accessed 7 Mar 2012
4. Benner SA, Hutter D (2002) Phosphates, DNA, and the search for nonterrean life: a second generation model for genetic molecules. Bioorg Chem 30:62–80. doi:10.1006/bioo.2001.1232
5. Crick FH (1968) The origin of the genetic code. J Mol Biol 38:367–379. doi:10.1016/0022-2836(68)90392-6
6. Orgel LE (1968) Evolution of the genetic apparatus. J Mol Biol 38:381–393. doi:10.1016/0022-2836(68)90393-8
7. Woese CR, Dugre DH, Dugre SA et al (1966) On the fundamental nature and evolution of the genetic code. Cold Spring Harb Symp Quant Biol 31:723–736
8. Guerrier-Takada C, Gardiner K, Marsh T et al (1983) The RNA moiety of ribonuclease P is the catalytic subunit of the enzyme. Cell 35:849–857. doi:10.1016/0092-8674(83)90117-4
9. Kruger K, Grabowski PJ, Zaug AJ et al (1982) Self-splicing RNA: autoexcision and autocyclization of the ribosomal RNA intervening sequence of Tetrahymena. Cell 31:147–157. doi:10.1016/0092-8674(82)90414-7
10. Cheng LKL, Unrau PJ (2010) Closing the circle: replicating RNA with RNA. Cold Spring Harb Perspect Biol 2:a002204. doi:10.1101/cshperspect.a002204
11. Ferré-D'Amaré AR, Scott WG (2010) Small self-cleaving ribozymes. Cold Spring Harb Perspect Biol 2:a003574. doi:10.1101/cshperspect.a003574
12. Joyce GF (2009) Evolution in an RNA world. Cold Spring Harb Symp Quant Biol 74:17–23. doi:10.1101/sqb.2009.74.004
13. Bartel DP, Szostak JW (1993) Isolation of new ribozymes from a large pool of random sequences. Science 261:1411–1418. doi:10.1126/science.7690155
14. Wochner A, Attwater J, Coulson A, Holliger P (2011) Ribozyme-catalyzed transcription of an active ribozyme. Science 332:209–212. doi:10.1126/science.1200752
15. Cech TR (1992) In: Frangsmyr T, Malmstrom BG (eds) Nobel lectures, chemistry 1981–1990. World Scientific Publishing Co., Singapore
16. Carothers JM, Oestreich SC, Davis JH, Szostak JW (2004) Informational complexity and functional activity of RNA structures. J Am Chem Soc 126:5130–5137. doi:10.1021/ja031504a
17. Kumar RK, Yarus M (2001) RNA-catalyzed amino acid activation. Biochemistry 40:6998–7004. doi:10.1021/bi010710x

18. Illangasekare M, Sanchez G, Nickles T, Yarus M (1995) Aminoacyl-RNA synthesis catalyzed by an RNA. Science 267:643–647. doi:10.1126/science.7530860
19. Lee N, Suga H (2001) A minihelix-loop RNA acts as a trans-aminoacylation catalyst. RNA 7:1043–1051. doi:10.1017/S1355838201010457
20. Turk RM, Chumachenko NV, Yarus M (2010) Multiple translational products from a five-nucleotide ribozyme. Proc Natl Acad Sci USA 107:4585–4589. doi:10.1073/pnas.0912895107
21. Yarus M, Widmann JJ, Knight R (2009) RNA-amino acid binding: a stereochemical era for the genetic code. J Mol Evol 69:406–429. doi:10.1007/s00239-009-9270-1
22. Ban N, Nissen P, Hansen J et al (2000) The complete atomic structure of the large ribosomal subunit at 2.4 A resolution. Science 289:905–920. doi:10.1126/science.289.5481.905
23. Nissen P, Hansen J, Ban N et al (2000) The structural basis of ribosome activity in peptide bond synthesis. Science 289:920–930. doi:10.1126/science.289.5481.920
24. Jencks WP (1969) Catalysis in chemistry and enzymology. McGraw Hill, New York
25. Joyce GF (1987) Nonenzymatic template-directed synthesis of informational macromolecules. Cold Spring Harb Symp Quant Biol 52:41–51
26. Zielinski WS, Orgel LE (1985) Oligomerization of activated derivatives of 3′-amino-3′-deoxyguanosine on poly(C) and poly(dC) templates. Nucleic Acids Res 13:2469–2484. doi:10.1093/nar/13.7.2469
27. Leu K, Obermayer B, Rajamani S, Gerland U, Chen IA (2011) The prebiotic evolutionary advantage of transferring genetic information from RNA to DNA. Nucleic Acids Res 39:8135–8147. doi:10.1093/nar/gkr525
28. Li X, Liu DR (2004) DNA-templated organic synthesis: nature's strategy for controlling chemical reactivity applied to synthetic molecules. Angew Chem Int Ed Engl 43:4848–4870. doi:10.1002/anie.200400656
29. Weber AL, Orgel LE (1978) The formation of peptides from the 2′(3′)-glycyl ester of a nucleotide. J Mol Evol 11:189–198. doi:10.1007/BF01734480
30. Weber AL, Orgel LE (1980) Poly(U)-directed peptide-bond formation from the 2′(3′)-glycyl esters of adenosine derivatives. J Mol Evol 16:1–10. doi:10.1007/BF01732065
31. Tamura K, Schimmel P (2001) Oligonucleotide-directed peptide synthesis in a ribosome- and ribozyme-free system. Proc Natl Acad Sci USA 98:1393–1397. doi:10.1073/pnas.98.4.1393
32. Tamura K, Schimmel P (2003) Peptide synthesis with a template-like RNA guide and aminoacyl phosphate adaptors. Proc Natl Acad Sci USA 100:8666–8669. doi:10.1073/pnas.1432909100
33. Tjivikua T, Ballester P, Rebek J (1990) Self-replicating system. J Am Chem Soc 112:1249–1250. doi:10.1021/ja00159a057
34. Lee DH, Granja JR, Martinez JA et al (1996) A self-replicating peptide. Nature 382:525–528. doi:10.1038/382525a0
35. Joshi PC, Aldersley MF, Delano JW, Ferris JP (2009) Mechanism of montmorillonite catalysis in the formation of RNA oligomers. J Am Chem Soc 131:13369–13374. doi:10.1021/ja9036516
36. Turk RM, Illangasekare M, Yarus M (2011) Catalyzed and spontaneous reactions on ribozyme ribose. J Am Chem Soc 133:6044–6050. doi:10.1021/ja200275h
37. Hammond GS (1955) A correlation of reaction rates. J Am ChemSoc 77:334–338. doi:10.1021/ja01607a027
38. Hilvert D, Carpenter SH, Nared KD, Auditor MT (1988) Catalysis of concerted reactions by antibodies: the Claisen rearrangement. Proc Natl Acad Sci USA 85:4953–4955. doi:10.1073/pnas.85.14.4953
39. Jacobsen JR, Schultz PG (1994) Antibody catalysis of peptide bond formation. Proc Natl Acad Sci USA 91:5888–5892. doi:10.1073/pnas.91.13.5888
40. Kirschner M, Gerhart J (1998) Evolvability. Proc Natl Acad Sci USA 95:8420–8427. doi:10.1073/pnas.95.15.8420

41. Eigen M (1971) Selforganization of matter and the evolution of biological macromolecules. Naturwissenschaften 58:465–523
42. Eigen M, McCaskill J, Schuster P (1988) Molecular quasi-species. J Phys Chem 92:6881–6891. doi:10.1021/j100335a010
43. Nowak MA (1992) What is a quasispecies? Trends Ecol Evol 7:118–121. doi:10.1016/0169-5347(92)90145-2
44. Nowak MA, Schuster P (1989) Error thresholds of replication in finite populations. Mutation frequencies and the onset of Muller's ratchet. J Theor Biol 137:375–395. doi:10.1016/S0022-5193(89)80036-0
45. Schuster P, Swetina J (1988) Stationary mutant distributions and evolutionary optimization. Bull Math Biol 50:635–660. doi:10.1007/BF02460094
46. Gago S, Elena SF, Flores R, Sanjuan R (2009) Extremely high mutation rate of a hammerhead viroid. Science 323:1308. doi:10.1126/science.1169202
47. Drake JW, Holland JJ (1999) Mutation rates among RNA viruses. Proc Natl Acad Sci USA 96:13910–13913. doi:10.1073/pnas.96.24.13910
48. Schuster P (2006) In: Klussmann S (ed) The aptamer handbook. Weinheim, Wiley-VCH
49. Gerland U, Hwa T (2002) On the selection and evolution of regulatory DNA motifs. J Mol Evol 55:386–400. doi:10.1007/s00239-002-0-2335-z
50. Hopfield JJ (1974) Kinetic proofreading: a new mechanism for reducing errors in biosynthetic processes requiring high specificity. Proc Natl Acad Sci USA 71:4135–4139. doi:10.1073/pnas.71.10.4135
51. Ninio J (1974) A semi-quantitative treatment of missense and nonsense suppression in the strA and ram ribosomal mutants of Escherichia coli. Evaluation of some molecular parameters of translation in vivo. J Mol Biol 84:297–313. doi:10.1016/0022-2836(74)90586-5
52. Johnson KA (1993) Conformational coupling in DNA polymerase fidelity. Annu Rev Biochem 62:685–713. doi:10.1146/annurev.biochem.62.1.685
53. Freier SM, Kierzek R, Jaeger JA, Sugimoto N, Caruthers MH, Neilson T, Turner DH (1986) Improved free-energy parameters for predictions of RNA duplex stability. Proc Natl Acad Sci USA 83:9373–9377. doi:10.1073/pnas.83.24.9373
54. Mathews DH, Sabina J, Zuker M, Turner DH (1999) Expanded sequence dependence of thermodynamic parameters improves prediction of RNA secondary structure. J Mol Biol 288:911–940. doi:10.1006/jmbi.1999.2700
55. SantaLucia J Jr (1998) A unified view of polymer, dumbbell, and oligonucleotide DNA nearest-neighbor thermodynamics. Proc Natl Acad Sci USA 95:1460–1465. doi:10.1073/pnas.95.4.1460
56. Schink S, Renner S, Alim K, Arnaut V, Simmel FC, Gerland U (2012) Quantitative analysis of the nanopore translocation dynamics of simple structured polynucleotides. Biophys J 102:85–95. doi:10.1016/j.bpj.2011.11.4011
57. Hamilton WD (1964) The genetical evolution of social behaviour. J Theor Biol 7:1–52. doi:10.1016/0022-5193(64)90038-4
58. Alcock J (1993) Animal behavior: an evolutionary approach. Sinauer Associates, Inc., Sunderland
59. Sherman PW (1977) Nepotism and the evolution of alarm calls. Science 197:1246–1253. doi:10.1126/science.197.4310.1246
60. Hamilton WD (1975) In: Fox R (ed) Biosocial anthropology. Wiley, New York
61. Sober E, Wilson DS (1998) Unto others: the evolution and psychology of unselfish behavior. Harvard University Press, Cambridge
62. Frank SA (1995) George Price's contributions to evolutionary genetics. J Theor Biol 175:373–388. doi:10.1006/jtbi.1995.0148
63. Wilson DS (1975) A theory of group selection. Proc Natl Acad Sci USA 72:143–146
64. Price GR (1972) Extension of covariance selection mathematics. Ann Hum Genet 35:485–490

65. Dugatkin LA (2002) Cooperation in animals: an evolutionary overview. Biol Philos 17:459–476. doi:10.1023/A:1020573415343
66. Rissing SW, Pollock GB, Higgins MR et al (1989) Foraging specialization without relatedness or dominance among co-founding ant queens. Nature 338:420. doi:10.1038/338420a0
67. Goodnight CJ, Stevens L (1997) Experimental studies of group selection: what do they tell us about group selection in nature? Am Nat 150(Suppl 1):S59–S79. doi:10.1086/286050
68. Wade MJ (1976) Group selections among laboratory populations of Tribolium. Proc Natl Acad Sci USA 73:4604–4607. doi:10.1073/pnas.73.12.4604
69. Buss LW (1987) The evolution of individuality. Princeton University Press, Princeton
70. Gould SJ (2002) The structure of evolutionary theory. Harvard University Press, Cambridge
71. Lewontin RC (1970) Units of selection. Annu Rev Ecol Syst 1:1–18
72. Smith JM, Szathmary E (1995) The major transitions in evolution. W.H Freeman, Oxford
73. Szathmáry E, Demeter L (1987) Group selection of early replicators and the origin of life. J Theor Biol 128:463–486. doi:10.1016/S0022-5193(87)80191-1
74. Cavalier-Smith T (2001) Obcells as proto-organisms: membrane heredity, lithophosphorylation, and the origins of the genetic code, the first cells, and photosynthesis. J Mol Evol 53:555–595. doi:10.1007/s002390010245
75. Koch AL (1984) Evolution vs the number of gene copies per primitive cell. J Mol Evol 20:71–76. doi:10.1007/BF02101988
76. Matsuura T, Yamaguchi M, Ko-Mitamura EP et al (2002) Importance of compartment formation for a self-encoding system. Proc Natl Acad Sci USA 99:7514–7517. doi:10.1073/pnas.062710399
77. Szostak JW, Bartel DP, Luisi PL (2001) Synthesizing life. Nature 409:387–390. doi:10.1038/35053176
78. Szabó P, Scheuring I, Czárán T, Szathmáry E (2002) In silico simulations reveal that replicators with limited dispersal evolve towards higher efficiency and fidelity. Nature 420:340–343. doi:10.1038/nature01187
79. Gebicki JM, Hicks M (1973) Ufasomes are stable particles surrounded by unsaturated fatty acid membranes. Nature 243:232–234. doi:10.1038/243232a0
80. Allen WV, Ponnamperuma C (1967) A possible prebiotic synthesis of monocarboxylic acids. Curr Mod Biol 1:24–28
81. Yuen GU, Lawless JG, Edelson EH (1981) Quantification of monocarboxylic acids from a spark discharge synthesis. J Mol Evol 17:43–47. doi:10.1007/BF01792423
82. McCollom TM, Ritter G, Simoneit BR (1999) Lipid synthesis under hydrothermal conditions by Fischer-Tropsch-type reactions. Orig Life Evol Biosph 29:153–166. doi:10.1023/A:1006592502746
83. Rushdi AI, Simoneit BR (2001) Lipid formation by aqueous Fischer-Tropsch-type synthesis over a temperature range of 100 to 400 degrees C. Orig Life Evol Biosph 31:103–118. doi:10.1023/A:1006702503954
84. Lawless JG, Yuen GU (1979) Quantification of monocarboxylic acids in the Murchison carbonaceous meteorite. Nature 282:396. doi:10.1038/282396a0
85. Naraoka H, Shimoyama A, Harada K (1999) Molecular distribution of monocarboxylic acids in Asuka carbonaceous chondrites from Antarctica. Orig Life Evol Biosph 29:187–201. doi:10.1023/A:1006547127028
86. Yuen G, Blair N, Des Marais DJ, Chang S (1984) Carbon isotope composition of low molecular weight hydrocarbons and monocarboxylic acids from Murchison meteorite. Nature 307:252–254. doi:10.1038/307252a0
87. Yuen GU, Kvenvolden KA (1973) Monocarboxylic acids in Murray and Murchison carbonaceous meteorites. Nature 246:301. doi:10.1038/246301a0
88. Deamer DW (1985) Boundary structures are formed by organic components of the Murchison carbonaceous chondrite. Nature 317:792. doi:10.1038/317792a0

89. Deamer DW, Pashley RM (1989) Amphiphilic components of the Murchison carbonaceous chondrite: surface properties and membrane formation. Orig Life Evol Biosph 19:21–38. doi:10.1007/BF01808285
90. Cistola DP, Hamilton JA, Jackson D, Small DM (1988) Ionization and phase behavior of fatty acids in water: application of the Gibbs phase rule. Biochemistry 27:1881–1888. doi:10.1021/bi00406a013
91. Small DM (1986) In: Small DM (ed) The physical chemistry of lipids: from alkanes to phospholipids. Plenum Press, New York
92. Apel CL, Deamer DW, Mautner MN (2002) Self-assembled vesicles of monocarboxylic acids and alcohols: conditions for stability and for the encapsulation of biopolymers. Biochim Biophys Acta – Biomembranes 1559:1–9. doi:10.1016/S0005-2736(01)00400-X
93. Walde P, Goto A, Monnard P-A et al (1994) Oparin's reactions revisited: enzymatic synthesis of poly(adenylic acid) in micelles and self-reproducing vesicles. J Am Chem Soc 116:7541–7547. doi:10.1021/ja00096a010
94. Grahame DC (1947) The electrical double layer and the theory of electrocapillarity. Chem Rev 41:441–501. doi:10.1021/cr60130a002
95. Hunter RJ (2001) Foundations of colloid science, 2nd edn. Oxford University Press, New York
96. Blandamer MJ, Cullis PM, Soldi LG et al (1995) Thermodynamics of micellar systems: comparison of mass action and phase equilibrium models for the calculation of standard Gibbs energies of micelle formation. Adv Colloid Interface Sci 58:171–209. doi:10.1016/0001-8686(95)00252-L
97. Israelachvili JN, Mitchell DJ, Ninham BW (1976) Theory of self-assembly of hydrocarbon amphiphiles into micelles and bilayers. J Chem Soc, Faraday Trans 2 72:1525–1568. doi:10.1039/F29767201525
98. Israelachvili JN, Mitchell DJ, Ninham BW (1977) Theory of self-assembly of lipid bilayers and vesicles. Biochim Biophys Acta 470:185–201. doi:10.1016/0005-2736(77)90099-2
99. Monnard P-A, Deamer DW (2003) Preparation of vesicles from nonphospholipid amphiphiles. Meth Enzymol 372:133–151. doi:10.1016/S0076-6879(03)72008-4
100. Walde P, Wick R, Fresta M et al (1994) Autopoietic self-reproduction of fatty acid vesicles. J Am Chem Soc 116:11649–11654. doi:10.1021/ja00105a004
101. Blöchliger E, Blocher M, Walde P, Luisi PL (1998) Matrix effect in the size distribution of fatty acid vesicles. J Phys Chem B 102:10383–10390. doi:10.1021/jp981234w
102. Berclaz N, Blöchliger E, Müller M, Luisi PL (2001) Matrix effect of vesicle formation as investigated by cryotransmission electron microscopy. J Phys Chem B 105:1065–1071. doi:10.1021/jp002151u
103. Allahyarov E, D'Amico I, Löwen H (1998) Attraction between like-charged macroions by Coulomb depletion. Phys Rev Lett 81:1334–1337. doi:10.1103/PhysRevLett.81.1334
104. Angelini TE, Liang H, Wriggers W, Wong GCL (2003) Like-charge attraction between polyelectrolytes induced by counterion charge density waves. Proc Natl Acad Sci USA 100:8634–8637. doi:10.1073/pnas.1533355100
105. Butler JC, Angelini T, Tang JX, Wong GCL (2003) Ion multivalence and like-charge polyelectrolyte attraction. Phys Rev Lett 91:028301. doi:10.1103/PhysRevLett.91.028301
106. Ha B-Y (2001) Modes of counterion density fluctuations and counterion-mediated attractions between like-charged fluid membranes. Phys Rev E 64:031507. doi:10.1103/PhysRevE.64.031507
107. Levin Y (1999) When do like charges attract? Physica A 265:432–439. doi:10.1016/S0378-4371(98)00552-4
108. Linse P, Lobaskin V (1999) Electrostatic attraction and phase separation in solutions of like-charged colloidal particles. Phys Rev Lett 83:4208–4211. doi:10.1103/PhysRevLett.83.4208
109. Dinsmore AD, Wong DT, Nelson P, Yodh AG (1998) Hard spheres in vesicles: curvature-induced forces and particle-induced curvature. Phys Rev Lett 80:409–412. doi:10.1103/PhysRevLett.80.409

110. Kaplan PD, Rouke JL, Yodh AG, Pine DJ (1994) Entropically driven surface phase separation in binary colloidal mixtures. Phys Rev Lett 72:582–585. doi:10.1103/PhysRevLett.72.582
111. Hanczyc MM, Fujikawa SM, Szostak JW (2003) Experimental models of primitive cellular compartments: encapsulation, growth, and division. Science 302:618–622. doi:10.1126/science.1089904
112. Hanczyc MM, Mansy SS, Szostak JW (2007) Mineral surface directed membrane assembly. Orig Life Evol Biosph 37:67–82. doi:10.1007/s11084-006-9018-5
113. Chen IA, Szostak JW (2004) A kinetic study of the growth of fatty acid vesicles. Biophys J 87:988–998. doi:10.1529/biophysj.104.039875
114. Berclaz N, Müller M, Walde P, Luisi PL (2001) Growth and transformation of vesicles studied by ferritin labeling and cryotransmission electron microscopy. J Phys Chem B 105:1056–1064. doi:10.1021/jp001298i
115. Zhu TF, Szostak JW (2009) Coupled growth and division of model protocell membranes. J Am Chem Soc 131:5705–5713. doi:10.1021/ja900919c
116. Bartlett JH, Kromhout RA (1952) The Donnan equilibrium. Bull Math Biophys 14:385–391
117. Tinoco I, Sauer K, Wang JC, Puglisi JD (2002) Physical chemistry: principles and applications in biological sciences, 4th edn. Prentice Hall, Upper Saddle River
118. Hallett FR, Marsh J, Nickel BG, Wood JM (1993) Mechanical properties of vesicles. II. A model for osmotic swelling and lysis. Biophys J 64:435–442. doi:10.1016/S0006-3495(93)81384-5
119. Mui BL, Cullis PR, Evans EA, Madden TD (1993) Osmotic properties of large unilamellar vesicles prepared by extrusion. Biophys J 64:443–453. doi:10.1016/S0006-3495(93)81385-7
120. Needham D, Nunn RS (1990) Elastic deformation and failure of lipid bilayer membranes containing cholesterol. Biophys J 58:997–1009. doi:10.1016/S0006-3495(90)82444-9
121. Shoemaker SD, Vanderlick TK (2002) Stress-induced leakage from phospholipid vesicles: effect of membrane composition. Ind Eng Chem Res 41:324–329. doi:10.1021/ie010049t
122. Chen IA, Roberts RW, Szostak JW (2004) The emergence of competition between model protocells. Science 305:1474–1476. doi:10.1126/science.1100757
123. Budin I, Szostak JW (2011) Physical effects underlying the transition from primitive to modern cell membranes. Proc Natl Acad Sci USA 108:5249–5254. doi:10.1073/pnas.1100498108

Further Reading

Deamer DW, Fleischacker GR (eds) (1994) Origins of life: the central concepts. Jones and Bartlett Publishers, Boston

Deamer D, Szostak JW (eds) (2010) The origins of life. Cold Spring Harbor Laboratory Press, New York

Gesteland RF, Atkins JF (eds) (1993) The RNA world, 3rd edn. Cold Spring Harbor Laboratory Press, New York

Jencks WP (1969) Catalysis in chemistry and enzymology. McGraw Hill, New York

Saenger W (1984) Principles of nucleic acid structure. Springer, New York

Chapter 10
Physical Chemistry: Extending the Boundaries

Sydney Leach

Abstract This chapter is conceived as a brief exposition of the content of the previous nine chapters, a commentary on them and added material, with the intent to enlarge reflection on the general theme, Physical Chemistry in Action. It can be considered as a guide to the book and, in its attempt to be syncretic, perhaps as a guide to the perplexed, confronted with the separate domains of physical chemistry, astrochemistry and astrobiology.

10.1 Introduction

Physical Chemistry was coined as a term in 1752 by the scientific polymath and poet Mikhail Lomonosov [1], but it was recognised and defined as a specific discipline only towards the end of the nineteenth century. Definitions and distinctions were discussed and exemplified by Jean Perrin in his magistral 'Traité de Chimie Physique' of 1903 [2] which clearly defined the overlap between the subjects now known as physical chemistry and chemical physics. In essence both of these fields involved the opening up of chemistry to the techniques and thought processes of physics. Extension of physical chemistry, and its relation to physics, to the sphere of biology became formally recognised 50 years later, as exemplified by the arguments and presentation of Cyril Hinshelwood in his popular tome 'The Structure of Physical Chemistry', published in 1951 [3].

In parallel to these developments there also occurred the extension of physical chemistry from Earthbound considerations to those of the Solar System, with the interpretation of spectroscopic observations of comets and planetary atmospheres, in the first half of the twentieth century, amplified later by satellite studies and space

S. Leach (✉)
Laboratoire d'Etude du Rayonnement et de la Matière en Astrophysique (Lerma)
Observatoire de Paris-Meudon, 5 place Jules-Janssen, Meudon 92195, France
e-mail: Sydney.Leach@obspm.fr

I.W.M. Smith et al. (eds.), *Astrochemistry and Astrobiology*, Physical Chemistry in Action, 307
DOI 10.1007/978-3-642-31730-9_10,

probes, and finally to the Universe as a whole with the discovery of interstellar molecules by optical and radiofrequency spectroscopies. The end of the twentieth century saw the first observations consistent with the existence of exoplanets, in far-off stellar systems, and this has led in recent years to determinations of the partial content and physical characteristics of exoplanet atmospheres, requiring physical chemistry knowledge for interpretation. These developments have also led to the birth of the field of astrobiology/exobiology which seeks to understand the origin of life and its possible existence in extraterrestrial sites. This new field is subject to much speculation, tempered by laboratory studies that often fall within the extensive boundaries of physical chemistry.

Accompanying these evolutionary developments of the intellectual and spatial boundaries of physical chemistry, and in large part being key contributors to these extensions, are a succession of new instrumental tools for the laboratory study and an enlarging capacity for computation and simulation of physical chemistry processes.

This chapter will include examples and lessons, drawn mainly from the previous nine chapters that illustrate the tenor of the above text. It will also point out gaps in these chapters and give additional material or comments in the context of the treatment of the overall subject of Physical Chemistry in Action: Astrochemistry and Astrobiology.

10.2 Physical Chemistry in Astrophysics and Astrobiology: The Basic Tools

Chapter 1, by Ian Smith, reviews fundamental aspects of Physical Chemistry that are essential for applications to astrophysics and astrobiology. The historical construction of the periodic table and of the concepts and determination of molecular structure constitute a paradigm for the development of our present day understanding of the structure of the universe. Both result from careful observation, in many cases by spectroscopy, the gradual winnowing of superfluous facts and incorrect concepts, the hardening of the principal concepts and processes through theoretical models and simulations. And in both cases there remain many unexplained observations, such as the Diffuse Interstellar Bands (DIBs), for which explanation is sought through attempts to correlate observed spectral properties with astrophysical sites and their physical conditions, with few valid results so far. Another area of uncertain attribution is that of the Unidentified Infrared Bands (UIBs), most often blithely assigned to polycyclic aromatic hydrocarbons (PAHs) or their avatars, but without a single firm identification. The difficulty is that observations have mainly been in the 3–15 μm infrared region where bands are specific to chemical bond vibrations but not to particular molecules. It is now hoped that observations in the Far InfraRed/THz region, home of the characteristic skeletal vibrations of large molecules (0.15–15 THz) will enable specific assignments to be made. Far InfraRed (FIR) observations are possible on the space telescope HERSCHEL, which has three spectral instruments: HIFI (0.48–1.25 THz, 1.41–1.91 THz), PACS (1.43–5.01 THz) and SPIRE (0.45–1.5 GHz). Interpretation of observations in the PAH trope will require preliminary building up a library of FIR spectra on PAH type molecules. At present there are few examples, and these are mainly restricted to FIR spectra of relatively small PAHs in the solid state. FIR observations with HERSCHEL have so far been restricted to small molecules such as water and related species. Interpretation of these observations has been based on laboratory studies in the sub-mm range and has helped to understand the water budget during star and planet formation.

Bonding properties, intermolecular forces and thermodynamics are fundamental to chemistry, along with the kinetics of reactions and these are adequately sketched in Chap. 1, along with appropriate examples. Apart from the equation of state of the universe, basically a thermodynamic concept, these aspects of physical chemistry are not usually associated with cosmology, but there have been exceptions in the past. One notable physical chemist whose work led to cosmological musings was Walther Nernst [4]. Who could be more of a physical chemist than Nernst, well known for his basic contributions to electrochemistry and to thermodynamics! A believer in the ether, seen by him as being rich in potential energy, he argued against the 'heat death' predicted for the universe of ever increasing entropy. In 1912 he proposed that the energy of the ether, in our more modern terms a sort of zero-point energy [5], would save the world. When quantum theory developed he postulated the existence of zero-point radiation energy and showed that, compressed in a container, it had the remarkable property of an invariant energy density. It is fascinating that, although zero-point radiation energy has fallen by the wayside, invariant energy density is a property of 'dark energy', one of the dark ladies of modern cosmology [6][1] and that the concept of all pervading ether(s) is perhaps disguised nowadays as that of field(s) [4].

Another excursion coupling physical chemistry and cosmology has been made, this time by a cosmologist, David Layzer [8] who discussed some numerical coincidences between astronomical and cosmological parameter values and those of importance in solid-state chemistry. He explained these coincidences by a cosmological theory that assumes the universe began to expand from a maximally dense initial state at zero temperature, i.e. a cold universe origin. In this model, chemical bonding plays a role in the formation of gravitationally bound systems. This is based on the coincidence between the binding energies of gravitationally bound systems and the cohesive energies of solids. In this cold model the cosmic background radiation is interpreted as thermalized starlight emitted by an earlier generation of massive stars. There are indeed plenty of cold and 'not so hot' (so-called 'ekpyrotic') early universe studies [9–12]. If a cold scenario survives, although the hot Big Bang is the favourite horse at present, the conclusion of Layzer that cosmology will become a discipline closely related to physical chemistry could perhaps be revived.

[1] Did Shakespeare have only one Dark Lady in his sonnet sequence [7]? Cosmology has certainly two Dark Ladies (Dark Matter, Dark Energy) whose identity is sought.

The section on surfaces, interfaces and catalysis is relevant to astrophysical processes, in particular to those occurring on interstellar grains and comets, and in general to objects subsumed under the notion of astrophysical ices. Observations by the infrared SPITZER space telescope have transformed this area of science. The high sensitivity of SPITZER has been capital in the observation of chemically interesting mantles formed on interstellar grains in sites previously difficult to observe, such as solar-type protostars, protoplanetary disks, star-less and star-forming clouds, and a range of extragalactic sources. The production of complex molecules observed in space has become a very active area of laboratory research and concerns, in particular, processes leading to molecule formation on surfaces, especially the fundamental one of H_2 formation, as well as physicochemical processes of adsorption and desorption. Surfaces, interfaces and catalysis intervene, of course, in a vast range of biological processes and phenomena, illustrated here by a discussion of enzyme-catalysed reactions and in detailed examples discussed in later, more biologically oriented chapters. Catalysis involves the formation of an appropriate reaction intermediate that helps to lower, or avoid, the energy hill between reactants and products. Nature has been extraordinarily inventive in creating catalysts in the biological domain.

Two further areas of physical chemistry that are important in both astrophysics and astrobiology are photochemistry and radiation chemistry. Singly or together they play important roles in the formation of key organic compounds in the ISM, in planetary and cometary atmospheres and in photosynthesis. Ionization by photon absorption or by transfer of energy on particle impact is important in many astrophysical and astrobiological contexts and this will be illustrated, albeit fragmentarily, in later chapters.

A word must also be said about energy sources. For us on Earth the Sun is a major source, both directly and indirectly. However, its luminosity and spectral distribution have evolved in time from a reduced luminosity at the birth of the Earth, but with a 1,000 times greater far ultraviolet component than at present, with important consequences in the evolution of the atmosphere [13, 14]. The luminosity and spectral distribution of stars found or suspected to have associated planets will depend on the evolutionary moment chosen and this will determine the possible occurrence and nature of photosynthesis on such exoplanets [15]. Many other energy sources have been considered as having possibly contributed to chemical synthesis on the Earth: these include lightning, coronal discharges, shocks due to infall of material such as meteorites, with various degrees of estimated efficiency [16]; their possible role in the formation of increasingly complex biomolecules has been questioned in view of a number of thermodynamic limitations and arguments given that favour UV radiation as the prime energy source [17]. Changing scale, we note that the most efficient energy sources in the universe are accretion disks. Those around black holes convert 7–40% of rest mass energy to radiation [18].

10.3 The ISM: A Chemical Cornucopia, or from Dust to Dust?

The molecular universe is the subject of Maryvonne Gérin's Chap. 2. It deals mainly with the interstellar medium (ISM) and leaves aside planets, dealt with in other chapters, except for early stages of their origin. How did molecules form, when did they form and where did they form are the questions addressed. The evolutionary path from the Big Bang until the formation of successive populations of stars is sketched out, based on observations and models. Recently, gaseous regions containing no elements heavier than helium have been observed in sites that correspond to about 2 billion years after the Big Bang [19]. These relics of the pristine Universe indicate that the distribution of 'metals' (the astrophysicists quaint name for elements heavier than helium) is inhomogeneous in the Universe. Metal-free stars, the so-called population III stars, could therefore still form much later than the early epochs assumed in current models.

Galaxy formation and distribution remain front-line questions under investigation. Observational evidence for molecules existing quite early on poses questions about the nature and the formation of dust grains as part of the molecule fabrication process in the early universe. Understanding the nature and physical properties of the different components of the ISM, summarised in Table 2.1, is mainly restricted to our own galaxy, the Milky Way. The role of supernovae explosions in the creation and scattering of molecular material is important and its understanding increasingly so. This chapter includes detailed discussion of our knowledge of the life cycle of the ISM and illustrates the profound interplay of gravity, turbulence and magnetic fields in its various physical phases and temporal stages. Molecules can preferentially lurk in dense regions of the ISM where they are protected from the debilitating and destructive effects of star-emitted far-UV and X-rays. But the cores of these dense regions are also the birthplaces of stars. Newly born stars, familiarly called YSO's (young stellar objects) emit energetic radiation ragingly, increase local temperatures in the surrounding molecular cloud, dissociate molecules, and yet release them by destroying ice mantles. A dynamic situation thus exists, monitored from the molecular viewpoint by mm and sub-mm radioastronomy over a variety of sites corresponding to different evolutionary stages, the dynamic aspects being expressed in computational simulations, discussed in later chapters.

One of the most aesthetically pleasing images in this evolutionary cycle is that of bipolar outflows and jets issuing from YSO's, schematically indicated in Fig. 2.2 but worth seeing in astronomical photographs or in an artist's view. These features play a role in the accretion of matter on the central object and have interesting chemical and molecular consequences. They also contribute to the outward transport of energy and angular momentum, which are fundamental physical effects permitting conservation of energy and angular momentum of the whole system by counterbalancing their losses due to the accretion process. The next evolutionary stage is the formation of circumstellar disks and it is here that planet embryos gradually build up, starting with the aggregation of small dust particles. Because of the relatively small size of circumstellar disks, the detailed study of their chemical composition can only be developed with the coming into action of high sensitivity and high angular resolution spectroscopic instruments such as ALMA. Some meteorites reaching Earth today contain relics of this circumstellar disk stage, as evidenced by their element abundances.

Radiative and particle energy impacting on the constituents of molecular clouds lead to photodissociation, photoionization and shock phenomena. The behaviour of photodissociation regions (PDRs) is the subject of much computational simulation, with improving developments towards full 3D treatments that should build up confidence in understanding the dynamics involved. Shocks of various origins and the influence of magnetic fields on their hydrodynamic aspects are also active areas of observation and modelling. Shocks can also be an easy way out in postulating reasons for ejection of matter in the ISM when no other obvious cause exists.

A very thorough presentation of the constituents of the interstellar medium, notably the atoms and molecules observed, and the physical conditions (density, temperature, degree of ionization, magnetic field) in their sites in the ISM, is given next. Here there is a lot of standard spectroscopy and its relevant applicable theory. Fifty years ago the ISM was thought to be a place where atoms and their ions reigned: molecules were thought to be rare and largely subject to destructive forces. Now well over 100 molecules have been observed in the ISM or in circumstellar regions; these include cations and anions, the latter with some surprise at the time of their initial observation, in view of their relatively small electron affinities and thus being susceptible to photoejection.

Radiofrequency astronomical spectroscopy often runs into the 'confusion limit', which occurs when the density of observed spectral lines is so great that picking out features characteristic of a particular molecule becomes impossible, except for the few high abundance species towering over their neighbours in spectral line intensity. The situation is improving and should improve further with the introduction of interferometric instruments (e.g. ALMA) with their high sensitivity. They will be able to observe much smaller regions of the sky (smaller aperture) than with the previous generation of radiotelescopes and, in principle, be able to observe regions containing fewer different types of molecules, thus diminishing the risks of a 'confusion limit'.

Determining the density, and energy distribution, of electrons at various sites is no easy matter. This information is essential for estimating collisional excitation effects due to electrons. A classical problem is in the determination of the extent of electron excitation of the CN radical, which is an important factor in measuring the Cosmic Background Radiation (CBR) temperature by CN spectroscopy, a technique usable for extragalactic sources and thus able to determine the CBR temperature in high redshift sources, i.e. at past epochs in cosmic evolution. This is in contrast to bolometric measurements such as those of COBE, which are limited to our region of the solar system and thus to the present [20].

Another datum that has proven elusive to pin down accurately, by theoretical calculations or by observations, is the ionisation rate due to cosmic rays impinging on molecules. It is now found that there is a distribution of ionisation rates in the ISM. This has been probed by H_3^+ in many lines-of-site in diffuse molecular clouds [21]. The mean rate is $\approx 3.5 \times 10^{-16}$ s^{-1}. Some of the highest inferred rates are 25 times larger than the lowest upper limits, indicating that there are variations in the underlying cosmic ray flux across the Galaxy. These variations are possibly related to the distance between an observed cloud and the nearest site of particle acceleration. The cosmic ray spectrum is apparently not uniform in the Galaxy. The ion H_3^+ is here considered as a tracer of astrophysical environments; several other tracers are discussed, for PDR regions, shocks, and the cold cores and hot cores existing within molecular clouds. This is a fast evolving area. As new molecular and ionic species are discovered with the latest telescopes, the number of tracer species and their diagnostic characteristics will increase bringing, for example, new knowledge about the phenomena of turbulence occurring in various regions of the ISM.

The subject of dust grains, a term that covers a large range of solid particles of various whereabouts, is briefly discussed from the viewpoint of their physical properties. Chemical aspects, notably their role as small chemical reactors in the building up and evolution of ices, are left mainly to other chapters. Future high spectral resolution observations by the Stratospheric Observatory for Infrared Astronomy (SOFIA) and the James Webb Space Telescope (JWST) will clarify many aspects of these processes in icy grain mantles. Element abundances in the gas phase, when smaller than those of the Sun, are indicators of the degree of condensation into solid structures. We note, however, as Snow and Witt have shown [22], that solar abundances of carbon and heavier elements are approximately a factor of 1.8 greater than for a large number of nearby stars. This led them to propose a revised list of cosmic abundances. Solar system abundances and condensation temperatures of the elements have been reviewed more recently by Lodders [23, 24]. The actual mechanisms of gas phase element depletion by condensation on grains are not always evident, given the low densities of matter in molecular clouds.

Since abundances generally decrease with galactocentric distance, coupled with the fact that there exist differences in abundances between our Galaxy and galaxies at high redshift, it is vain to consider that there is a generic cosmic composition valid for all cosmic systems [24].

A useful list of different types of interstellar dust grains and their characteristic infrared bands is presented by Gérin (Table 2.4). It is remarkable that spectra of novae change, in just a few days, from a gas spectrum to a spectrum largely characteristic of dust [25]. Photoelectric effects on dust grains lead them to retain charge and to contribute to local heating in molecular clouds. Furthermore, redox reactions on the surface of interstellar dust grains could possibly contribute to the synthesis of molecules, the chemical driving force being the electrostatic charging of grains [26–28].

Besides contributing to specific bands in the IR, dust grains are also responsible for very broad absorption, culminating in a peak at 217 nm, and continuously rising beyond into the far UV, observed until the Lyman limit at 91.2 nm in the interstellar extinction curve of our Galaxy. Other galaxies have variants of this extinction curve, including several without the 217 nm feature, indicating that the nature and/or size range of dust grains are not universal. Furthermore, it is not clear what fraction of the dust grains arises from evolved stars and supernovae and what fraction is grown in the ISM. With so few specific spectral characteristics, it has been an interesting and not too difficult game, played quite brilliantly, to invent suitable mixtures of species able to simulate the extinction curve. These have sometimes provided pegs to hang on amusing, occasionally very exotic, theories concerning the nature and origin of the species involved [29, 30].

One class of species, polycyclic aromatic hydrocarbons (PAHs), has a 25-year old pedigree in astrophysics [31]. They are thought to be the principal carriers of the Unidentified Infrared Bands (UIB) mentioned previously but, despite careful experimental and theoretical laboratory studies, as stated earlier, no definitive assignments have been made. Indeed, the assignment that the UIB are due to pure aromatic carriers has been disputed and aspects of the somewhat consecrated single photon excitation, followed by relaxation, mechanism have been argued to be invalid for PAHs [32]. These bands appear to be characteristic of mixtures of aliphatic and aromatic molecules and they are assigned to amorphous organic solids with such a mixture, similar to that in meteorites.

On the other hand, fullerenes, long searched for in the ISM, and occasionally assigned, in particular in ionic form, by subtle forms of wishful thinking, have finally really been observed through their infrared spectra [33, 34]. The ISM observation of neutral fullerenes in the 200–800 nm spectral range has been investigated without success so far [35]. Based on laboratory studies [36, 37] it has been shown to be worth attempting, even if difficult to carry out. Fullerenes are potentially important as denizens of space since they are electron acceptors and have properties intermediate between those of molecules and those of bulk materials. On the basis of laboratory studies of fullerene formation in carbon sources these species were long expected to be formed only around carbon stars especially poor in hydrogen such as Coronae Borealis stars [38]. They have been observed (C_{60}, C_{70}) in planetary nebulae [33, 34] including, however, hydrogen rich circumstellar environments [39], which suggests that they are formed in the destruction of hydrogenated amorphous carbon grains (HAC) [40]. Very recently SPITZER IR observations of the binary star system XX Oph show evidence for the possible presence of *solid* C_{60}; it is suggested that solid C_{60} would be excited by stars having effective temperatures in the range 15,000–30,000 K [41].

This chapter concludes with a thorough presentation of the spectroscopic and imaging instruments, techniques and methods used for measuring and determining various physical properties of the ISM and in identifying objects therein. Ground-based as well as space-borne instruments are discussed and their capabilities are well delineated. Technical progress in this area is rapid, but the cost of realisation of new ground and space-borne instruments is becoming of major economic, and thus political, concern.

10.4 Reactants into Products: How Fast?

After this survey telling us about the existence and nature of molecules in the ISM and the physical conditions of their formation and existence we enter into chemical territory in the chapter on chemical processes in the ISM by Michael Pilling. Its scope concerns the rates of chemical reactions, in particular those necessary for devising and exploiting chemical networks, discussed in detail in Chap. 4, that conceivably lead to the formation and abundances of interstellar molecules at various astrophysical sites. The majority of gas-phase reaction rate studies have been carried out at temperatures far above those existing in the ISM, so that techniques for measurement at low temperatures, although rare, are of capital importance. These measurements are not only necessary per se but also to provide benchmark values for assisting and qualifying theoretical computations of reaction rates at low temperatures.

Present day databases for astrochemical networks involve several thousand elementary reactions, of which only a few have been measured, even fewer at low temperatures. Thus estimations, using well-worn simple theory or more sophisticated reaction rate theory, have been necessary for many cases. Reactions of interest include those involving different types of reactant: atoms, molecules, free radicals, ions. The appropriate potential energy surfaces and reaction pathways have distinctive qualities in the various cases and usually require quite distinct experimental techniques for their measurement. The methods used in reactions between an ion and a neutral species are described and exemplified, and the few low temperature techniques emphasised, in particular the CRESU supersonic isentropic flow technique with which it has been possible to measure down to 8 K. The appropriate theory valid for ion-neutral reactions is given, the nature of the interaction between the two species being dependent on whether the neutral species is dipolar, fixed or induced, or quadrupolar. Various reformations of the theory, in particular by the introduction of trajectory calculations, or phase space considerations which assume an isotropic potential between the reactants, are discussed and exemplified by a number of specific reactions. Neutral-neutral reactions, including those involving free radicals, have mainly been studied using variants of laser flash photolysis, with a low temperature limit of 80 K, and to lower temperatures by suitable adaptations of the CRESU technique. Some other, more traditional, techniques are also discussed. A new technique for studying cold ion-neutral reactive collisions in ion traps has recently been developed in which the molecular ions can be selected in particular rovibrational states [42, 43].

How, using theory, does one picture and follow a bimolecular reaction? A reaction involves rearrangements of atoms, so that the timescale of the chemical reaction is that for the motion of atoms, 10^{-9}–10^{-12} s. During the rearrangement process there is a reorganisation of the electronic structure, but this dance is choreographed by the atomic rearrangement, so that the timescale of electron redistribution is that of the nuclei whenever the Born-Oppenheimer approximation is valid, as it is in most bimolecular reactions. Probing this redistribution can be done on the attosecond scale [44]. The electron reorganisation can create an energy barrier, whether the reaction is exothermic or endothermic. The rate of the reaction reflects the effort necessary to pass over this barrier. Because of low thermal energies in the ISM, reactions that have small energy barriers will be favoured. Transition state theory and its avatars have been developed for calculating the rate coefficient for a bimolecular reaction in which an energy barrier must be mastered. Chemical reactions are described in terms of the coordinates of the atoms participating in bond rupture and formation. The activated complex exists at the saddle point on the potential energy surface corresponding to the barrier height maximum. The minimum energy path in the reaction will usually pass through this saddle point. For the small barrier low temperature reactions occurring in the ISM the transition state is considered to be located at the maximum of the Gibbs free energy. Its determination requires variational transition state theory, whose canonical and micro-canonical forms are adaptable to different types of bimolecular reactions. In some cases there exist more than one transition state, created by the counterplay of long and short-range interactions that depend on the nature of the local binding forces. Switching between transition states can occur and is particularly important in some reactions, including a number of those that may occur in the ISM.

Ways of avoiding cumbersome or impracticable calculations necessary for evaluating whether a reaction will be of significant importance in the ISM are being explored. These include the use of sensitivity analysis to identify key reactions in cosmochemical schemes, as discussed in Chap. 4. However, the picture given above of the importance of the transition state barrier needs serious modification in some monomolecular dissociation cases. It has been found that some reactions, for example the photodissociation of NO_3 into $NO + O_2$, important in atmospheric chemistry, proceed without a transition state [45]. The saddle point on the potential energy surface is by-passed. Instead, an incomplete bond cleavage leaves part of the molecule unable to escape, so that it orbits the other molecular remnant until reaching a reactive site, usually involving a conical intersection of potential energy surfaces in photochemical cases, leading to the products being formed after internal conversion. This 'roaming' reaction involves large excursions on the potential energy surfaces and situations where the Born-Oppenheimer approximation stumbles. Roaming reaction pathways have also been invoked in thermal reactions, ion molecule reactions and in shock tube studies [46]. They rarely constitute an exclusive pathway, as is the case for NO_3 photodissociation, but can occur alongside, and thus be competitive with, transition state channels.

Association reactions need to remove surplus energy either by an internal relaxation process, such as in radiative association, or by third body interaction. This is also true for electron-ion recombination processes. Aspects of these cases are briefly considered in this chapter, which concludes with detailed discussions of the experimental and theoretical aspects of some neutral-neutral, radical-radical and radical-molecule reactions and surface reactions, notably the formation of H_2 on surfaces, which is the paramount way of associating two H atoms to form the molecule in the ISM.

Investigation of the chemistry occurring on the surfaces of interstellar grains is a very active, but challenging, area of research. From laboratory experiments we can reasonably assume that molecular hydrogen can be formed efficiently on cold grains, although determining the amount of internal and kinetic energy released in H_2 leaving the surface is a delicate problem. It has been shown that other surface reactions can occur, forming species such as methane, ammonia, water and even methanol. A basic problem in application of laboratory obtained information to astrophysical models is that the nature, structure and other physical properties of interstellar grains is not known with sufficient exactitude. They are surely not uniform in composition or in density.

In spite of these drawbacks there has been much effort and some progress, both in experimental and in theoretical determination of the rates of physico-chemical processes occurring in surface reactions relevant to cosmochemistry. Application to cosmochemical models is still full of ambiguities. Chemical models, in which surface reactions were treated with rate equations, are now subject to more accurate treatments, in which the stochastic nature of the process is dealt with by different approaches. But detailed gas-grain models require knowledge or reasonable estimation of grain size distribution, the nature of the grain surface and characteristics and extent of its inhomogeneities which can play an important role in affecting reaction rates, catalysing or restraining reaction processes. These features and parameters are essential to clarify for the development of valid models.

Assessing the feasibility of a reaction requires accurate data on thermodynamic quantities. Knowledge of $\Delta_f H$ (M), the heat of formation (enthalpy) of a molecular species M, is necessary for quantitative understanding of chemical equilibria and reactions. There is an extensive literature on the heats of formation of molecules determined experimentally, as reported for example in the NIST [47] and Lias et al. [48] compilations, or as derived using empirical, or semi-empirical methods based on the parameterisation of structural similarities [49]. The NIST compilation [47] often quotes the early experimental $\Delta_f H$ data of Lemoult [50, 51], either exclusively or in comparison with other data including group additivity scheme values. Regularities in the values of the heats of formation of homologous series of molecules were early recognised by Lemoult [52] and were extended in his further studies [50, 51]. Later parameterisation methods, developed since 1940, and reviewed by Cox and Pilcher [53], culminated in the development of many additivity schemes, in particular that of Group Additivity [49, 54, 55] for calculating the heat of formation of *neutral* molecules. These methods are mainly based on the concept that the heat of formation of a molecule in the gas state is equal to the sum of contributions from sub-structural components within the molecule. Extensive comparisons between experimental heats of formation and the values determined using these various parameterisation schemes are given in books by Cox and Pilcher [53] and Pedley, Naylor and Kirby [56]. The neutral molecule heats of formation reported in the compilations of Lias et al. [48] and of the NIST [47] are often those based on estimations tabled in these two books. Quantum mechanical methods for calculating molecular heats of formation have recently been critically reviewed [57]. In some developments they include group contribution methods.

For molecular *ions*, reported heats of formation [47, 48] have very often been determined by one or other of two principal methods:

1. For a molecular ion M^+, $\Delta_fH(M^+)$ is given by $\Delta_fH\,(M^+) = IE_{ad}(M) + \Delta_fH(M)$, i.e. the sum of $IE_{ad}(M)$, the adiabatic ionization energy of the molecule M, and $\Delta_fH(M)$, the heat of formation of M.
2. For a molecular ion $(MH)^+$, the heat of formation $\Delta_fH(MH)^+ = \Delta_fH(M) + \Delta_fH(H^+) - PA(M)$, where PA is the proton affinity (PA) of molecule M.

The critical values, $IE_{ad}(M)$ and PA(M), are often not known to better than 10 or 20 kJ mol^{-1}. For example, in the NIST [47] and Lias et al. [48] compilations, many IE values are given with an error of ±0.1 or 0.2 eV, often on the basis of estimations of onset energies in a photoelectron spectrum first band, a notoriously subjective task. In very many cases, quoted IE values are obtained by subtracting $\Delta_fH(MH)$ from $\Delta_fH(MH)^+$, the latter being obtained via the proton affinity of $\Delta_fH(M)$, whose measurement can be difficult to perform and interpret and which, in the past, has not always been known with great accuracy [47, 48, 58–60].

There are a small, but increasing number of molecules whose $IE_{ad}(M)$ is now known to very high accuracy, via various photoionization techniques involving lasers or synchrotron radiation, coupled with coincidence detection techniques [61]. Thus an accurate value of the cation heat of formation would provide the opportunity of obtaining a value of the heat of formation of the neutral species. The latter is also not always well known. For example, for ethyl formate, the NIST compilation cites two values that differ by 36 kJ/mol, $\Delta_fH = -361.7$ kJ/mol and -398 kJ/mol [47].

Some reported Δ_fH values of cations were derived from appearance energy (AE) measurements in dissociative ionization processes (10.1), either by photon impact or by electron impact. In their determination it is usual to follow the approximation [61], that two heat capacity terms in the full energy balance equation [62] cancel each other so that the 298 K Δ_fH values of fragment ions ${m_1}^+$ can be obtained from measured AE's using (10.2):

$$M + h\nu/e^- \rightarrow M^+ + e^-/2e^- \rightarrow {m_1}^+ + m_i \tag{10.1}$$

$$AE\ + \Delta_fH_{gas}(M) - \sum\left[\Delta_fH_{gas}(m_i)\right] = \Delta_fH_{gas}({m_1}^+) \tag{10.2}$$

There have been a few attempts to correlate $\Delta_fH(M^+)$ in a simple additive manner. Franklin devised a group equivalents method, with application to gaseous free radicals and carbonium ions [63]. His results are not conclusive as to the validity of this method. Holmes et al. [64] developed an additivity scheme for estimating the heats of formation of gas-phase organic ions, based on data for unbranched homologous ions but additivity methods are not generally successful for ions, mainly because ionization energies are not simply proportional to the number of added sub-units.

We emphasize that particular care must be exercised in the use of the standard compilations of heats of formation of molecular species [47, 48], especially of cations and anions; careful examination of the sources of information and a critical cast of mind in their respect are essential in evaluating the validity of the data.

10.5 Chemical Networking in the ISM and Elsewhere

The next chapter, by Wakelam, Coppen and Herbst, builds on the previous chapters in presenting models that enable one to simulate the chemical composition of molecular clouds under a variety of physical conditions and at various stages in their temporal evolution. These models are conceived for comparison with spectral observations of the abundances in appropriate astrophysical sites, in particular where reliable estimates have been made of the age and initial composition of the cloud. Initially limited to reactions occurring solely in the gas phase, the revelations of observations and a gradual awakening to the shortcomings of these purely gas-phase models have driven modellers towards ever greater integration of reactions occurring between dust and gases as well as those specific to the solid phase. The case of the formation of the major molecular constituent of the ISM, H_2, is a leading exemplar of the situation.

This activity is in an area of the first stirrings of interest in astrobiology, since the molecular clouds in the ISM contain the raw material for the formation of stars and their planetary systems and the subsequent development of life, at least on our Earth and possibly elsewhere. Cosmochemical models, which include a number of parameters very sensitive to their outcome, are in practice limited to certain portions of the star-to-star life-cycle since their stages and their coupling lack definitive understanding. These parameters, such as the geometry, temperature and density of the cloud or of a circumstellar envelope, and their temporal evolutions, are exemplified and the characteristics of appropriate models, including those involving turbulence, are sketched out.

It is here that the various processes creating the chemistry of the ISM are delineated and discussed in detail: effects of cosmic rays and UV photons, bimolecular reactions in the gas phase, and processes involving dust grains. The rates of these interactions and their dependence on particular cloud conditions are considered. The shortcomings of present-day databases of reaction rates, mainly because of lack of reliable data relevant to the low temperature conditions of the ISM, as mentioned earlier, is a grave handicap in applying current chemical networks in astrochemical modelling. Estimates of model error bars and the use of sensitivity analysis can be employed to reduce uncertainties in the models and to highlight key reactions [65]. Very large chemical networks, cumbersome and costly for computation, can be used in reduced form but may require a succession of analyses if applied to a large range of physical conditions.

An important section in this chapter concerns gas-grain interactions and surface reactions, exemplifying and going deeper into many of the relevant experimental and theoretical points already mentioned above. The coupling of gas phase and grain surface reactions, via the use of rate equations, allows computational possibilities for models embracing a variety of different physical conditions. The disadvantages of this method are also discussed, a serious one of which is the limited and coarse-grained view of the grain surface and the interacting species. Stochastic and other methods to overcome some of these disadvantages are discussed. A particularly difficult task is how to differentiate and take into account the variation in reaction behaviour on the grain surface and in successive strata of the grain interior. The processes of diffusion of various species, their subsequent reactions and product desorption are studied both experimentally and theoretically. The formation and chemistry of molecular oxygen in the ISM is examined in detail as an example of the efforts made and problems inherent in obtaining adequate observational, laboratory and modelling data and the difficulties in integrating these aspects into a coherent picture valid for different regions of the ISM. As emphasised earlier, the advent of interferometric telescopes should much improve the quality and relevance of observations to the understanding of the chemistry in the ISM and thus stimulate improvements in the design and computational aspects of theoretical models, including the provision of databases that adequately include gas-grain reaction aspects.

As the chemical reaction networks become more complex, patterns in their organisation become more evident and are of interest in mastering their applications. These networks constitute complex systems composed of many interacting units or parts, simultaneously or in temporal succession. The collective integrated behaviour of a chemical network will display organisational and emergent behaviour resulting, in the cases of interest here, in an evaluation of the abundance of a molecular species in a specific site at a specific time. The topology of these complex networks, i.e. recognising structure in the interaction of individual reactions, can yield useful information. For example, the mechanisms of certain interstellar chemical reactions have been proposed on the basis of graph-theoretical methods [66]. The large-scale topology of chemical networks in the ISM and those involved in planetary atmospheres has been studied [67]. For the ISM it displays patterns consistent with an equilibrium state whereas the structures found for the Earth's atmosphere reveal features similar to those in cellular networks relevant to biological evolution and is evidence of strong, nonlinear coupling between atmosphere and biosphere.

Biochemical and astrochemical networks exhibit different scaling behaviours and have both some similarities but also significant differences as is discussed by Jolley and Douglas [68]. This was shown in their examination of gas phase reactions in dark clouds in the ISM and the biochemical networks involved in the metabolism of *E coli*. In the astrochemical sphere, the chemical networks are distinguished by the atomic or molecular species acting primarily as reactants or products whereas in metabolic networks in the biological realm most species are found to contribute as both reactants and products. These distinct topological and functional features may be of fundamental relevance to our understanding of the origin and evolution of life on Earth as well as in its possible existence elsewhere in the solar system or beyond. However, in these considerations one must be aware that there is a whole range of differences between astrochemical and metabolic reaction networks beyond the presence or absence of biological organisation.

10.6 Inferring Life in Exoplanets

The atmospheres of the planets and their satellites in the Solar System have been investigated by Earthbound and space-borne telescope spectroscopy and more directly by a succession of space missions to these planets. Much information has been gathered by the full resources of spectroscopy ranging from the X-ray region to that of radiofrequencies and, in space missions, by mass spectroscopy and a host of other analytical instruments. The harvest has been extensive and has provided matter for applications of physicochemical concepts, laws and properties in rationalising the observations. Detailed pictures have been obtained not only of the composition and dynamic characteristics of these atmospheres but also some insight into the composition of planetary interiors.

Excellent reviews have been published concerning the atmospheres of solar system bodies [69–72]. Many aspects of physical chemistry and chemical physics are involved in modelling planetary atmospheres, whether they are in the Solar System or are of exoplanets: proving or disproving characteristics of radiative equilibrium, hydrostatic equilibrium (the balance between gravitational and atmospheric pressure forces), chemical equilibrium, radiative transfer (absorption, emission, diffusion), atomic and molecular opacities, Raleigh diffusion in gases, Mie diffusion in particles, etc. Solar System studies thus provide a background of information that can be exploited in investigating the atmospheres and structures of exoplanets and can be used in modelling the chemistry/photochemistry of exoplanet atmospheres, including effects of charged particles (stellar winds), and in searching for spectroscopic biomarkers (a controversial area).

The search for exoplanets is intimately linked to the search for extraterrestrial life. Since the only life we know occurs in our solar system, it is logical to search for exoplanets in similar systems. No extrasolar systems with close similarities to our own have yet been discovered, due in part to observational bias, through lack of adequate technical means. Furthermore, present ideas as to the mechanisms of planet formation are in a state of flux. Thus the search for extrasolar systems similar to our own is an exciting and evolving pursuit. Our solar system contains not only planets but also comets, asteroids, meteorites, which could be sources for the molecular building blocks of life. In analogous circumstances they could penetrate whatever atmosphere exists on an exoplanet and thus deliver these key molecules to the atmosphere, the surface or any liquid haven. Material can also be transferred from one planet to another, as illustrated by the existence of Martian and Lunar meteorites on Earth. In addition, cosmochemistry in the ISM can be a source of prebiotic molecules that can eventually be deposited on planetary sites.

Chapter 5 by Lisa Kaltenegger examines our present knowledge of exoplanet atmospheres and future observational prospects, within the context of searching for biomarkers of extraterrestrial life. This is done on the assumption that physicochemical and biochemical characteristics of our only known life and their effects on planetary atmospheres constitute the search model. She first lists the exoplanet physical and chemical properties that have been measured so far by observational techniques, mainly on the basis of planet-star mutual gravitational effects: radial velocity (stellar wobbling via Doppler shifted spectra), micro-lensing (caustic of interposed gravitational lens gives magnified image), transits (hide and seek and eclipses) and astrometry (stellar wobbling via precise star position measurement), some of which are briefly explained in various stages in this chapter. There have also been some direct imaging observations of a few exoplanets, mainly using coronographic techniques to block out the parent star's bright light. The various exoplanet detection methods are described in detail, with emphasis on their respective sensitivity to exoplanet masses, in an early review article by Perryman [73]. Recent developments in exoplanet detection techniques are to be found in symposia reports, e.g. [74]. To date, several hundred exoplanets have been confirmed as observed, while many others, detected by the Kepler space observatory, are awaiting confirmation; discoveries are continuously monitored and reported in a database [75]. A rough division of exoplanets into two groups, rocky and gaseous, depends on the exoplanet mass and radius, with an uncertain quantitative boundary between the two classes. A sophisticated taxonomy of exoplanets has recently been proposed [76].

The effect of an exoplanet atmosphere on the analysed radiation that gives clues to the exoplanet mass and radius is discussed and the difficulties of interpretation are mentioned in this chapter. The spectrum of a planet is composed of two principal components: the scattered incident radiation from the planet's star and the thermal emitted flux from the planet itself. These two components are well separated in wavelength in solar system planets but can overlap in hot exoplanets. For a transiting planet with a known radius, the thermal emission spectrum is often equated, for convenience, to the thermal emission from a blackbody, but this is an approximation and a wavelength dependent "brightness" temperature can be defined. However, except for conditions of isothermal atmosphere, the brightness temperature is not a measure of physical or effective temperature. Hot Jupiters, i.e. large gaseous exoplanets that orbit close to their central star, also exhibit auroral emission. Atomic and molecular species observed in exoplanet atmospheres include H, Na, CO, CO_2, H_2O, CH_4 [77, 78] as well as suspected aerosols of VO and TiO [79].

Photochemistry and heated stratospheres have profound effects on exoplanet atmospheres. Clouds also play an important role since they scatter incident light back into space and trap condensed species. In a planetary atmosphere energy is transported upwards by convection until the atmosphere becomes optically thin to thermal radiation. At higher altitudes the outgoing energy is transported by radiation. The pressure level of this radiative-convective boundary depends on a number of factors: the atmospheric composition, the opacity of its major constituents, temperature and gravitational forces. The general structure of an irradiated atmosphere is a function of the depth at which incident photon and charged particle energy is absorbed. This will depend on how close the planet is to its central star. In our solar system the energy incident on Jupiter is absorbed fairly deep within the atmosphere, in fact below the level where the atmosphere becomes optically thick in the thermal IR. The absorbed energy then contributes to the internal energy being transported outwards by convection. For hot Jupiter exoplanets the large incident flux is absorbed above the radiative-convective boundary, and the global temperature distribution then becomes inhomogenous. This results in the formation of a significant temperature gradient between the equator and the poles of the exoplanet.

There is an increasing effort to model the processes and effects of photochemistry in exoplanet atmospheres [79–82]. The absorption of UV and VUV radiation will lead to dissociation of atmospheric molecules; this will occur high in the atmosphere before most incident UV is scattered back into space. The dissociation products will participate in complex reaction schemes. For hot Jupiters, photochemistry will likely be more complex than for our solar system Jupiter. Molecules that are condensed below the Jovian clouds (e.g. H_2O, H_2S, NH_3) and are thus protected from photodissociation, will be gaseous in the atmospheres of hot Jupiters. Sulphur and nitrogen compounds may play an important role in hot Jupiter photochemistry and also perhaps in haze production [83].

A central issue discussed by Kaltenegger is how to characterise a habitable planet, including the definition and determination of a stellar system's habitable zone, here linked essentially to the presence of liquid water, at least on an exoplanet's surface, but also subject to effects of an atmosphere. The Earth is used as a model planet in this investigation and considerable sections of this chapter are devoted to characterising the spectra of the Earth observable from far-off, as a proxy exoplanet. Observed spectra using Earthshine as a background spectral source show, besides the presence of water, direct or indirect characteristics of biotic activity such as molecular oxygen, ozone and methane as well as the red chlorophyll abrupt spectral edge in the 700–750 nm region, harbinger of the presence of vegetation on the Earth. Simulations of the photochemistry of planetary atmospheres in an Earth-like planet orbiting different types of star suggest that those orbiting G- and K-type stars would be the best candidates for observing the ozone signature. Other possible biosignature gases could be created by microbial life from redox reaction by-products or generated from secondary metabolism processes (cf. Chap. 7), as has been recently reviewed, using the Earth as an exoplanet analogue [84].

The evolution of the Earth's atmospheric composition over geological times is modelled in an important section of this chapter and the resulting simulated spectra corresponding to various epochs are presented. These provide possible benchmark spectra for comparison with habitable zone exoplanet spectra that can be used to estimate the evolutionary stage of an exoplanet in the context of the development of life forms.

One important player in the atmospheres of solar system planets, affecting their chemistry and content, is the solar wind [85]. The presence or absence of a planetary magnetic field determines many effects of charge particle deposition in these atmospheres. The corresponding effects of stellar winds on exoplanet atmospheres are not considered in this chapter. Many observed exoplanets are relatively close to their central star and are thus subjected to intense energetic photon irradiation and plasma fluxes which can lead to heating the atmosphere to high temperatures and high energy chemistry. Furthermore, due to their proximity to the star, these exoplanet atmospheres can be distorted by strong tidal forces [86]. All strongly magnetized planets in the solar system generate non-thermal radio emission caused by the cyclotron maser instability. The power emitted in Jupiter's hectometric emission is strongly correlated with solar wind plasma parameters. The operation of the Low Frequency Array (LOFAR),[2] which has a sensitivity of about 10^{-3} Jy (1 Jansky = 10^{-26} Wm^{-2} Hz^{-1}) in the radio emission decametre range should lead to improved knowledge of the magnetospheres of exoplanets [86].

10.7 Water, Water Everywhere?

Water is indeed everywhere in the Universe, but where did it come from? It is originally formed in the ISM as H_2O^+, via the very reactive $H_3{}^+$ ion (see Chap. 2); H_2O^+ reacts with H_2 to form H_3O^+ that recombines with free electrons to form the OH radical and neutral H_2O (recombination of the H_2O^+ ion with electrons leads essentially to dissociation to OH + H). In solid or gaseous form water has been found in a variety of astrophysical sites besides the ISM: planets, satellites, comets, circumstellar disks, other galaxies and in our Sun and on our Moon. It also forms a matrix for trapping gases, as clathrates in which guest molecules are trapped within polyhedral water cages; the most prominent example is that of methane hydrates which occur on ocean floors and in permafrost.

Liquid water, so conducive to the development of life, is expected to be found at the surface of rocky planets when the temperature is suitable, our Earth being the prime example, as well as in the interior of planetary bodies at high temperatures and pressures. There are many proposed sources for water on the Earth [87]; the geological study of zircons suggests the existence of water on the Earth as long ago as 4.3 Ga [88]. Ongoing questions concerning the early Earth are how did water become partitioned between interior and surface reservoirs? Was water originally contained in the interior and outgassed via volcanism? Or was it mostly in the atmosphere and slowly outgassed via processes such as dissolution into an originally molten planet and, later, subduction through a solid one? Was the deep water cycle established together with plate tectonics, or did one come first, perhaps enabling the other? Estimations of the relative content of water in the Earth's mantle to that on its surface range from a factor of 0.1 to 2.5 [89].

The concept of a water world or ocean planet, in which the surface of the planet is entirely covered by a deep layer of water, has arisen in the context of the search for exoplanets [90]. At least one exoplanet thought to correspond to such a structure, having nearly 50% of its mass as water, has been observed [91].

[2] LOFAR, recently built in Holland, is a new radio interferometric array consisting of many low-cost antennae, organised in stations arranged in an area of 100 km diameter as well as several international stations and operating between 10 and 250 MHz.

Hydrothermal vents would not be expected on ocean planets, which might reduce their biological potential.

Philip Ball (Chap. 6) examines the role of water in sustaining life, as it is on Earth (but as it is in the heavens?). He discusses the unique properties of water and whether they exclude other possible molecular rivals for its biotic functions. In a brief historical recall, he opens the question of the possibly unique fitness of water for this role, which extends beyond its capacity as a mere solvent to embrace its manipulative effects on structure and reactivity in interacting with biochemical solutes. Ball discusses the various physical and interactive qualities of water, in which hydrogen-bonding plays a capital role both as a structure determinant and in a dynamic solvent network. The shortcomings of the properties of water are also mentioned and this brings up the question of the adaptation of life to water. The action of water in the living cell is evoked and the question is raised as to the possible increase in its viscosity in this crowded confined space bounded by a permeable membrane.

The hydrophobic effect, which we will later see to play an important role in membrane structure, is here discussed in terms of the general tendency of hydrophobic species to aggregate or modify their structure in aqueous solution. Examples of affected species are proteins and some of these are shown to be able to undergo dewetting transitions that enhance interactions between hydrophobic surfaces. A more general biochemical role of dewetting is sketched. This is followed by a detailed discussion of protein stability and denaturation. It emphasises the role that water plays in protein folding in determining the path within conformational space that leads to the stable protein configuration. The question of denaturation is complex in that it can be initiated in various ways; the roles of a solvent, and of solutes that modify the hydration shell, are important and are considered.

The subject of protein folding and stability has been highlighted by the discovery of protein-misfolding diseases. In this context it has led to speculation on the complex role of hydration and, in particular, to considerations on the sensitivity of the energy surfaces of proteins to the degree of hydration, noting that rearrangements of the folded state are necessary for carrying out some protein functions. In protein-substrate binding there is a role for water in mediating the transfer of information from one biochemical entity to another or others, whether directly or indirectly; thermodynamic constraints govern these processes. A mechanistic role in function is also evoked, in the important example of the allosteric regulation of oxygen binding to haemoglobin; examples are also given of the direct participation of water in reactions occurring at binding sites where it can act as a nucleophile or as a proton source. Protons can indeed be transported rapidly through a chain of water molecules that form a ‘water wire’, by a Grotthuss-type mechanism [92] which, in modern terms, involves rapid dynamic rearrangements of hydrogen bonds. A critical evaluation of the classical Grotthuss mechanism and its modern guise has been made by Cuikerman [93]. Biological examples of the functioning of these water wires include photon energy conversion systems, transport of water through cell membranes and control mechanisms concerning membrane permeability to other substances.

Concerning the mechanisms in play in water wires inside protein cavities, it is sometimes difficult to decide whether protons are being transferred in one direction or the hydroxyl anion OH^- in the opposite direction. The difficulty lies in the specific and unknown physicochemical nature of the H^+ (or OH^-) pathways in a protein cavity that is obviously distinct from bulk water. In this context, it is difficult to define what proton transfer really means. Are protons transferred in a sequence of hopping steps between waters and (de)protonatable polar residues, or is it also possible that a protonated water cluster diffuses in a protein cavity for a relatively short distance? What and where is the rate limiting step for H^+ transfer? Are protons just transferred between water molecules in the water wire? Can polar residues shuttle protons between water molecules? These questions have been asked and are essential to the full understanding of proton transfer processes in this context [93].

In his discussion of dynamical aspects of protein hydration, Philip Ball remarks on the driver role of the solvent in determining the dynamics of the protein but concludes that the protein and its hydration water really function as a single dynamic entity within cells. Monitoring the dynamic aspects requires experimental techniques covering several orders of magnitude in their timescales; these techniques are rapidly developing from classical spectroscopic and analytic forms to ultramodern variants. As an example, ultrafast IR spectroscopy has shown that orientational relaxation times are considerably increased when water interacts with a variety of interfaces or a large molecule [94]; the presence of an interface appears to be more important in slowing hydrogen bond dynamics than the chemical nature or geometry of the interface.

Whether water is the only useful solvent for proteins is a question arising in this chapter. Ball deplores the aquacentric prejudice that has limited imaginative discussion on possible non-terrestrial biochemistries. He goes on to discuss the role of water in determining the structure of nucleic acids in specific biotic sites and the effects of various amounts of water on their conformations and on their information transmitting capabilities arising through qualities of self-recognition and self-organisation. Are these properties limited to water as a solvent? This is an open question on which only a limited amount of research has been done at the present time.

Another aspect of these problems is the effect of ions existing in the electrolytic solutions that constitute the fluid component of the cytoplasm. These ions can affect the conformation, interactions and biochemical functions of molecules in the cell. The Hofmeister series, which was first noted in 1888 [95], is invoked in this more modern context. It ranks the relative influence of ions on the physical behaviour of a wide variety of aqueous processes ranging from colloidal assembly to protein folding. The influence of an ion on the properties of macromolecules was initially thought to arise, at least in part, from its capacity of modifying bulk water structure. However, recent time-resolved and thermodynamic studies of water molecules in salt solutions show that bulk water structure is not central to the Hofmeister effect. Models are now being developed that take into account direct interactions between ion and macromolecule, and the interactions with water molecules that are operative in the first hydration shell of the macromolecule.

This chapter ends with a recapitulation of the role of water as a biomolecule and then addresses the astrobiological question as to whether non-aqueous solvents could play an analogous role in living organisms. This is discussed in detail by Benner et al. [96]. Philip Ball concludes that it would be difficult to find a solvent that possessed the same versatility, sensitivity and responsiveness that water exhibits in the essential biochemical processes of living organisms. But living entities other than those we have encountered on Earth may be full of biochemical surprises in this respect! Some of these possibilities are vigorously discussed by Benner et al. in speculations that include life forms in a number of surprising places: the interstellar space vacuum, solids in the Oort cloud, Venusian clouds, Titan and the habitable zones of the giant planets [96].

10.8 On Limits of Life

Life? So what do I think about the origin of life? First of all, what is life? This is a good moment in the book to muse on the subject. Is life just a philosophical concept, a property of matter fulfilling a set of defined qualities? But in order to conceive of these qualities there must be an a priori concept of life. Where does that come from? From experience, subtly insinuated into the conscious brain. Experience = collection of material for a set which delimits a concept. To collect is to apprehend.

What is in my collection?

At first glance:

Birth = multiplication of the living entity in a similar form. Reproduction by a procedure technically known as replication. Leads to more of approximately the same, the similar.

Growth, necessitating transformation of matter, acted on actively by uptake from the surroundings. Can lead to more of the same or similar. Functions through metabolism.

Death, but this also includes transformation of matter, acted on passively, by reaction from surroundings. Leads to less of the same or similar.

Implications at a higher level: (1) Uptake of matter from surroundings requires not only matter that can be transformed but also energy available for the work of transformation. (2) This implies obedience to laws of thermodynamics, chemical reactivity, mechanisms for uptake of energy.

The origin of life, as a subject for scientific study, involves examination of these two implications, the search for model experimental or theoretical systems that lead from inert matter to matter fitting the "collection". Of course, a universally accepted definition of life would be a great help in this quest. This is a notoriously slippery subject, as can be gathered from my reflections above; more professional discussions on possible definitions of life have been given and reviewed [97–99]. In the chapter by Cockell and Nixon, the origin of life is not examined directly but can be detected as a shadow subject in their discussion of the boundary conditions for the existence and persistence of life.

First a few more words about energy sources. The stars are universal sources of energy, created in nuclear reactions, and as photon energy. Although nuclear energy is on a scale unsuitable for biochemical reactions, the existence in the early Earth of many radioisotopes such as those of uranium, thorium and potassium, created in past supernovae explosions, may have given rise to some radiation chemistry reactions playing a role in the origin or the development of life. These radioisotopes are not in a situation of thermodynamic equilibrium and they can contribute to the creation of disequilibrium in the environment. Decay of the nuclei can drive tectonics and volcanism on Earth and, prospectively, on other planetary-type bodies whether in our solar system or beyond, and even conceivably provide energy to create and sustain life independent of the presence of a star.

Photon energy from the sun, and of all stars, although emitted over a large spectral range, includes that useful in chemical transformations. Getting useful hold of this energy requires a collection system and links to an energy conversion system. These involve mainly chemical physics, in photon collection processes known as light harvesting, and physical chemistry in energy transformation processes. This constitutes photosynthesis, which is largely at the origin of our food system as well as that for a host of living entities. Drawing a supply of energy from food requires organic chemistry reactions in the biochemical systems that constitute the processes of metabolism. Energy is also produced in respiratory processes. For us this involves the intake of oxygen and the release of carbon dioxide from the oxidation of complex organic substances. Once energy is gathered or produced, almost invariably in the form of ATP (adenosine triphosphate), its biochemical use must be thermodynamically possible and this is considered in detail later in Chap. 7.

There are thus two main sources of energy for creating and sustaining life: initially via photons, then followed by energy sources based on food, via life itself. But there are many creatures that live in the ocean's depths, in quasi-total darkness and that have both organic and inorganic sources of energy. And there are also bacteria that live deep underground, in the dark and which extract energy from rocks, feeding on inorganic, not organic, matter. This may correspond to the most usual situation in the solar system. Future exploration of the solar system will test these concepts in strange far away places.

Strange places, not so far away, the homes of extremophiles on Earth, are the sites for the discussion of the boundaries of life by Cockell and Nixon (Chap. 7). The three main parameters affecting the functioning of cellular biology are temperature, acidity and salinity. We, as human beings, in our threescore years and ten, live in environments with restricted ranges of temperature, air pressure, water availability, pH and access to energy sources. But there are many creatures that are born and live in environments that far transgress our limited range of these parameters. Are these extremophiles models or templates for organisms that can potentially exist and thrive in extraterrestrial sites? This question is one of the underlying motifs in this chapter.

In the discussion it is clear that the existence of life itself in particular extreme physical conditions may be passive in form, in the sense of the creature's metabolic and/or replication processes being restricted or absent. There are many cases of differential and competitive effects of physicochemical and other parameters on these key functions associated with life under extreme conditions. Incompleteness in the action of these functions makes it difficult to assign clear boundaries between life and non-life. Furthermore, the rates of biochemical reactions may become so low in extreme conditions of temperature and/or other parameters, that simply being able to measure replication and/or metabolic rates, in particular of microbial species, takes agonisingly long times, not always conterminous with our own everyday time scales. The examples discussed involve low (permafrost) and high (hydrothermal vent) temperatures, high pressure (deep sea) environments, scarce water situations and extremes of pH. Experimental procedures for metabolic and replication rate measurements, and the practical difficulties in doing so, are expertly presented and analysed for a variety of microorganisms, physicochemical conditions and geographical sites, with an emphasis on explicating the physical chemistry involved in the way organisms tolerate and thrive in extreme conditions.

This is followed by a section on water as a solvent and on other possible solvents compatible with biochemistry, which complement certain aspects of the presentation on these subjects given in Chap. 6. It is agreed that the properties of water seem fine-tuned for life but equally possible that life on Earth has evolved to be fine-tuned to a watery environment. The electron configurational virtues of carbon on which chemical bonding in life structures depends are then compared and contrasted with those of silicon in their ability to form biochemically valid molecules and structures: proteins, informational molecules, membranes. Silicon is a loser in this competition, although it must be said that fascinating and somewhat outlandish suggestions have been made concerning particular silicon-based materials as possible components in novel forms of life yet to be discovered. Nevertheless, silicon must not be completely dismissed as a contributor to life, since silicates have been shown to mediate the formose reaction [100], in which sugars are produced from simple aldehydes, a well-known hobbyhorse of origin of life theories.

Life is exercised within a thermodynamic framework and it is this that sets real limits to its possible extreme physicochemical conditions. Energy extraction and conversion in biological entities often rely on the existence of appropriate redox reactions that facilitate these processes, illustrated here by the actions of microorganisms which are to be found in many extreme environments. Gibbs free energy is the dominant factor. The Gibbs free energy (the IUPAC recommended name is Gibbs energy or Gibbs function) is a thermodynamic potential that measures the process-initiating work obtainable from a thermodynamic system under specific physical conditions. It is also the chemical potential that is minimized when a system reaches equilibrium at constant pressure and temperature. The derivative of the Gibbs energy as a function of the reaction coordinate of the system is zero at the equilibrium point. This is a good marker indicating the feasibility of a chemical process proceeding under well-defined conditions, although it does not indicate whether the reaction rate will be measurable; kinetic factors are required to this end.

Methods for determining the change in Gibbs energy (ΔG) occurring in the formation of a chemical compound are discussed and illustrated with examples, including a number involving microbially-mediated redox reactions. Particular care must be taken in these determinations of ΔG when the physical conditions are very different from the standard conditions used in defining the Gibbs energy; this is often the case in the extreme conditions under which extremophiles live and thrive and the consequences are discussed and illustrated for temperature, pressure and other physical or physicochemical excursions from standard conditions.

There are other applications of thermodynamics to understanding metabolic pathways, both in nature and in laboratory-inspired studies. Cockell and Nixon cite studies on microbial assemblages in marine hydrothermal systems, in which reaction energetics have been modelled for a large number of redox reactions under the prevailing physical conditions, and the energy yields estimated. The energetic potential depends strongly on the geochemical composition of the environment in which the microbes live. Price and Sowers [101] have used experimental data on communities of microbes to stress that there are three separable modes of metabolic processes in which this energy is consumed: growth, maintenance, and survival. Organisms in maintenance mode can conduct basic cellular functions but lack sufficient energy for growth; those in survival mode use energy solely at the rate required to repair macromolecular damage. Thermodynamic considerations will obviously be of importance in determining possible habitats for life, whether on Earth or elsewhere. This chapter includes some further considerations on the possible existence of life governed by other forms of biochemistry than the familiar one dominated by carbon compounds and aqueous solutions defined here as our ‘biospace’.

10.9 Energy, the Essential Primer of Self-Organisation

In Chap. 8 the energy-involved processes essential to life are analysed by Robert Pascal in terms of their thermodynamic and kinetic aspects, from standpoints complementary to those of previous chapters and in somewhat more detail. The emphasis here is on the physical chemistry of self-organisation in the context of metabolism.

Questioning the energy source of the first living organisms is usually characterised by the possibility of dividing living organisms into two categories, those that acquire their nutrients directly from the environment and those that make their own, i.e. only indirectly from the external environment. Organisms are thus considered to be either heterotrophic or autotrophic. A burning question, much discussed, and to which an answer is still sought, concerns which of these two forms of life is simpler and can thus be expected to have arisen first on the early Earth. Because of its supposed simplicity heterotrophy has often been viewed as more likely. In this case, where a cell of an organism is unable to produce its own nutrients, the implication is that the environment harbours complex substances that supply the nutrients of the organism. Many possible avenues of prebiotic synthesis of such nutrients have been explored but these have not convinced the sceptics. In the words of Morowitz [102], 'early life could not have survived off of a free lunch'. He considers that heterotrophic cells would not have persisted since they would have quickly exhausted the nutrients available in their surroundings, an improbability argument. This favours the view that autotrophic organisms were the original inhabitants of the early Earth.

Morowitz claims that the study of the metabolism of present day cellular organisms can provide important clues to the origins of cellular life. He considers that prebiotic processes are expected to leave conspicuous evidence in contemporary biochemistry, a molecular phylogeny. The closure of a phospholipid bilayer is considered to be the critical event in the origin of cellular entities distinct from their environment. This differs from the central role played by the development of catalytic polypeptides and of a coding system for their perpetuation as preferred in other early life schemes. Thus Morowitz sees the presence of phospholipids and of simple energy-transducing systems for their further production, as primary to the development of protein catalysts and of their nucleic acid coding systems. Here Robert Pascal stresses use of the findings and laws of physical chemistry as an alternative to the molecular phylogeny approach to the emergence of life. Introducing kinetics into such schemes [103] has persuaded Pross [104] that replication must have preceded metabolism. An excellent critical review of theories concerning the place of metabolism in the origin of life has been made by Anet [105].

The opposition between the autotrophic and heterotrophic approaches to metabolism has resulted in different emphases on research avenues. The autotrophic school is enthralled with geochemical cycles and their biochemical implications, whereas the heterotrophic school concentrates more on the emergence of self-replicating systems. In effect, the heterotrophy versus autotrophy question becomes engulfed in the fundamental quarrel between the proponents of replication-first versus metabolism-first processes in the emergence of life on the early Earth. Measured aspects of this confrontation are found in Chaps. 8 and 9. It should also be said that there are attempts to bridge the gap between these two viewpoints by devising a metabolism-driven replication scheme [17].

Robert Pascal sets his scene in a background of evolutionary principles introduced into physical chemistry in attempts to elucidate the origin of life. This is applied to the qualitative and quantitative aspects of metabolic processes in living organisms and their natural emergence and development into enzymatic mediated systems. Matter, energy and thermodynamic concepts are reviewed and their natures illuminated by inclusion of the historical contexts of their conception. A section on self-organisation and kinetics contains a subtle discussion of the means by which living systems achieve and maintain their necessary excursions from thermodynamic equilibrium. Kinetic barriers are the key to this situation, and the processes in which they are involved are, of necessity, chemically selective in their nature and mediated by operative catalysis. Transition state theory is evoked in these processes, which are considered to demonstrate dynamic kinetic stability in their tendency to achieve a maximisation of the equilibration rate. Eschenmoser proposed that this could be achieved by chemical self-organisation, through the generation of autocatalytic dissipative structures along gradients of increasing rates of the environment's overall free energy dissipation [106]. He considers that the circumvention of kinetic barriers is fundamental to chemical self-organisation. This provides a key for the operation of chance and that of necessity in the emergence and propagation of self-replicating autocatalytic systems capable of evolving into a living system.

The emergence of metabolism involves sequences of reactions that include, somewhere along the line, an irreversible step. Increasingly complex autocatalyic chemical cycles have been suggested, some of which have been proposed to occur on mineral surfaces. Reaction topologies can vary in complexity and can be affected by modifications of physicochemical conditions. Biochemical catalysts, ribozymes, behave under fine-tuned conditions and on operationally valid timescales in order to achieve specific and varied chemical outcomes. It is here that evolutionary effects on a biochemical level and its consequences can occur, involving both chance and necessity. Life certainly did not have the opportunity to completely explore the whole vast universe of chemical possibilities; our own biochemistry must result in part from some contingent choices, not necessarily optimal, occurring by chance. In this context new molecules and functions develop by co-optation and/or modification of pre-existing ones. And sometimes these leave long-living traces, such as is manifested by the similarity in the sodium content of the sea and that of the internal liquids of multicellular animals, which possibly harks back to the emergence of the first multicellular organisms in sea waters. Another example is the fact that the redox potential of the cell cytoplasm is very low, less than 0 mV, as compared to that for the redox state of oxygenated environments, >600 mV [17], which is consistent with the first cells having evolved on Earth before the oxygenated atmosphere was established. Thus organisms in our present atmosphere spend part of their resources to maintain a large redox gap with respect to the environment. They are victims of features of biochemistry that emerged under selective pressures that no longer exist. Taking into effect selective pressures and their evolution over long time spans is a challenge for evaluation of the possibilities and nature of life in exotic and spatially far-off places.

The requirements of a source of energy for proto-metabolic processes is examined in detail by Robert Pascal and evaluated for different physical and chemical conditions. These sources include those mentioned earlier in the present chapter, light, heat, redox reactions, etc. Their capabilities and efficiencies in driving reactions are considered in terms of physicochemical and molecular parameters and processes. The composition of the atmosphere and the geological state of the early Earth provide constraints on the types of nutrients that could be synthesised and take part in metabolic processes at life's infancy. One must remember that current biochemistry was then non-existent but that organic chemistry and physical chemistry were identical with today's versions. Various scenarios for the emergence of metabolism and the realisation of its biochemical role in the functioning of cells and in the synthesis of organic matter are discussed; the role of a chemical natural selection is stressed. These considerations are primordial in the possibility of sustaining life and their relevance to possible habitats for life on Earth and on exoplanets is briefly mentioned. The chapter ends with a final trill on the fundamental role of quantum concepts in the processes we recognise as inherent to life.

10.10 The Machinery of Biological Repetition

In tackling the issue of replication, in Chap. 9, Turk-MacLeod, Gerland and Chen analyse the components and physicochemical basis of this important stage in the emergence of life, without worrying as to whether this was anterior or posterior to metabolism in the early Earth. Two principal aspects are discussed, the molecular carriers of information and the lipid membranes that encapsulate them along with other species active in membrane-bounded cells. The structures and components of the nucleic acids DNA and RNA, informational molecules, are described and their functions elaborated. This discussion assumes the essential nature for life of the DNA-RNA-protein biopolymer system. The geometrical lability of the nucleic acids, enabling them to fold into complex structures, and the discovery of catalytic RNA molecules, ribozymes, were essential steps in the development of the concept of a 'RNA world' existing prior to the DNA world of today.

Biological catalysis by RNA was indeed a sensational discovery at the beginning of the 1980s. The biochemical paradigm had it that only proteins were capable of functioning as biocatalysts. However, Altman studied the enzyme ribonuclease P, which is composed of RNA and protein subunits, and found conditions under which the RNA component could catalyse the formation of mature tRNA in the absence of protein [107]. Furthermore, Cech discovered the phenomenon of self-splicing rRNA in *Tetrahymena thermophila*. This rRNA catalyses the consecutive transesterification of specific phosphodiester groups in its nucleotide sequence and thereby the excision of an intervening sequence and ligation of the remaining rRNA molecule [108]. The discoveries of Altman and Cech enabled an RNA world to be envisaged, in which RNA is the sole genetically encoded component of biological catalysts [109]. Thus was born a new conceptual stage in the origin of life. Its real existence in the past is subject to debate; there is also a peptides-first hypothesis which proposes that proteins were the first catalysts in life and indeed aminoacids and peptides can more easily be formed than nucleic acids in abiotic conditions. However, there do exist some features of our present day biology that might be the fossilised traces of a long gone RNA world. The main objection to an RNA-first world is in the difficulty of divining how a molecule as complex as RNA could be assembled spontaneously. A possible answer is provided by the work of Powner et al. [110] who have proposed and demonstrated a mode of pyrimidine ribonucleotide synthesis in which the sugar and the nucleobase emerge from a common precursor under prebiotically plausible conditions.

Communication between molecules is a matter of specific interactions, mainly occurring through electromagnetic forces. How does this operate in replication in which the genetic information is carried by nucleic acids? It is here that hydrogen bonds between the nitrogenous bases, and the phosphate groups supporting the structural backbone, play a role. Phosphate groups, which participate widely in biological chemistry, are important to the functioning of DNA and RNA. The polyanion backbone gives DNA the capability of replication, following simple rules, and to be capable of evolving. The anionic nature of the backbone helps to prevent the nucleic acids from folding, so enabling them to act as templates for replication and polymerization. The interaction between two strands to form a duplex occurs in ways that provide simple rules involving base pair relations to guide molecular recognition. A polyanion or polycation is probably required for the establishment of a self-sustaining chemical system capable of Darwinian evolution; its structure may therefore well be a universal signature of life.

Turk-MacLeod et al. maintain that substrate positioning is a key factor in the ability to build complementary strands and consider that, in principle, this can occur in the absence of enzymes, thus providing a possible scheme for copying DNA strands in the period before protein enzymes had evolved. The proximity of interacting units is ordained and ensured by hydrogen-bond effects and/or covalent bonding. Nucleic acids acting as templates can facilitate the synthesis of some compounds by modifying the effective concentration of reactants or rendering inoperative the formation of unwanted products. This is a possible pathway for peptide formation in the early Earth. In laboratory experiments it has been shown that ribosomes and ribozymes can be avoided for peptide synthesis by using highly reactive aminoacyl phosphate nucleotides bound to RNA guide sequences in which the aminoacyl groups mimic chemical processes found in modern biosynthesis. It has also been shown that peptides can form self-replicating systems. Nucleic acids are apparently the best in achieving template chemical synthesis and self-replication but other molecules are possible and one can imagine alternative life-forming processes.

The somewhat unexpected efficiency of a small RNA enzyme, only five nucleotides in length, suggests that many different reactions could be accelerated by small RNAs and even be operative after the most primitive ribonucleotide polymerisation had occurred [111]. The evolution of small entities into very specific and efficient RNA enzymes is considered in terms of effects on the transition states of reactions. There are many RNAs that bind amino acids and this may be the origin of catalysers of peptide bond formation.

The capacity for evolution of an organism depends in part on the fidelity of copying of its replication processes. The fidelity criteria are discussed by Turk-MacLeod et al. in terms of thermodynamic limits. A critical factor is the error threshold for replication, which is the critical copying fidelity below which the fittest genotype deterministically disappears, being vanquished by spontaneous decay, i.e. by some irreversible process that destroys, for example, the DNA [112]. The error threshold limits the length of the genome (number of nucleotides) that can be maintained by selection, as was initially remarked by Eigen [113]. Turk-MacLeod et al. consider that in a prebiotic context the concept of error threshold is not one of competition, as it is in the Eigen formulation, but refers to the limit of the process that generates one functional copy per core replication molecule before the template is destroyed. Nature's methods of verifying the copying fidelity are discussed and are related to the possible error limits at different evolutionary epochs including that of the prebiotic world. A thermodynamic bound on copying fidelity involves the processes of matching, and mismatching, of nucleosides and thus the thermodynamics of base-pairing. The thermodynamic based models are shown to exhibit the same qualitative trends as experimentally determined error ratios for non-enzymatic template-directed polymerization, with DNA and RNA templates and primers. The much higher copying fidelity of DNA polymerization as compared with RNA polymerization is argued as a fundamental reason for the passage from an RNA to a DNA world.

The section on lipid membranes starts with a description of amphiphilic molecules, containing both polar and nonpolar domains, often based on fatty acids, and explains the *modus operandi* of the hydrophobic effect. These molecules can aggregate in sheets or pseudo-spheres, micelles and vesicles, by lining up their nonpolar domains inward and polar domains outward, the latter interfacing water. The lipid membrane that encloses every living cell is essentially a lipid sheet formed into a bubble. Membranes overcome the dissipative disadvantages of free solutions in the carrying out of complex biochemistry. Besides their role as confining containers, membranes can also act as semi-permeable barriers that mediate the flux of molecules entering and undergoing biochemical reactions. Some bilayer-membrane vesicles can exhibit morphological changes that can be characterised as growth, fusion, division by budding, internal synthesis of new vesicles; vesicle-surface interactions as a mode of mediating vesicle reproduction has also been suggested [114]. Self-reproduction, through growth followed by division, and cell-like properties, emerge from the molecules forming the vesicles via thermodynamic and kinetic constraints [115]. Interactions between membrane proteins can create a form of steric pressure that may influence the biological processes that drive vesicle formation [116].

The preferential interactions between molecules due to cell confinement are considered by Turk-MacLeod et al. as examples of groupings, for example of genes or of individual living organisms, that are widespread in nature and which can be treated in terms of altruistic and non-altruistic traits, and thus related to Darwinian evolution.

The formation of the encapsulating membranes is discussed by Turk-MacLeod et al.; the operative strictures of thermodynamics in these processes and in the functional role of cell membranes are elaborated. The competition between vesicles that encapsulate RNA and those incapable of doing so, considered as model protocells, and its relation to the evolutionary fitness of replicator functions, is considered at length in terms of the driving forces of thermodynamics. It is noted that membrane stabilization is a key objective in this competition but this results also in a reduction of permeability, thus diminishing the ability of the protocell to use nutrients. Further evolution of the membrane and its constituents is necessary to overcome this restriction in function. In this respect it is of interest that model protocell membranes composed of particular mixtures of amphiphiles have superior properties, with respect to growth and division, than membranes composed of single amphiphiles [117]. In the protocell the amphiphiles must mainly have been fatty acids, which are also present in modern phospholipids. But the latter have the disadvantage of being relatively impermeable to polar solutes. However, a selective advantage of phospholipids is their facilitation of vesicle growth by absorbing fatty acids from neighbouring vesicles; phospholipid membranes emerged as the winners in Darwinian evolution.

A present day cell constitutes a thermodynamic open system that exchanges matter and energy with its environment. Its complexity is far greater than that of the protocells conceived to have existed in the early Earth. We have seen that reactivity and catalytic functions are necessary for the development of structural and dynamic chemical complexity; indeed one can consider that life is a naturally emergent property of a molecular system when an appropriate degree of complexity is attained.

The logical steps in self-reproduction have been analysed, beginning with the work of von Neumann on self-reproducing machines, a subject very much alive [118] and of interest to space exploration. It has been extended to the protocell domain [119] and its general concepts have recently been discussed by Nurse in terms of a computational picture of the logic of living systems, focussing on how information is managed in these systems and how this creates higher-level biological phenomena [120]. In this context it is necessary to improve our knowledge as to how molecules interact to generate logic modules in the living system and how these modules function in biochemical networks. Nurse presents a practical programme to this end. This is a renewal of themes that were in the forefront of science 50 years ago, in the early days of molecular biology. It requires, among other things, sophisticated polyvalent databases and the development of new experimental techniques to improve in vivo analysis of living systems, using advanced imaging techniques for real-time experiments.

Finally, we recall that the discussion in this chapter is based on the assumption that the DNA-RNA-protein biopolymer system is essential for life. However, the possible construction or existence of other forms of life is hovering on the horizon. Six alternative genetic polymers, based on simple nucleic acid architectures not found in nature, and capable of heredity and of Darwinian evolution, have recently been described [121]. DNA and RNA are therefore not functionally unique as genetic materials and so the question as to whether extraterrestrial life has a basically different genetic structure than that found on Earth remains open. The work of Pinheiro et al. [121] thus leaves intact a fundamental conundrum of astrobiology and weighs on the design strategies of life detection space missions.

References

1. Shiltsev V (2012) Mikhail Lomonosov and the dawn of Russian science. Phys Today 64:40–46

2. Perrin J (1903) Traité de Chimie Physique I: Les Principes. Gauthier-Villars, Paris
3. Hinshelwood C (1951) The structure of physical chemistry. Clarendon, Oxford
4. Bartels H-G, Huebener R (2007) Walther Nernst: pioneer of physics and chemistry. World Scientific, Singapore
5. Wilcek F (1999) The persistence of ether. Phys Today 52:11–13
6. Kragh H (2012) Walther Nernst: grandfather of dark energy. Astron Geophys 53:1.24–1.26
7. Shakespeare W (2002) Sonnets and poems. Oxford University Press, Oxford
8. Layzer D (1993) Chemistry and cosmology. J Phys Chem 97:2395–2399
9. Canuto V (1978) On the origin of Hawking mini black-holes and the cold early universe. Mon Not R Astron Soc 184:721–725
10. Aguirre AN (1999) Cold big bang nucleogenesis. Astrophys J 521:17–29; (2000) The cosmic background radiation in a cold big bang. Astrophys J 533:1–18
11. Khoury J, Ovrut BA, Steinhardt PJ, Turok N (2001) Ekpyrotic universe: colliding branes and the origin of the hot big bang. Phys Rev D 64:123522–123523
12. Martin J, Peter P (2004) On the "causality argument" in bouncing cosmologies. Phys Rev Lett 92:061301–061304
13. Sagan C, Chyba C (1997) The faint Sun paradox: organic shielding of ultraviolet-labile greenhouse gases. Science 276:1217–1221
14. Ribas I, Guinan EF, Güdel M, Audard M (2005) Evolution of the solar activity over time and effects on planetary atmospheres. I. High-energy irradiances (1–1700 Å). Astrophys J 622:680–694
15. O'Malley-James JT, Raven JA, Cockell CS, Greaves JS (2012) Life and light: exotic photosynthesis in binary and multiple-star systems. Astrobiology 12:115–124
16. Chyba C, Sagan C (1992) Endogenous production, exogenous delivery and impact-shock synthesis of organic molecules: an inventory for the origins of life. Nature 355:125–132
17. Mulkidjanian AY, Galperin MY (2007) Physicochemical and evolutionary constraints for the formation and selection of first biopolymers: towards the concensus paradigm of the abiogenic origin of life. Chem Divers 4:2003–2015
18. Luminet J-P (2011) Black holes. Cambridge Univesity Press, Cambridge
19. Fumagalli M, O'Meara JM, Prochaska JX (2011) Detection of pristine gas two billion years after the big bang. Science 334:1245–1249
20. Leach S (2012) Why *COBE* and CN spectroscopy cosmic background radiation temperature measurements differ, and a remedy. Mon Not R Astron Soc 421:1325–1330
21. Indriolo N, McCall BJ (2012) Investigating the cosmic-ray ionization rate in the galactic diffuse interstellar medium through observation of H_3^+. Astrophys J 745:91-1-17
22. Snow TP, Witt AN (1996) Interstellar depletions updated: where all the atoms went. Astrophys J Lett 468:L65–L68
23. Lodders K (2003) Solar system abundances and condensation temperatures of the elements. Astrophys J 591:1220–1247
24. Lodders K (2010) Solar system abundances of the elements. In: Goswami A, Reddy BE (eds) Principles and perspectives in cosmochemistry, Astrophysics and space science proceedings. Springer, New York, pp 379–417
25. Ney EP, Hatfield BF (1978) The isothermal dust condensation of Nova Vulpeculae 1976. Astrophys J Lett 219:L111–L115
26. Duley WW (1980) Redox reactions and the optical properties of interstellar grains. Astrophys J 240:950–955
27. Field D (2000) H_2 formation in space: a negative ion route ? Astron Astrophys 362:774–779
28. Caruana DJ, Holt KB (2010) Astroelectrochemistry: the role of redox reactions in cosmic dust chemistry. Phys Chem Chem Phys 12:3072–3079
29. Hoyle F, Wickramasinghe NC (1979) On the nature of interstellar grains. Astrophys Space Sci 66:77–90
30. Hoyle F, Wickramasinghe NC, Al-Mufti S (1985) The ultraviolet absorbance of presumably interstellar bacteria and related matters. Astrophys Space Sci 111:65–78

31. Léger A, d'Hendecourt L, Boccara N (eds) (1987) Polycyclic aromatic hydrocarbons and astrophysics. Reidel, Dordrecht
32. Kwok S, Zhang Y (2011) Mixed aromatic-aliphatic organic nanoparticles as carriers of unidentified emission features. Nature 479:80–83
33. Cami J, Bernard-Salas J, Peeters E, Malek SE (2010) Detection of C_{60} and C_{70} in a young planetary nebula. Science 329:1180–1182
34. Zhang Y, Kwok S (2011) Detection of C_{60} in the protoplanetary nebula IRAS 01005+7910. Astrophys J 730:126-1-5
35. Herbig GH (2000) The search for interstellar C_{60}. Astrophys J 542:334–343
36. Leach S, Vervloet M, Desprès A, Bréhéret E, Hare JP, Dennis TJ, Kroto HW, Taylor R, Walton DRM (1992) Electronic spectra and transitions of the fullerene C_{60}. Chem Phys 160:451–466
37. Sassara A, Zerza G, Chergui M, Leach S (2001) Absorption wavelengths and bandwidths for interstellar searches of C_{60} in the 2400–4100 Å region. Astrophys J Suppl 135:263–273
38. Goeres A, Sedlmayr E (1992) The envelopes of R Coronae Borealis stars I. A physical model of the decline events due to dust formation. Astron Astrophys 265:216–236
39. García-Hernández DA, Iglesias-Groth S, Acosta-Pulido A, Manchado A, García-Lario P, Stanghellini L, Villaver E, Shaw RA, Cataldo F (2011) The formation of fullerenes: clues from new C_{60}, C_{70}, and (possible) planar C_{24} detections in the Magellanic cloud planetary nebulae. Astrophys J Lett 737:L30-1-7
40. Duley WW, Hu A (2012) Fullerenes and proto-fullerenes in interstellar carbon dust. Astrophys J Lett 745:L11-1-4
41. Evans A, van Loon JT, Woodward CE, Gehrz RD, Clayton GC, Helton LA, Rushton MT, Eyres SPS, Krautter J, Starrfield S, Wagner RM (2012) Solid-phase C60 in the peculiar binary XX Oph? Mon Not R Astron Soc 421:L92–L96
42. Tong X, Winney AH, Willitsch S (2010) Sympathetic cooling of molecular ions in selected rotational and vibrational states produced by threshold photoionization. Phys Rev Lett 105:143001-1-4
43. Hall FJ, Aymar M, Bouloufa-Maafa N, Dulieu O, Wilitsch S (2011) Light-assisted ion-neutral reactive processes in the cold regime: radiative molecule formation versus charge exchange. Phys Rev Lett 107:243202-1-5
44. Goulielmakis E, Loh Z-H, Wirth A, Santra R, Rohringer N et al (2010) Real-time observation of valence electron motion. Nature 466:739–743
45. Grubb M, Warter ML, Xiao H, Maeda S, Morokuma K, North SW (2012) No straight path: roaming in both ground- and excited-state photolytic channels of $NO_3 \text{---} > NO + O_2$. Science 335:1075–1078
46. Bowman JM, Schneider BC (2011) Roaming radicals. Annu Rev Phys Chem 62:531–553
47. NIST Chemistry Webbook (June 2005) National Institute of Standards and Technology Reference Database. Available from http://webbook.nist.gov (current 2010)
48. Lias SG, Bartmess JE, Libman JF, Holmes JL, Levin RD, Mallard WG (1988) Gas-phase ion and neutral thermochemistry. J Phys Chem Ref Data 17(supplNo.1)
49. Cohen N, Benson SW (1983) Estimation of heats of formation of organic compounds by additivity methods. Chem Rev 93:2419–2438
50. Lemoult P (1907) Recherches théoriques et expérimentales sur les chaleurs de combustion et de formation des composés organiques. 1. Amines primaires, secondaires et tertiaires. Ann Chim Phys 8e série:395–432
51. Lemoult P (1908) Recherches théoriques et expérimentales sur les chaleurs de combustion et de formation des composés organiques. 2. Composés hydrazoiques. Ann Chim Phys 8e série:562–574
52. Lemoult P (1905) Relations générales entre la chaleur de combustion des composés organiques et leur formule de constitution. Calcul des chaleurs de combustion. Ann Chim Phys 8e série: 5–70

53. Cox JD, Pilcher G (1970) Thermochemistry of organic and organometallic compounds. Academic, New York
54. Benson SW, Buss JH (1958) Additivity rules for the estimation of Molecular properties. Thermodynamic properties. J Chem Phys 29:546–573
55. Benson SW (1976) Thermochemical kinetics, 2nd edn. Wiley, New York
56. Pedley JB, Naylor RD, Kirby SP (1986) Thermochemical data of organic compounds, 2nd edn. Chapman and Hall, London
57. van Speybroek V, Gani R, Meier RJ (2010) The calculation of thermodynamic properties of molecules. Chem Soc Rev 39:1764–1779
58. Holmes JL, Lossing FP (1989) Bond strengths in even-electron ions and the proton affinities of free radicals. Int J Mass Spectrom Ion Processes 92:111–122
59. Meot-Ner Mautner M, Sieck LW (1991) Proton affinity ladders from variable-temperature equilibrium measurements. 1. A reevaluation of the upper proton affinity range. J Am Chem Soc 113:4448–4460
60. Czakó G, Mátyus E, Simmonnett AG, Császár G, Schaefer HF III, Allen WD (2008) Anchoring the absolute proton affinity scale. J Chem Theory Comput 4:1220–1229
61. Lias SG, Bartmess JE (2005) Gas-phase ion thermochemistry. NIST Chemistry Webbook, http://webbook.nist.gov
62. Traeger JC, McLoughlin RG (1981) Absolute heats of formation for gas-phase cations. J Am Chem Soc 103:3637–3652
63. Franklin JL (1953) Calculation of the heats of formation of gaseous free radicals and ions. J Chem Phys 21:2029–2034
64. Holmes JL, Fingas M, Lossing FP (1981) Towards a general scheme for estimating the heats of formation of organic ions in the gas phase. Part 1. Odd-electron ions. Can J Chem 59:80–93
65. Vasyunin AI, Semenov D, Henning Th, Wakelam V, Herbst E, Sobolev AM (2008) Chemistry in protoplanetary disks: a sensitivity analysis. Astrophys J 672:629–641
66. Patra SM, Mishra RK, Mishra BK (1997) Graph-theoretic study of certain interstellar reactions. Int J Quantum Chem 62:495–508
67. Solé RV, Munteanu A (2004) The large-scale organization of chemical networks in astrophysics. Europhys Lett 68:170–176
68. Jolley C, Douglas T (2012) Topological signatures: large-scale structure of chemical networks from biology and astrochemistry. Astrobiology 12:29–39
69. Wayne RP (2000) Chemistry of atmospheres. An introduction to the chemistry of the atmospheres of earth, the planets, and their satellites, 3rd edn. Oxford University Press, Oxford
70. Taylor FW (2010) Planetary atmospheres. Oxford University Press, Oxford
71. Pierrehumbert RT (2010) Principles of planetary climate. Cambridge University Press, Cambridge
72. Lellouch E (2011) The composition of planetary atmospheres: an historical perspective. In: Beaulieu J-P, Dieters S, Tinetti G (eds) Molecules in the atmospheres of extrasolar planets, ASP conference series, Paris, vol 450, pp 3–18
73. Perryman MAC (2000) Extra-solar planets. Rep Prog Phys 63:1209–1272
74. Sozzetti MT, Lattanzi MG, Boss AP (eds) (2011) The astrophysics of planetary systems: formation, structure, and dynamical evolution. Proceedings IAU symposium, 276 Torino
75. Schneider J, Dedieu C, Le Sidaner P, Savalle R, Zolotukhin I (2011) Defining and cataloging exoplanets: the exoplanet.eu data base. Astron Astrophys 532:A79-1-13
76. Plavalova E (2012) Taxonomy of the extrasolar planet. Astrobiology 12:361–369
77. Seager S (2010) Exoplanet atmospheres: a theoretical outlook. In: Sozzetti MT, Lattanzi MG, Boss AP (eds) The astrophysics of planetary systems: formation, structure, and dynamical evolution. Proceedings IAU symposium. Torino, 276, pp 198–207
78. Seager S, Deming D (2010) Exoplanet atmospheres. Annu Rev Astron Astrophys 48:631–672
79. Burrows A, Budaj J, Hubeny I (2008) Theoretical spectra and light curves of close-in extrasolar giant planets and comparison with data. Astrophys J 678:1436–1457

80. Liang M-C, Seager S, Parkinson C, Lee AY-L, Yung YL (2004) On the insignificance of photochemical hydrocarbon aerosols in the atmospheres of close-in extrasolar giant planets. Astrophys J Lett 605:L61–L64
81. Line MR, Vasisht G, Chen P, Angerhausen D, Yung YL (2011) Thermochemical and photochemical kinetics in cooler hydrogen-dominated extrasolar planets: a methane-poor GJ4336b? Astrophys J 738:32-1-14
82. Miller-Ricci Kempton E, Zahnle K, Fortney JJ (2012) The atmospheric chemistry of GJ 1214b: photochemistry and clouds. Astrophys J 745:3-1-13
83. Marley MS, Fortney J, Seager S, Barman T (2007) Atmospheres of extrasolar giant planets. In: Reipurth B, Jewitt D, Keil K (eds) Protostars and planets V. University of Arizona Press, Tucson, pp 733–747
84. Seager S, Schrenk M, Bains W (2012) An astrophysical view of earth-based metabolic biosignature gases. Astrobiology 12:61–82
85. Fox JL, Galand MI, Johnson RE (2008) Energy deposition in planetary atmospheres by charged particles and solar photons. Space Sci Rev 139:3–62
86. Yelle R, Lammer H, Ip W-H (2008) Aeronomy of extra-solar giant planets. Space Sci Rev 139:437–451
87. Lunine JI (2005) Astrobiology: a multidisciplinary approach. Addison Wesley, San Francisco
88. Trail D, Mojzsis SJ, Harrison TM, Schmitt AK, Watson EB, Young ED (2007) Constraints on Hadean zircon protoliths from oxygen isotopes, Ti-thermometry, and rare earth elements. Geochem Geophys Geosystems 8:Q06014-1-22
89. Hirschmann M, Kohlstedt D (2012) Water in Earth's mantle. Phys Today 65:40–45
90. Léger A, Selsis F, Sotin C, Guillot T, Despois D et al (2004) A new family of planets? "Ocean Planets". Icarus 169:499–504
91. Marcy G (2009) Water world larger than Earth. Nature 462:853–854
92. de Grotthuss CJT (1806) Sur la décomposition de l'eau et des corps qu'elle tient en dissolution à l'aide de l'électricité galvanique. Ann Chim (Paris) 58:54–74
93. Cuikerman S (2006) Et tu, Grotthuss! and other unfinished stories. Biochim Biophys Acta 1757:876–885
94. Fayer MD (2012) Dynamics of water interacting with interfaces, molecules, and ions. Acc Chem Res 45:3–14
95. Zhang Y, Cremer PS (2006) Interactions between macromolecules and ions: the Hofmeister series. Curr Opin Chem Biol 10:658–663
96. Benner S, Ricardo A, Carrigan MA (2004) Is there a common chemical model for life in the universe? Curr Opin Chem Biol 8:672–689
97. Cleland CE, Chyba CF (2002) Defining 'life'. Origins Life Evol B 32:387–393
98. Ruiz-Mirazo K, Pereto J, Moreno A (2004) A universal definition of life: autonomy and open-ended evolution. Origins Life Evol B 34:323–346
99. Deamer D (2010) Special collection of essays: what is life? Astrobiology 10:1001–1002
100. Lambert JB, Gurusamy-Thangavelu SA, Ma K (2010) The silicate-mediated formose reaction: bottom-up synthesis of sugar silicates. Science 327:984–986
101. Price PB, Sowers T (2004) Temperature dependence of metabolic rates for microbial growth, maintenance and survival. Proc Natl Acad Sci 101:4631–4636
102. Morowitz HJ (1992) Beginnings of cellular life: metabolism recapitulates biogenesis. Yale University Press, New Haven/London
103. Pross A (2003) The driving force for life's emergence. Kinetic and thermodynamic considerations. J Theor Biol 220:393–406
104. Pross A (2004) Causation and the origin of life. Metabolism or replication first? Origins Life Evol B 34:307–321
105. Anet FAL (2004) The place of metabolism in the origin of life. Curr Opin Chem Biol 8:654–659
106. Eschenmoser A (1994) Chemistry of potentially prebiological natural products. Origins Life Evol B 24:389–423

107. Altman S, Baer MF, Bartkiewicz M, Gold H, Guerrier-Takada C, Kirsebom LA, Lumelsky N, Peck K (1989) Catalyses by the RNA subunit of RNase P - a minireview. Gene 82:63–64
108. Cech TR, Zaug AJ, Grabowski PJ (1981) In vitro splicing of the ribosomal RNA precursor of tetrahymena: involvement of a guanosine nucleotide in the excision of the intervening sequence. Cell 27:487–496
109. Cech TR (1993) The efficiency and versatility of catalytic RNA: implications for an RNA world. Gene 135:33–36
110. Powner MW, Gerland B, Sutherland J (2009) Synthesis of activated pyrimidine ribonucleotides in prebiotically plausible conditions. Nature 459:239–242
111. Turk RM, Chumachenko NV, Yarus M (2010) Multiple translational products from a five-nucleotide ribozyme. Proc Natl Acad Sci 107:4585–4589
112. Szathmáry E (2006) The origin of replicators and reproducers. Philos Trans R Soc B 361:1761–1776
113. Eigen M (1971) Self-organization of matter and the evolution of biological molecules. Naturwissenschaften 58:465–523
114. Hancyzc MM, Szostak JW (2004) Replicating vesicles as models of primitive cell growth and division. Curr Opin Chem Biol 8:660–664
115. Stano P, Luisi PL (2010) Achievements and open questions in the self-reproduction of vesicles and synthetic minimal cells. Chem Commun 46:3639–3653
116. Čopič A, Latham CF, Horlbeck MA, D'Arcangelo JG, Miller EA (2012) ER cargo properties specify a requirement for COPII coat rigidity mediated by Sec13p. Science 335:1359–1362
117. Szostak JW (2011) An optimal degree of physical and chemical heterogeneity for the origin of life. Philos Trans R Soc B 366:2894–2901
118. Zykov V, Mytilinaios E, Adams B, Lipson H (2005) Self-reproducing machines. Nature 435:163–164
119. Solé RV (2009) Evolution and self-assembly of protocells. Int J Biochem Cell Biol 41:274–284
120. Nurse P (2008) Life, logic and information. Nature 454:424–426
121. Pinheiro VB, Taylor AI, Cozens C, Abramov M, Renders M et al (2012) Synthetic genetic polymers capable of heredity and evolution. Science 336:341–344

Index

I.W.M. Smith et al. (eds.), *Astrochemistry and Astrobiology*, Physical Chemistry in Action,
DOI 10.1007/978-3-642-31730-9, © Springer-Verlag Berlin Heidelberg 2013